PRACTICAL SIGNAL PROCESSING

The principles of signal processing are fundamental to the operation of everyday devices such as digital cameras, mobile telephones and digital audio players. This book introduces the basic theory of digital signal processing, placing a strong emphasis on the use of techniques in real-world applications. The author uses intuitive arguments rather than mathematical ones wherever possible, reinforced by practical examples and diagrams.

The first part of the book covers sampling, quantisation, the Fourier transform, filters, Bayesian methods and numerical considerations. These ideas are then developed in the second part, illustrating how they are used in audio, image, and video processing and compression, and in communications. The book concludes with methods for the efficient implementation of algorithms in hardware and software. Throughout, links between various signal processing techniques are stressed and real-world examples showing the advantages and disadvantages of the different approaches are presented, enabling the reader to choose the best solution to a given problem.

With over 200 illustrations and over 130 exercises (including solutions), this book will appeal to practitioners working in any branch of signal processing, as well as to undergraduate students of electrical and computer engineering.

MARK OWEN received his Ph.D. in Speech Recognition from Cambridge University in 1992, after which he has worked in industry on digital signal processing applications in video, audio and radar. He is currently a freelance consultant in this, and related, fields.

PRACTICAL SIGNAL PROCESSING

MARK OWEN

CAMBRIDGE UNIVERSITY PRESS
Cambridge, New York, Melbourne, Madrid, Cape Town,
Singapore, São Paulo, Delhi, Mexico City

Cambridge University Press
The Edinburgh Building, Cambridge CB2 8RU, UK

Published in the United States of America by Cambridge University Press, New York

www.cambridge.org
Information on this title: www.cambridge.org/9781107411821

First published 2007
First paperback edition 2012

A catalogue record for this publication is available from the British Library

ISBN 978-0-521-85478-8 Hardback
ISBN 978-1-107-41182-1 Paperback

Contents

Preface

This is a book you can read in the park, on the beach, at the bus stop – or even in the bath.

The book is in two parts. The first part takes you step-by-step through the fundamental ideas of digital signal processing, while the second part shows how these ideas are used in a wide range of practical situations. My aim is that by the end of the book you will understand many of the signal processing algorithms and techniques that are essential to everyday devices such as digital cameras, modems, digital set-top boxes, mobile telephones and digital audio players. I have used examples drawn from the operation of such devices to help explain points in the text.

You do not need to know any calculus to understand any of the ideas discussed. A basic understanding of trigonometry and of arithmetic on complex numbers is necessary, however; and a very basic knowledge of the principles of electronic circuits is helpful, but by no means essential.

If you are a student, I hope that the approach this book takes will give you a more concrete and more intuitive grasp of the principles of digital signal processing than a purer mathematical treatment would. If you are a practising engineer or programmer with a particular problem to solve, I hope that the book helps you understand the problem and decide on the right way to tackle it. And if you are just interested in the subject for its own sake, I hope you enjoy the book.

There are exercises at the end of each chapter. Some of them you can probably do in your head; for some you might need pencil and paper; and for some you will need to write a short program. Some are slightly more ambitious programming projects. A few ask you to criticise inappropriate solutions to signal processing problems suggested by a hypothetical friend who has clearly not read this book: if you have any friends in this position you can remedy the situation by buying them a copy. Please try the easier exercises and at least think about how you would go about the harder ones; and please don't take your computer into the bath.

1

Introduction

1.1 What is a signal?

A *signal* is a varying quantity whose value can be measured and which conveys information.

For example, we can consider temperature to be a signal. It can vary over time, we can measure it using a thermometer, and it conveys information: knowing the temperature outside will inform our decision as to which clothes to wear.

In a digital signal processing system we represent a signal as a sequence of numbers either on a computer or in digital hardware. For example, we could store the temperature at various times of the day as a sequence of numbers in an array on a computer: each number might be a temperature reading in Celsius.

Digital signal processing involves transforming one signal into another signal, represented digitally throughout. The transformation is achieved using simple operations on the numbers representing the signal. For example, we might want to know the average temperature over a day: we could calculate this by adding up the elements in the array of temperature data and dividing the total by the size of the array.

1.2 Domain and range of a signal

Temperature is a function of a single real-valued variable, time: see Figure 1.1. We say that the *domain* of the signal is one-dimensional. Some signals are functions of more than one variable. For example, a black-and-white photograph can be regarded as a signal: the brightness u of a point on the photograph is a function of two variables, the x and y coordinates of the point on the photograph: see Figure 1.2. In this case the domain of the signal is two-dimensional.

In a black-and-white photograph, the brightness of a point on the photograph can be represented as a single real number, and so we call it a real-valued signal, and say that the *range* of the signal is one-dimensional. In a colour photograph,

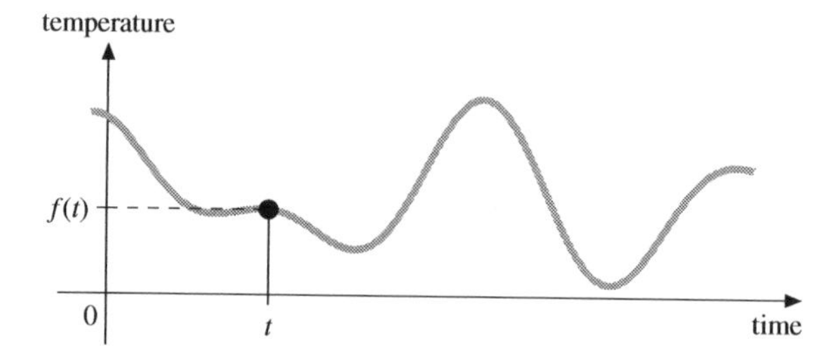

Figure 1.1 A signal with one-dimensional range and one-dimensional domain.

Figure 1.2 A signal with one-dimensional range and two-dimensional domain.

however, one real number will not do. The colour of a point can be expressed as three real numbers, separately giving the amount of red, green and blue that go to make up the colour. In this case, we would say that the range of the signal is three-dimensional.

Now let us consider a colour movie. The colour of a point on the screen with given x and y coordinates can be expressed as three real numbers for the amounts of red, green and blue as described above. However, the picture also changes with time, which adds an extra dimension to the domain of the signal. In total we therefore have three dimensions in the domain (x, y and time) and three in the range (red, green and blue).

1.3 Converting signals from one form to another

A device that converts a signal from one form to another is called a *transducer*. Often the signal on one side of the conversion will be an electrical one.

symbols. The symbols for the three main types of filter, known as 'low-pass', 'band-pass', and 'high-pass', are shown in Figure 1.5(f) to (h).

Exercises

1.1 At the beginning of the chapter we used outside temperature of an example of a signal, considering it only as a function of time. A weather forecast, however, will talk about temperature not only as a function of time, but as a function of location in the country: how many dimensions does the domain of this signal have?

An aircraft pilot whose hobby is signal processing thinks of the outside temperature as a signal. How many dimensions would you imagine he thinks its domain has?

1.2 Consider a black-and-white movie as a signal. How many dimensions do its domain and range have?

1.3 Use the following procedure to determine roughly how accurately you can represent real numbers on your computer. Set a floating-point variable `a` to 1. Set a variable `b` to $1 + 10^{-k}$ for various integer values of k, and compare `a` for equality with `b`. Find the largest value of k for which your computer reports that `a` is not equal to `b`. The machine precision is then roughly one part in 10^k.

A capacitor is manufactured using metal plates of length 1 millimetre, and the capacitance is directly proportional to this length. If the capacitor is to be made to the same precision as the numbers you can represent on your computer, what variation can we allow in the length of the metal plates? If an atom is 10^{-10} metres across, how many atoms does this correspond to?

1.4 Five hikers with varying levels of expertise at reading a compass attempt to measure the bearing of a landmark (i.e., the angle between a line pointing due north from where they are standing and a line to the landmark, measured clockwise when viewed from above). Each produces an answer in degrees from 0° to 359° inclusive. Using complex numbers, devise a procedure to take the five readings and average them to produce a more reliable result. Try it on the following sets of bearings: (a) 12°; 15°; 13°; 9°; 16°; (b) 358°; 1°; 359°; 355°; 2°; (c) 210°; 290°; 10°; 90°; 170°.

Suppose you have more confidence in the compass-reading skills of some of the hikers than others. Describe a simple extension to your procedure to give a different weight to each reading.

2
Sampling

In Chapter 1 we looked at various types of signal. We now start to examine how signals can be processed digitally.

2.1 Regular sampling

The first step is to reduce a continuous signal to a finite number of values, in a process called *sampling*. For example, if we have a signal that varies with time, such as an audio signal, then we normally measure its value at equal temporal intervals. The audio signal is then represented by these regularly spaced measurements or *samples* as shown in Figure 2.1. The sample values (+4.7, +3.3, +4.2, etc.) would typically be stored in consecutive elements of a one-dimensional array. The sampling process is similar to the way a movie camera takes a series of pictures of a scene, regularly spaced in time; or to the way a dieter weighs himself every morning.

We can sample a two-dimensional signal in a similar way. For example, a still image can be sampled spatially by overlaying it with a rectangular grid of points. A greyscale image is represented by a two-dimensional array of values giving the intensity at each sample point. Usually the value 0.0 is used to represent black, and 1.0 is used to represent white, although other scales are also used.

Note that sampling the intensity of an image at a single point is not quite the same thing as averaging it over a small area around that point, which is what typical scanners and cameras do. We shall return to this distinction in Section 3.5.

In a colour image, three numbers may be used to represent each sample point, the numbers being proportional to the amount of each of the three primary colours (red, green and blue) present in the colour at that point.

When sampling occurs at regular intervals of time, as in the case of an audio signal, we usually speak of its *sample rate*, measured in units such as samples per second, or Hertz (Hz). If sampling is spatial, as in the case of the still image, we normally talk of its *resolution*, measured in units such as pixels per millimetre or dots per inch (dpi).

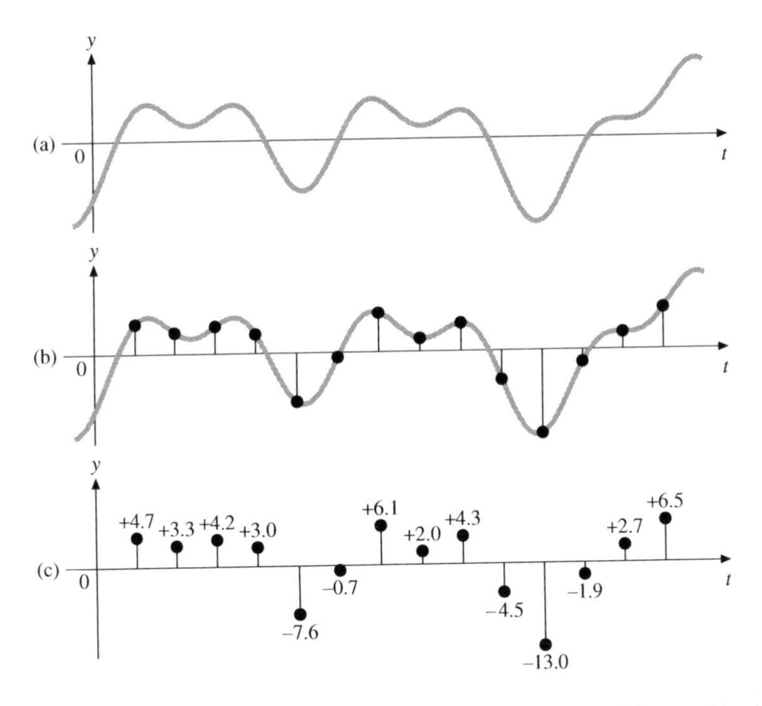

Figure 2.1 A typical audio waveform: (a) the original signal; (b) sample points on the waveform; (c) sampled representation.

The process of sampling discards some information about the signal. If an event occurs in the moment between the frames taken by a movie camera, it will not appear on the film. This can be seen in coverage of (especially outdoor) sporting events, when, because the lighting is very bright, the camera's shutter only opens very briefly for each frame. A fast-moving ball will appear to jump from position to position across the screen. Of course, the ball does not really jump; rather, the fact that the ball was between those positions was just never captured on the film.

In the same way, if an audio signal changes very quickly over time compared with the rate at which we take samples, or if an image includes features which are small compared with the sampling resolution, then we will not catch all the variations in the original. We will now make this idea more precise.

2.2 What is lost in sampling?

Figure 2.2 shows two audio waveforms superimposed. The first is a sine wave of frequency 200 Hz, and the second is a sine wave of frequency 800 Hz. The waveforms have been sampled at 1 kHz. At the points where the samples have been taken, shown by the dark dots, the two waveforms always have the same value. This means that if we are using a sample rate of 1 kHz then, given just the

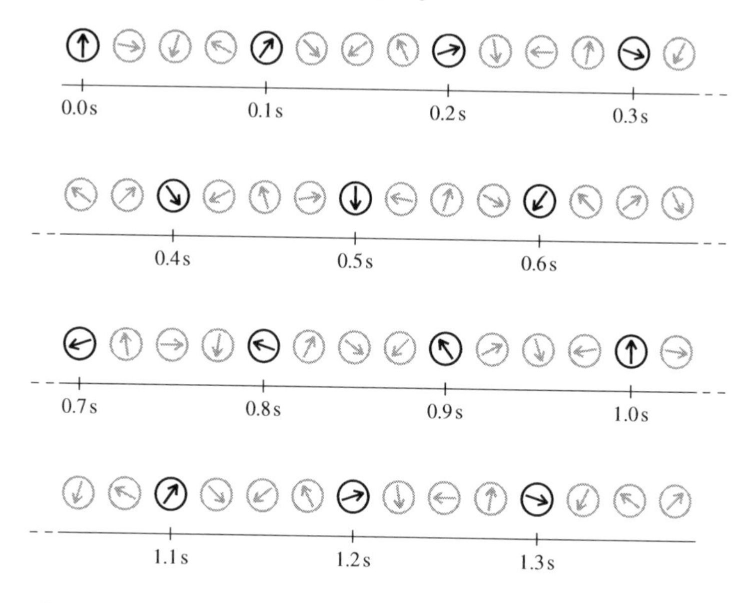

Figure 2.4 A wheel turning at 11 revolutions per second: the black images show samples taken every 0.1 s.

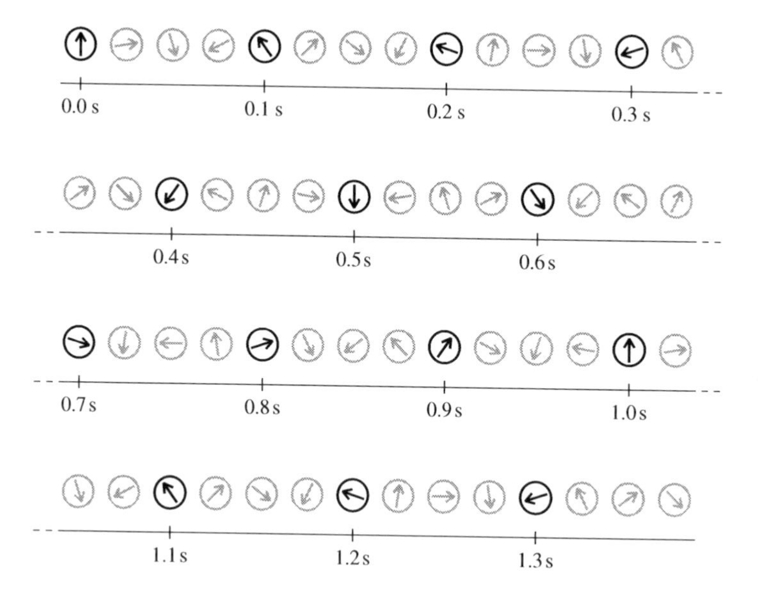

Figure 2.5 A wheel turning at 9 revolutions per second.

In these last two examples things are in fact not quite as simple as they might appear. Consider what happens when the wheel in Figure 2.4 rotates clockwise at $f = 9$ revolutions per second. How does this appear when sampled at $f_s = 10\,\text{Hz}$? As we can see from Figure 2.5, where the samples are again shown in black, the wheel again appears to rotate at one revolution per second, but this time the apparent motion is backwards.

The situation here is therefore different from what we saw in the case of sine waves above. In this case, although we are still not able to distinguish between f and $f_s + f$, sampling has not destroyed the distinction between frequencies f and $f_s - f$. We investigate this difference in more detail in the next section.

2.4 Negative frequencies

What is the difference between a movie of a swinging pendulum and one of a rotating wheel? Why is it that the first movie looks the same when played backwards, whereas the second looks different?

To understand what is happening, imagine a 'conical pendulum': a bob, suspended by a string from a fixed point, moving in a circle in a horizontal plane. The string sweeps out a cone as the bob moves. Viewed edge-on, the bob appears to move from side to side like an ordinary pendulum, and we cannot tell whether the underlying circular motion is clockwise or anticlockwise. However, if we move our point of view up or down so that we observe it at an angle, we suddenly gain extra information about its motion and can determine its direction of rotation.

In the edge-on view, we can only see one component of the motion of the bob – the left-to-right component. Viewed at an angle, we can also see the front-to-back component of the motion. Being able to see this component lets us distinguish clockwise from anticlockwise motion.

Now let us formulate this idea mathematically.

Consider a point z lying on the circle of radius 1 centred at the origin in the complex plane: see Figure 2.6. The line from z to the origin makes an angle θ with the real axis. The real part of z is $\mathrm{Re}(z) = \cos\theta$, and the imaginary part of z is $\mathrm{Im}(z) = \sin\theta$. If z moves around the circle at constant speed, then $\theta = 2\pi f t$, where f is the frequency of the motion in revolutions per unit time, and t is time, chosen so that $\theta = 0$ when $t = 0$. If f is positive, the motion is anticlockwise (the direction of increasing θ); if f is negative, the motion is clockwise.

The real and imaginary parts of z together determine its position unambiguously: if we can observe them both, then we can reconstruct the motion of z perfectly, and in particular we can tell if the motion is clockwise or anticlockwise. But what if we are looking at our conical pendulum edge-on – in other words, if we can only observe the real part of z, $\mathrm{Re}(z) = \cos\theta$? Since $\cos\theta \equiv \cos(-\theta)$, our observations will be the same if $\theta = 2\pi f t$ or if $\theta = -2\pi f t$; or, equivalently, we cannot distinguish a rotation of frequency f from one of frequency $-f$. We can determine the magnitude of the frequency of the motion, but not its sign.

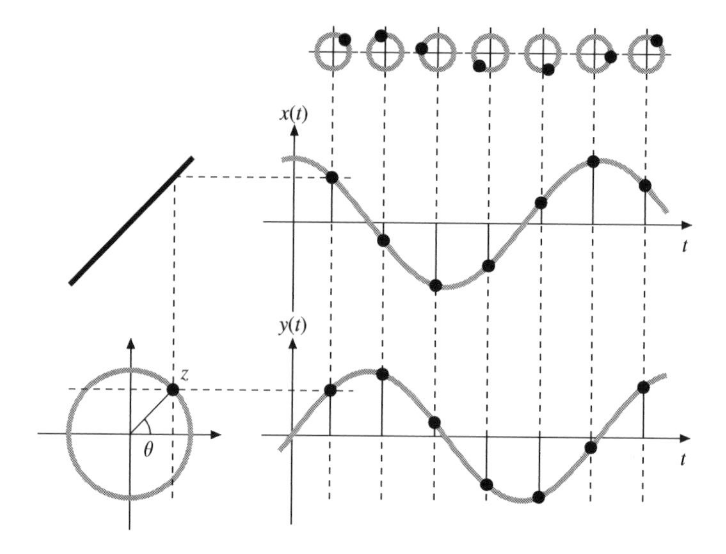

Figure 2.6 z moves in a circle in the complex plane.

If our signal is real-valued, then negative frequencies are indistinguishable from positive ones. If our signal is complex-valued, then positive and negative frequencies can be distinguished.

In the light of this, let us now revisit the phenomenon of aliasing. We saw that, for a sound waveform sampled at a frequency f_S, a sine wave at a frequency f is indistinguishable from sine waves at frequencies $kf_S \pm f$ for any integer k. Note, however, that our sound waveform samples are real-valued, which *by itself* means that we cannot distinguish between positive and negative frequencies. Comparing this with the example of the wheel of Figure 2.4 and Figure 2.5, we come to the following conclusions:

(i) Sine waves at frequencies $kf_S + f$ for any integer k, positive or negative, are indistinguishable from one another when sampled at frequency f_S.
(ii) When samples are real-valued, as in the case of an audio signal, a sine wave at frequency f is indistinguishable from one at frequency $-f$.

Exercise 2.2 shows that these conclusions are just a different way of stating those we reached at the end of Section 2.2.

2.5 The Nyquist limit

Normally a sampling rate is chosen that is fast enough to capture the quickest changes in the signal of interest. With a 1 kHz sample rate, as in our example above, we would normally work on the assumption that if a set of samples corresponds to a 200 Hz sine wave, then that is what the signal was. We would ignore the possibility that the signal might have been a sine wave at some alias frequency.

So what is the highest frequency we can work with if we have a sample rate f_s of 1 kHz? If we see real-valued samples of a sine wave with a frequency f of 499 Hz, we would recognise it as such rather than as the 501 Hz ($f_s - f$) sine wave that it might equally well have been; but if we see samples of a 501 Hz sine wave, we will mistakenly assume that the frequency was 499 Hz. The two candidate frequencies are f and $f_s - f$, and we recognise the frequency correctly when f is the smaller of these alternatives. This means that we need $f \leq f_s - f$, or equivalently $f \leq f_s/2$.

This upper bound of $f_s/2$ on the frequencies that can be processed using a sample rate of f_s is called the *Nyquist limit*.

As we noted earlier, if the samples are real-valued, we cannot distinguish negative frequencies in the range from 0 down to $-f_s/2$ from their positive counterparts in the range from 0 up to $f_s/2$. The available range of frequencies is thus $f_s/2$. However, if the samples are complex-valued, the range of distinguishable frequencies runs all the way from $-f_s/2$ to $f_s/2$, giving a total span of f_s. This is what you might expect: after all, the set of complex-valued samples contains twice as many numbers as the set of real-valued samples, and so it is reasonable that they should be able to represent twice the range of frequencies.

2.6 Irregular sampling

So far we have assumed that the sampling process is under our control, and that we will therefore be able to choose the points at which we wish to sample a signal. The obvious approach is then to sample at regular intervals, which makes analysis relatively straightforward. Indeed, most of the techniques described in this book can only be used on signals that are regularly sampled. Also, regularly spaced samples can simply be stored in an array in order: there is no need to tag each with the time (or position) at which it was taken.

Sometimes, however, we do not have direct control over the sampling process, and we are presented with samples taken at irregular intervals. We can consider two distinct cases: where there is an underlying regular sample rate, but some of the data points are missing; and where there is no such underlying regular process.

For an example of the first case, consider analysing the changes in a share price as a signal. We might have one sample of share price information per day, but there would be gaps in our data corresponding to weekends and holidays.

Another example might be a signal that has been sampled regularly, but where the sample data have been transmitted over an unreliable connection or stored on an unreliable medium, and some of the samples have been lost.

An example of the second case is land height data. Ideally, we would like to place a regular sampling grid over the area of interest and measure the height of the land at each grid point. However, this may not be feasible in practice – if a sample point happens to fall in the middle of a building, for instance – and we may have to make do with data samples taken from wherever it is possible to obtain them.

Working with irregular samples

There are three main methods available for dealing with irregular samples.

Ignore the problem In the stock market data example above, it may be satisfactory simply to ignore the missing data points and close up the gaps. This is arguably the correct approach if there were no trading on the days corresponding to the gaps. However, we would encounter difficulties if we wanted to compare sets of data which had gaps in different places, such as stock market data from exchanges that observe different holidays.

Convert to regular samples by interpolation This is the commonest approach. A regular sampling grid is laid over the irregular samples, and a value for each of the new sample points is computed from the samples that are available. As Figure 2.7 shows, the computation may be as simple as copying the value from the nearest available sample point (so-called 'zero-order interpolation'), or may involve linear ('first-order') interpolation between the two nearest points; more sophisticated techniques can also be used.

Techniques like this are used in digital cameras. It is prohibitively expensive to make image sensors that have millions of pixels without some of the pixels being faulty. Faulty pixels are identified during the manufacturing process, and their locations are programmed into the camera. The camera can then discard the samples from these locations and replace them by averages of values from adjacent (working) pixels.

Likelihood methods These are the most powerful techniques for dealing with irregular samples. They seek to determine the 'most likely' original signal that gave rise to the samples collected, whether regularly spaced or not. Of course, in order to do this, we need to have some idea – a *model* – of what constitutes a plausible signal and what does not. Likelihood methods are used in the restoration of audio from old records, where likely sounding segments of audio waveform are constructed to replace sections destroyed by scratches.

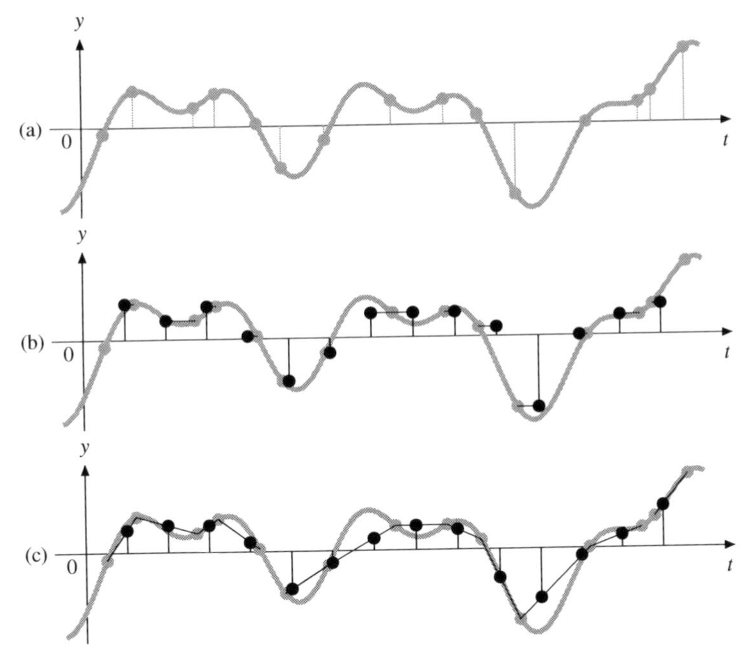

Figure 2.7 Irregularly sampled waveform: (a) original samples, in grey; (b) regular samples (black) derived by copying nearest sample; (c) regular samples (black) derived by linear interpolation.

Non-rectangular grids

A special case that often arises is where two-dimensional data are sampled on a regular, but not a rectangular, grid. Sometimes it is possible to convert the data directly into samples on a rectangular grid.

For example, an image may have been sampled by a sensor on a hexagonal grid as shown in Figure 2.8. Observe that if we could move the sample points in alternate columns (the black points in the figure) down by half a pixel then the samples would all line up horizontally, the resulting sample grid would be rectangular, and our problem would be solved. We have therefore reduced the problem to one of resampling, or interpolating, the columns of black samples, each of which is a one-dimensional, regularly sampled, signal. Techniques for doing that are described in Chapter 5.

In other cases, it is possible to ignore the fact that the sampling grid is not rectangular. In simple radar systems, an antenna revolves slowly, regularly emitting brief, tightly focussed pulses of radio waves as it does so. A pulse reflected back by an object is converted into a voltage at the antenna, now being used as a receiver: the time delay between transmitting the original pulse and receiving the reflection lets us work out the distance to the reflecting object and back. If the

Reduce the effective sampling rate to 2 kHz (or a convenient nearby value), and play back the resulting sound. You should be able to hear a high-pitched alias of the original 400 Hz signal. What is the frequency of this first alias?

Now reduce the effective sample rate to 500 Hz (or as near as you can manage). Work out what you are expecting to hear and then play the sound back. What is the frequency of the lowest alias?

Find a sampled piece of music, or, if you have the facilities to do so, sample a piece of music yourself. Make sure the original sample rate is at least 20 kHz, and repeat the above experiment of reducing the effective sample rate. The experiment works best with 'tonal' music, i.e., not music that consists only or mostly of percussion. At what sample rate does the sound become unacceptable? Repeat the experiment with a speech waveform. At what sample rate does the speech become clearly distorted? At what sample rate does the speech become unintelligible?

Note that simulating a lower sample rate by zeroing samples is not the same as telling the computer to use a different sample rate. Computer sound systems usually contain 'anti-aliasing filters' which remove frequencies above the Nyquist limit for the actual sample rate being used, and so you will probably not be able to produce an audible alias by simply using a different sample rate directly.

2.4 The highest frequency the human ear can hear is about 20 kHz: it varies from person to person and reduces with age. What is the minimum sample rate we can use to record a sound waveform so that it is indistinguishable from the original when played back? How does this compare with the sample rate used on compact discs, which is 44.1 kHz? With that used for NICAM (near instantaneous companded audio multiplex) stereo television sound, which is 32 kHz?

2.5 A Western includes a shot of a wagon whose wheels have circumference 4 m and which have eight equally spaced spokes, being otherwise perfectly circularly symmetrical. The movie is shot at 24 frames per second, and although the wagon is moving forwards, its wheels appear to be turning backwards at 1 revolution per second. What are the possible speeds at which the wagon may be moving? (**Hint:** Why do you need to know the number of spokes?)

2.6 A tuning fork whose pitch is known to lie between 262 Hz and 524 Hz (the octave above 'middle C' on a musical instrument) is struck and held in front of a bright CRT (cathode ray tube) monitor whose refresh rate is 75 Hz. The fork appears to oscillate at 10 Hz. What are the possible pitches of the tuning fork?

2.7 This exercise looks at the effectiveness of a couple of simple techniques for concealing faulty pixels in digital camera sensors. You will need to find out how to manipulate and display a photographic image on the computer.

Obtain a photographic image and load it into an array. Try to select something with a fair amount of detail: buildings are better than seascapes. The picture should ideally have a resolution of about 250 pixels by 250 pixels, but this is not particularly critical, and the image does not need to be square. You will find it simpler to work with a greyscale image, but you can use a colour one if you wish.

Set 1% of the pixels in the image, randomly selected, to black. Compare with the original image, preferably side-by-side, at a normal viewing distance. Is the image noticeably degraded? Adjust the percentage of affected pixels up or down to determine the point at which the effect just becomes noticeable.

Method (a) Return to the original image, and select 1% of the pixels at random. Set each of these pixels to the same value as its immediate neighbour to the left. (For simplicity, make sure your 'random selection' doesn't include any pixels in the leftmost column.) Is the image noticeably degraded when compared with the original? Again, experiment with adjusting the percentage of pixels you modify up or down to determine the point at which the effect just becomes noticeable.

Method (b) Return to the original image, and again select 1% of the pixels at random. Set each of the selected pixels to the average value of its four nearest neighbours: the pixels immediately above it, to its left, to its right and below it. (Again, make sure your 'random selection' doesn't include any pixels from any of the edges of the image.) If you are working with a colour image, you will have to ensure that you calculate the averages of the red, green and blue components of each pixel separately. Is the image noticeably degraded when compared with the original? Once more, adjust the percentage up or down to determine the point at which the effect just becomes noticeable.

Which of the two methods, (a) and (b), is more effective at concealing faulty pixels?

2.8 This exercise looks at *sub-Nyquist sampling*, an approach to dealing with signals that occupy only a narrow range of frequencies. Although this exercise deals with audio frequencies (so that the numbers are easy to deal with), the techniques are most relevant in radio applications.

(a) Suppose you have an audio signal that you know is definitely a sine wave with a frequency between 120 Hz and 130 Hz. Armed with this knowledge, would you be able to reconstruct the original signal from real-valued samples taken at 300 Hz? At 200 Hz? At 100 Hz? At 30 Hz? What is the lowest sample rate you could use and still be able to reconstruct the signal?

(b) In (a), what would happen if you tried to use a sample rate of 25 Hz?

(c) Now suppose you know that the audio signal is a sine wave with a frequency that might be in one of two ranges, either between 120 Hz and 130 Hz or between 150 Hz and 170 Hz. What is the lowest sample rate you could use, and still be able to reconstruct the signal from (real-valued) samples? What is the answer if the second band extends from 150 Hz to 172 Hz?

3

Conversion between analogue and digital

In Chapter 2 we looked at the effect of representing a signal by a series of samples. In this chapter we will look at how we can represent the individual sample values digitally.

3.1 A simple digital signal processing system

Imagine a digital system designed to process sound as shown in Figure 3.1.

The incoming sound (which is just a variation in air pressure) is converted into a voltage using a microphone. This voltage is then sampled at regular intervals and measured using a device called an *analogue-to-digital converter*, or *A/D converter* or *ADC* for short. The output from the analogue-to-digital converter is a sequence of numbers which represents the original sound pressure at the times the samples were taken.

The sequence of numbers is then processed using some algorithm to produce a new sequence of numbers. This new sequence is passed to a *digital-to-analogue converter* (or *D/A converter* or *DAC*), whose output is a varying voltage corresponding to the sequence of numbers. This voltage can now be converted back into a sound using a loudspeaker.

Unless we know in advance that the highest sound frequency will be less than half the sample rate (see Section 2.5 on the Nyquist limit), an *anti-aliasing filter* needs to be included between the microphone and the analogue-to-digital converter. This filter removes any frequencies above the Nyquist limit. Similarly, a filter has to be included between the output of the digital-to-analogue converter and the loudspeaker to ensure that the correct continuous waveform is reconstructed from the samples: we want the one that contains only frequencies below the Nyquist limit, rather than any of the higher frequencies consistent with the same sequence of samples. The output filter is sometimes called a *reconstruction filter*.

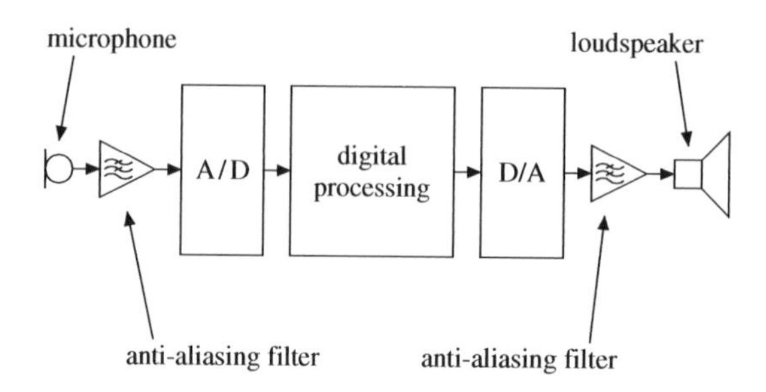

Figure 3.1 A simple digital signal processing system.

The input and output filters work on analogue signals. We will not discuss their design in detail here. For now we shall work on the assumption that they do a perfect job of removing aliases; in Chapter 5 we will see how to characterise filters more accurately.

Linear quantisation

A continuous analogue quantity such as a voltage can represent an infinite number of different sample values. However, the output of an analogue-to-digital converter, being digital, can only take on one of a finite number of different values. The voltage will normally be rounded, or *quantised*, to the nearest representable value. Figure 3.2 illustrates the commonest case, *linear quantisation*. Here the representable values are equally spaced over the range of possible sample values.

Figure 3.3 shows how the output code of an ideal eight-bit analogue-to-digital converter depends on its input voltage V_i. This plot is called the *transfer function* of the converter. An eight-bit binary number can take on 2^8 different values, and here we have assumed that the quantisation levels are at intervals of 0.01 V ranging from 0 V to 2.55 V.

Figure 3.4 shows the effect of linear quantisation on (a) a large-amplitude sine wave and (b) a small-amplitude sine wave. As you can see, the shape of the large-amplitude sine wave is reasonably well preserved, while the small-amplitude

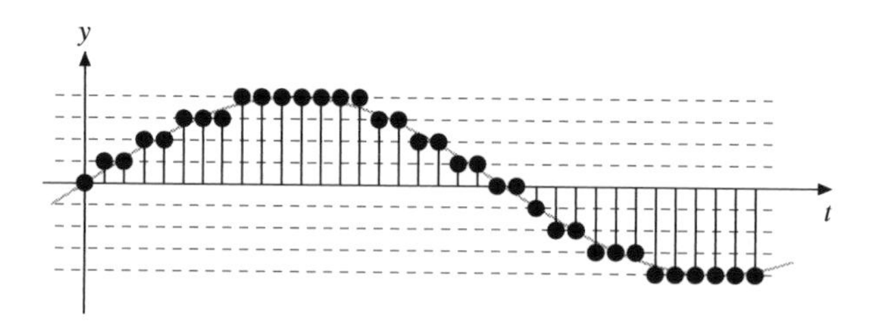

Figure 3.2 Linear quantisation.

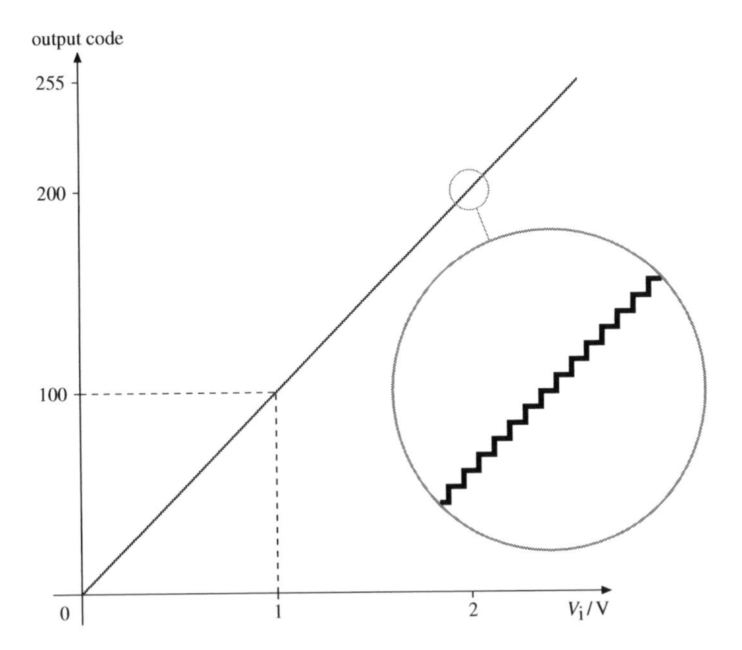

Figure 3.3 Transfer function of an ideal 8-bit linear analogue-to-digital converter.

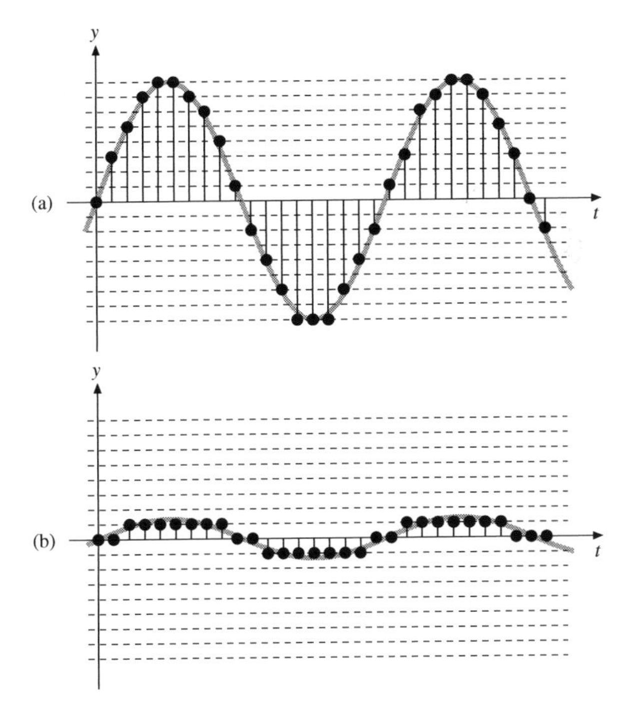

Figure 3.4 Effect of linear quantisation on (a) high-amplitude, and (b) low-amplitude signals.

sine wave has been distorted to the extent that it is barely distinguishable from a square wave. The human ear is very sensitive to distortion of this kind.

3.2 Non-linear quantisation

It is not essential to have the representable values equally spaced over the range of possible sample values. In audio applications *logarithmic quantisation* is very common.

Figure 3.5 shows the dependence of output code on input voltage for an eight-bit analogue-to-digital converter which operates according to the 'A-law' quantisation scheme used for digitising telephone audio in Europe. (The United States and Japan use a scheme called 'mu-law' which differs only in minor details.) First, the scheme is symmetrical with respect to positive and negative values: one bit of the converter's output represents the sign of the input voltage. The remaining seven bits represent the magnitude of the voltage. The codes are more densely packed at lower input amplitudes than at higher ones, which means that quiet sounds can be reproduced with good fidelity, although at the expense of a slight reduction in fidelity for louder sounds. Figure 3.6 shows the effect of an exaggerated form of A-law quantisation on a large-amplitude and a small-amplitude sine wave: the waveform shape is preserved equally well in the two cases. Compare this with Figure 3.4.

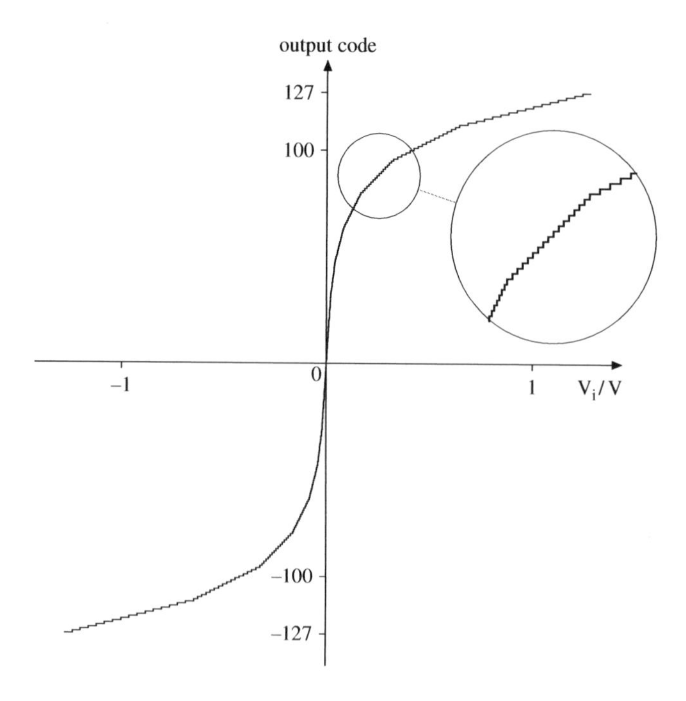

Figure 3.5 A-law transfer function.

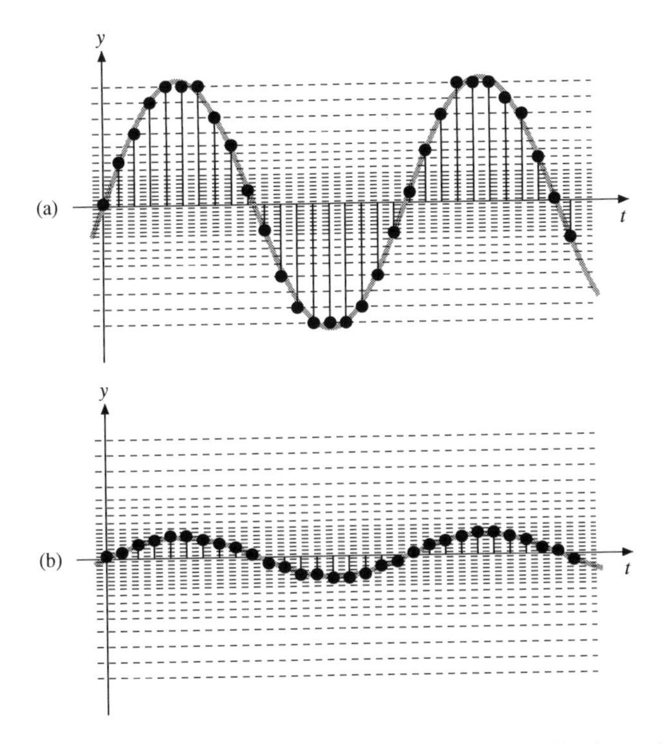

Figure 3.6 Effect of non-linear quantisation on (a) high-amplitude, and (b) low-amplitude signals.

In practice it is difficult to build analogue-to-digital converters with non-linear quantisation. A-law quantisation is usually done using a linear converter (a 13-bit converter is required in this case) followed by a further quantisation to the A-law levels in digital hardware or in software. Most of the signal processing techniques we will be describing in this book are also more easily applied to linearly quantised than to logarithmically quantised data. For this reason, logarithmic quantisation is usually postponed until immediately before a signal is to be stored or transmitted.

The logarithmic mapping of sample values to a smaller set of values is called *compression* and the device that does it a *compressor*. A device that undoes the mapping is called, naturally enough, an *expander*, and the system as a whole is called a *compander*. (Note that the term 'compression' is also used more generally to describe a process where a data file is reduced in size by exploiting patterns within it; in this case, the reverse process is called *decompression*.)

3.3 How many bits do we need?

It is clear from the example in Figure 3.4 that quantisation introduces distortion into a signal, or, equivalently, discards information about it. How precisely do we need to represent the samples in the digital part of our system (see Figure 3.1) to ensure that the signal is not distorted unacceptably? Of course, this depends on what you decide is 'unacceptable'; the exercises at the end of the chapter invite you to investigate for yourself. As a guide, compact discs are recorded using linear quantisation with 16 bits per sample, which is enough for audio in all but the most demanding applications (in a recording studio, 24 bits might be used). In telephony, eight bits per sample are normally used with logarithmic quantisation, giving the same range as a 13-bit (in the case of A-law quantisation) or 14-bit (mu-law) linear conversion would.

Consumer video and imaging devices use typically 8 or 10 bits per sample in each of the red, green and blue channels, which is enough for ordinary digitised images.

In some cases, such as in medical imaging applications, rather finer quantisation is used. An X-ray might be digitised using 16-bit quantisation so that it can subsequently be enhanced by artificially increasing the contrast between nearby shades in the image, as we shall describe in Chapter 9. If coarser quantisation had been used, the distinction between these shades might have been irretrievably lost in the digitisation process.

It is important to ensure that the input range of an analogue-to-digital converter is well matched to the signal level that is expected. If excessive range is provided, the situation is similar to that illustrated in Figure 3.4(b), and much of the precision of the converter is lost; if insufficient range is allowed, the converter will normally 'clip' the signal to fit within its range. We will look at the undesirable effects of clipping in Chapter 7.

Conversely, if the range of the input signal is not accurately known in advance, extra precision will have to be provided in the analogue-to-digital converter and subsequent processing.

3.4 Dither

If you look closely at a greyscale photograph printed in a newspaper (you might need a magnifying glass) you will see that areas of what appear to be grey in the image are actually made up of small black dots. Darker areas of grey are made up of bigger dots with smaller white spaces in between; lighter areas consist of smaller black dots with larger white spaces. The printing process is only capable of output with one bit of quantisation – either there is ink on the page at a particular point or there is not – and yet the illusion of a continuous range of tones is created. A similar effect can be seen in the output of an ordinary inkjet printer, especially when rendering colour photographs. The printer can exert only coarse control over the amount of ink it deposits to make a dot on the page, but can exert very fine control over where it puts the dots. The dot patterns produced by inkjet printers tend to be more intricate than those used in printing newspapers, and so you may need a more powerful magnifying glass to see them.

In signal processing, this technique is called *dither*. The idea is to use coarsely quantised samples at a high resolution or sample rate to represent more finely quantised samples at a lower resolution or sample rate.

A simple way to dither a signal is to add random numbers (or *noise*) to the sample values before quantisation.

Figure 3.7(a) shows an image to which one-bit quantisation has been applied: greys darker than a certain shade are rendered as black, those lighter than that shade appear as white. Figure 3.7(b) shows the same image where noise has been added before quantisation. Without the addition of noise, a light grey sample value has zero chance of producing a black pixel on the page and so a large area of light grey appears completely white. With the addition of noise, each pixel has a small chance of producing a black pixel, and so a light grey area will appear as a sparse random pattern of black dots with an appropriate overall density. As you can see – particularly if you hold the page at a distance – the visual effect is considerably better. An exercise at the end of the chapter illustrates the effect of dither on audio signals.

(a)

(b)

Figure 3.7 Image (a) quantised to one bit; (b) quantised to one bit after the addition of noise (hold the page at a distance for best effect).

We shall look at more sophisticated dithering techniques and their application to audio in Section 8.6.

3.5 Non-ideal conversion

Even when processing a perfectly clean signal, real-world analogue-to-digital conversion systems suffer from two significant sources of error.

Aperture

Ideally, the analogue-to-digital converter shown in Figure 3.1 would take an instantaneous snapshot of the level at its input and turn that value into a number. In practice, however, the process is not instantaneous; a reasonable approximation is to assume that the converter averages the input over a short period, called the *aperture* or *aperture time*. If the aperture time approaches the time between samples (in other words, if the converter effectively averages over most of a sample period rather than making a point measurement) and the input signal has a high frequency, the accuracy of the conversion results can be impaired: Figure 3.8 shows an example of a sine wave of amplitude 1 being sampled with averaging over most of the sample time, as indicated by the shading in the upper plots. For sine wave frequencies near the Nyquist limit, as in Figure 3.8(b), the amplitude of the sine wave represented by the samples is somewhat lower than the amplitude

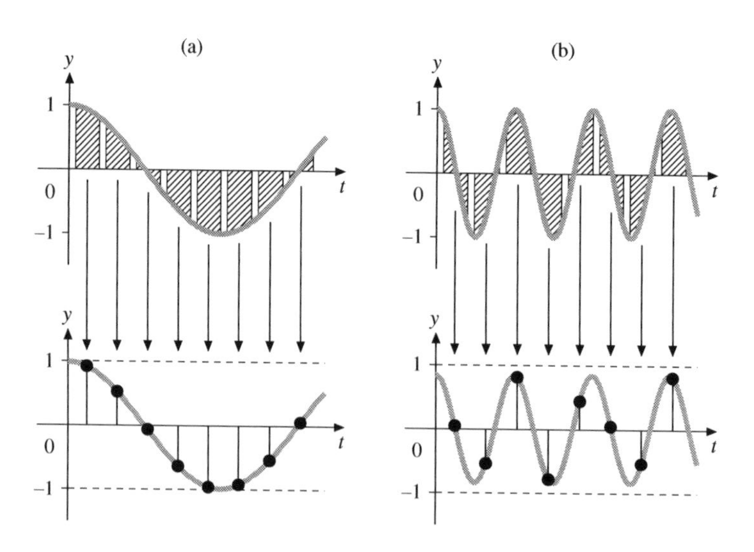

Figure 3.8 Effect of aperture on (a) low-frequency, and (b) high-frequency signals; shaded regions indicate aperture time over which signal is averaged.

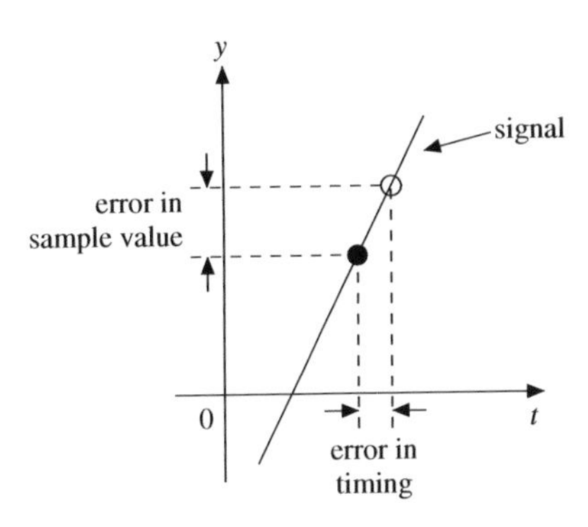

Figure 3.9 Effect of timing jitter: filled circle is ideal sample, open circle is sample affected by jitter.

of the original. The reduction in amplitude is frequency-dependent; it is possible to use the techniques we describe in Chapter 5 to characterise the effect precisely.

Jitter

The second main cause of error arises when samples are not taken at perfectly regular intervals. If the signal is changing rapidly, even a tiny error in timing can lead to a large error in the measured sample: see Figure 3.9. Jitter can affect digital-to-analogue converters in a similar way.

In systems where the conversion timing is controlled by hardware, jitter is usually negligible. In embedded systems, however, where conversions are initiated under software control – in an interrupt service routine for example – jitter can be a serious problem.

Exercises

3.1 This exercise illustrates linear quantisation of audio signals.

Record a vector containing several seconds of speech using a sample rate of at least 32 kHz and 16-bit samples. Select a sentence with several sibilant sounds such as ‘s’ or ‘sh’: ‘she sells sea shells on the sea shore’ or ‘the sixth sheik’s sixth sheep’s sick’. Play it back and verify that the sound is clean; make a new recording if necessary. Examine the data and find the maximum and minimum sample values: call these $s_{\max}$ and $s_{\min}$ respectively. Calculate $r = s_{\max} - s_{\min}$, the range of values.

Now we will simulate the effect of linear quantisation on this waveform.

To simulate quantisation to 512 different levels (i.e., 9-bit quantisation, because $2^9 = 512$) proceed as follows:

Step 1 Multiply all the values in the vector by $512/r$: this scales the samples to make the overall range equal to 512. **Step 2** Take the integer part of each scaled sample. (The function you need might be called something like `int` or `floor`.) **Step 3** Multiply all the values in this new vector by $r/512$. This restores the original range, so the volume of the sound will be about the same.

Play back the new vector. Is the sound acceptable? Repeat the experiment with five-bit quantisation (use 32 instead of 512 in the instructions above), three-bit quantisation (use 8 instead of 512) and one-bit quantisation (2 instead of 512). Make sure you start from the original recording each time. Which sounds are worst affected by quantisation? How many bits of quantisation result in what you would consider to be 'telephone quality' speech? Is the speech still intelligible after one-bit quantisation?

Repeat the experiment using music instead of speech. If you can, use a segment that includes both loud and quiet parts. At how many bits of quantisation does distortion become noticeable?

Save the quantised vectors that you have produced for comparison with the results of later exercises.

3.2 This exercise illustrates logarithmic quantisation of audio signals. We will use the same recorded speech vector as in the previous exercise.

Mu-law quantisation converts samples which range in value from -8191 to $+8191$ into eight-bit values. The sign of the sample is stored in one bit of the result: 1 for positive and 0 for negative. From now on we will only consider the magnitude of the sample value.

The remaining seven bits are divided into a three-bit 'chord' number c, $0 \leq c \leq 7$, and a four-bit 'step' number s, $0 \leq s \leq 15$. The quantisation level associated with given values of c and s is $(2s+33) \cdot 2^c - 33$.

Write a program to convert an eight-bit number into the corresponding mu-law quantisation level. Obtain the sign bit from the most-significant bit (bit 7) of the input, the chord number from the next most significant bits (bits 6 down to 4) and the step number from the four least significant bits of the input (bits 3 down to 0). Generate a table of the 256 quantisation levels.

Now write a program that will take a vector as input and for each element compare it against each entry in your table to find the nearest one. Generate a quantised vector as output where each element of the input has been replaced by the nearest quantisation level. (It doesn't make much difference which alternative you choose when two levels are equally close as long as you are consistent.)

Run your program on the speech vector and play back the output. Is it distinguishable from the original? (You will need to ensure that the range of sample values in the speech vector is compatible with the range of values in the quantisation table. You may have to multiply each element of the speech vector by a scale factor, and possibly also add or subtract an offset.)

Describe a faster algorithm for mu-law quantisation than the exhaustive table search method you have just implemented.

3.3 Invent a 5-bit logarithmic quantisation scheme along the lines of the eight-bit mu-law system described in Exercise 3.2. Try it on the speech vector. Is the quality as good as 9-bit linear quantisation? Is it better than 5-bit linear quantisation? Is 3-bit logarithmic quantisation a reasonable thing to try?

3.4 Obtain a digitised black-and-white still image. Experiment with five-bit linear and logarithmic quantisation of the sample values. Is linear or logarithmic quantisation more suitable for use with images?

3.5 Modify the procedure of Exercise 3.1 by inserting the following between steps 1 and 2:

Step 1A Add a random real number in the range 0 to 1 to each element of the vector. Make sure you generate a new random number for each vector element.

Repeat the exercise using three- and one-bit quantisation. You should be able to hear the noise you have added as a background 'hiss'; is the sound itself less distorted overall than when you used the same level of quantisation but without adding noise?

3.6 A friend is planning to use an analogue-to-digital converter to sample a 100 MHz sine wave using a sample rate of 30 MHz ('sub-Nyquist sampling' – see Exercise 2.8). The converter has an aperture time of 10 ns. Why is his plan probably not a good one?

3.7 Generate a vector $x(t)$ containing several seconds of a 1 kHz sine wave sampled at 8 kHz using the formula $x = \sin 2\pi t/8$ with suitable scaling; play it back and check that it sounds clean. (If you cannot use an 8 kHz sample rate choose a suitable nearby alternative value, but use the same formula.)

Now replace every third sample in your vector by a sample computed using the formula $x = \sin 2\pi(t+0.1)/8$. This simulates a jitter of 10 % of the sample period. Listen to the result: you should hear that the original signal has been slightly distorted. Adjust the amount of jitter and determine the minimum amount of jitter that leads to perceptible distortion.

3.8 A CD stores 16-bit linearly quantised samples with a sample rate of 44.1 kHz. Suppose that the recorded signal is a 20 kHz sine wave with maximum possible amplitude. How much jitter (in nanoseconds) can be tolerated before the resulting error is comparable to an error in the least-significant bit of a sample? (**Hint:** The worst case is when the signal goes through zero.) What is the answer for a 200 Hz sine wave?

If the samples read from the disc were passed directly to the digital-to-analogue converter in the player, it would be very difficult indeed to control the rotation of the disc accurately enough to keep the jitter below this value. What would you include in the design of a CD player to get around this problem?

4

The frequency domain

This chapter describes how to express a signal as a sum of cosine and sine functions. Such an expression is called a *frequency domain* representation, the original form of the signal being called its *time domain* representation. The process which converts a signal from its time domain representation to its frequency domain representation is called a *Fourier transform*.

We will initially only consider *periodic* signals, i.e., those that consist of a short segment that is repeated over and over again at a fixed rate. Later we will see how the ideas can be applied to non-periodic signals.

4.1 Measuring rotational speed

Imagine a movie showing a point moving in a circle of radius 1 about the origin in the x-y plane, making three anticlockwise revolutions per second. Figure 4.1 shows the first sixteen frames of this movie, taken at sixteen frames per second. Because the point completes a whole number of revolutions in one second, these sixteen frames will simply repeat forever. We can think of the positions of the point in these frames as samples of a periodic signal with a two-dimensional range. Alternatively, thinking of the point as lying in the complex plane, we have a periodic sequence of complex-valued samples.

If we were given the sixteen pairs of x and y coordinates of the point, one pair corresponding to each of these frames – i.e., the samples of this signal – how would we go about writing a program to determine the speed of rotation of the point? A simple approach would be to count the number of times the point goes upwards past the x axis (for example) in one second; however, we are going to try a slightly less obvious algorithm which will lead to some useful results.

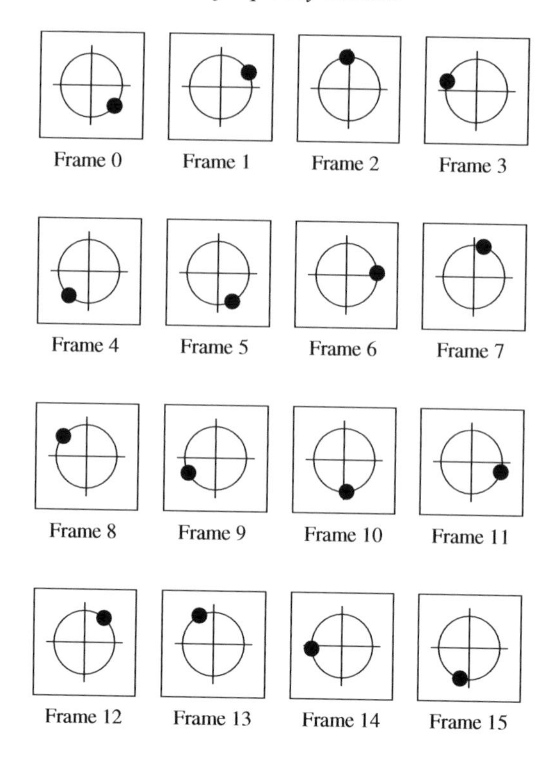

Figure 4.1 Sixteen frames from a movie of a point moving in a circle.

Figure 4.2(a) shows the same frames as Figure 4.1, but stacked along the time axis. The spiral line joins the point in one frame to the point in the next: as you can see, the spiral twists a total of three times around the time axis. This suggests our algorithm: suppose we try to untwist the spiral by some number of turns f. If the result is straight, we have exactly undone the motion of the point, and hence its rotation speed is f; if the result is still twisted, we have not found the correct speed. Figure 4.2(b) to (e) show the result of untwisting the motion of the point by $f = 1$, 2, 3 and 4 rotations per second. In Figure 4.2(d) the result is straight, and so the answer is that the rotation rate is three revolutions per second.

The only step we have not explained in the paragraph above is precisely how we determine whether we have succeeded in straightening the spiral out. Look at Figure 4.3(a) to (e): these correspond to Figure 4.2(a) to (e) and show in each case the sixteen frames superimposed. If you prefer, you can think of these as pictures of the point taken by a rotating camera. In all cases except Figure 4.3(d) the resulting points are uniformly spread around a circle, and hence their *average* x and y coordinates (or, equivalently, the x and y coordinates of the centre of gravity of the points) are both zero. In the case of Figure 4.3(d) the point remains fixed away from the origin, and so its average position is not zero.

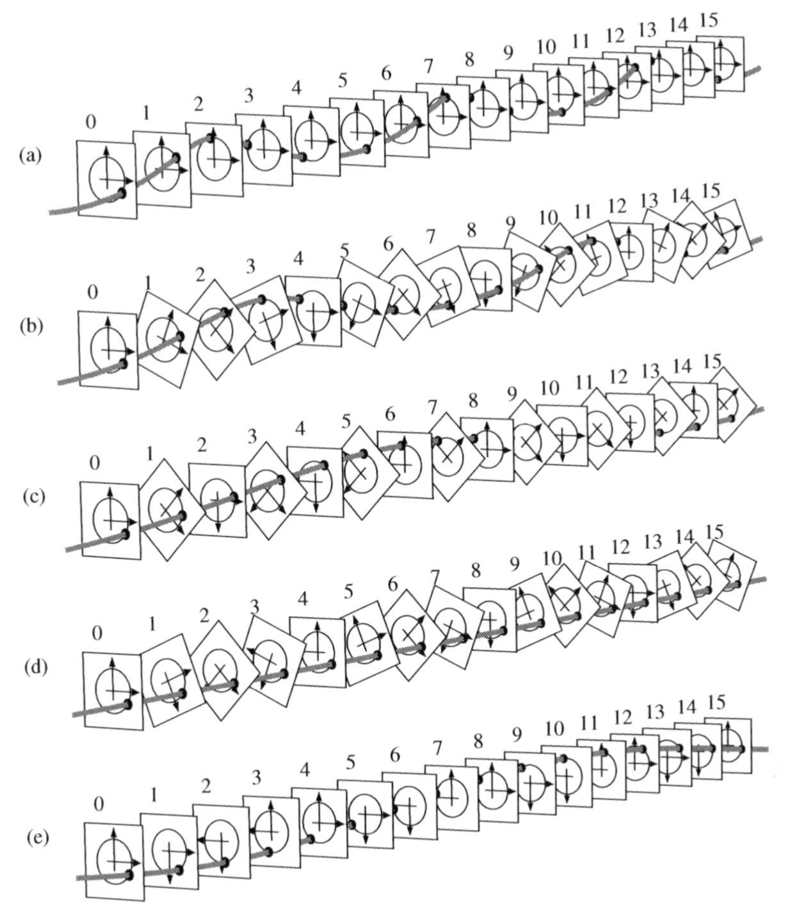

Figure 4.2 The frames of the movie stacked in time (a) with no untwisting; (b) with one untwisting rotation added over the sixteen frames; (c) with two rotations; (d) with three rotations; (e) with four rotations.

4.2 More complicated motion

Now imagine a point tracing out the more complicated path shown in Figure 4.4, making one complete circuit per second. It looks like the motion is the sum of two circular motions, one large overall motion at one revolution per second with a smaller, faster motion added to it – like the way the moon moves around the earth as the earth orbits the sun.

What happens if we apply our untwisting algorithm to this motion? Figure 4.5 shows the results obtained by untwisting at rates from $f = 0$ to $f = 11$ rotations per second, with the average position of the point in each case marked by a cross. The cross is at the origin except where $f = 1$ and $f = 7$: our algorithm is telling us

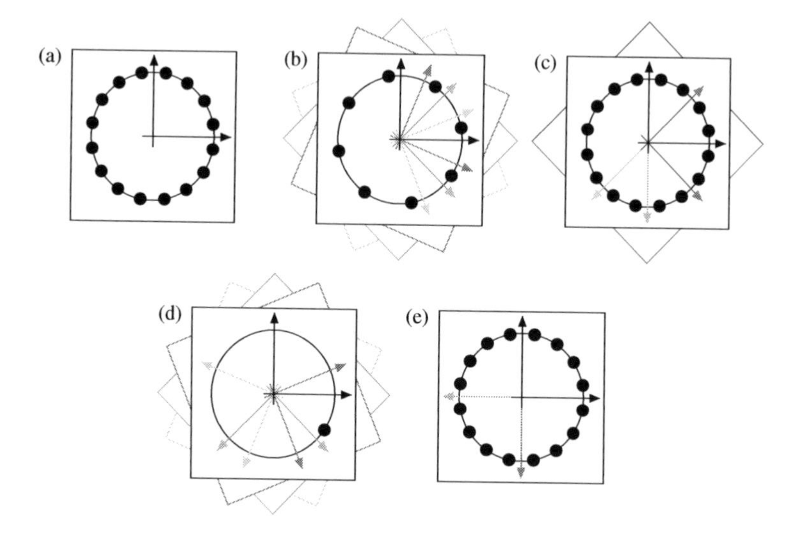

Figure 4.3 The frames of the movie superimposed with (a) zero; (b) one; (c) two; (d) three; and (e) four rotations of untwisting applied.

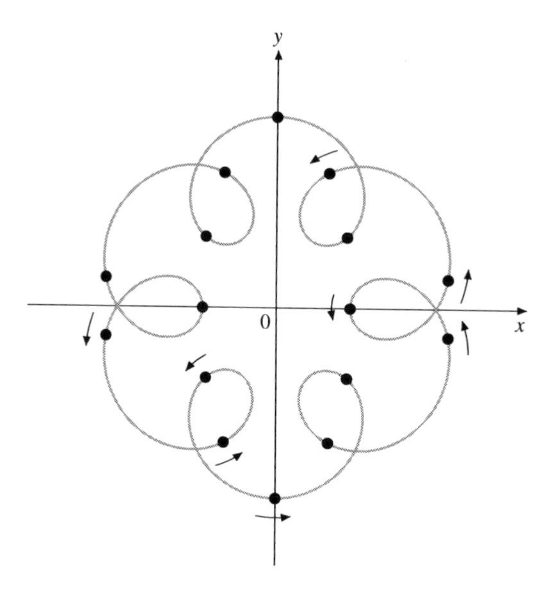

Figure 4.4 The sum of two circular motions.

that there are circular motions at one revolution per second and seven revolutions per second.

Our algorithm has thus managed to analyse a compound motion into its constituent parts, something we could not have done with the naïve method of counting the crossings of the x axis.

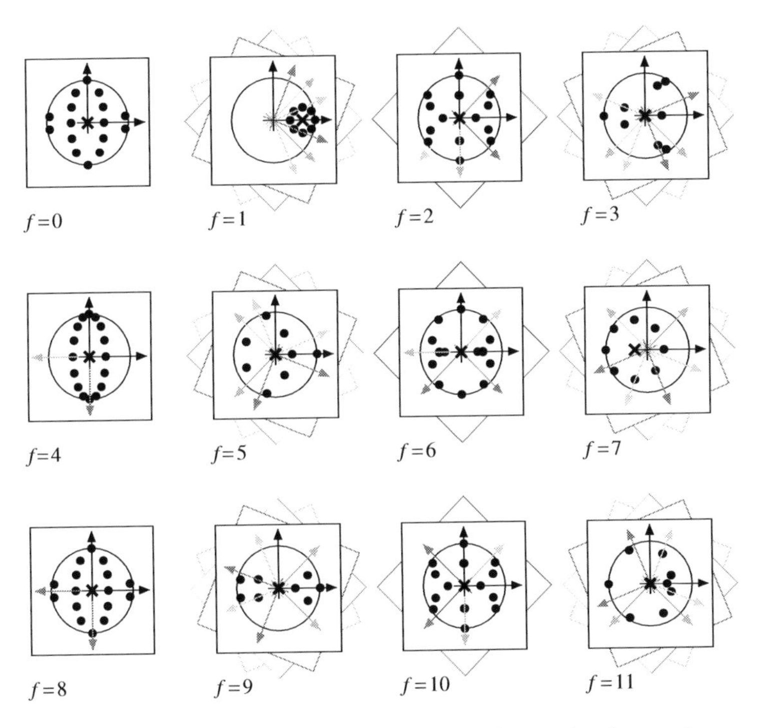

Figure 4.5 Untwisting applied to the sum of two circular motions.

Let us now write our algorithm down in mathematical terms. If we have a point with coordinates (x, y) and rotate it anticlockwise through an angle θ, the new coordinates (x', y') are given by

$$x' = x\cos\theta - y\sin\theta$$

and

$$y' = x\sin\theta + y\cos\theta.$$

If we are untwisting at f revolutions per second and we are sampling the position of the point N times per second then each frame is rotated by an angle of $-2\pi f/N$ more than the previous one, and so the kth frame is rotated by an angle of $\theta = -2\pi fk/N$: the minus signs are there because we are *un*twisting. Writing $x(k)$ and $y(k)$ for the coordinates of the point in the kth frame, the coordinates of the rotated point are

$$x'(k) = x(k)\cos(-2\pi fk/N) - y(k)\sin(-2\pi fk/N)$$

and

$$y'(k) = x(k)\sin(-2\pi fk/N) + y(k)\cos(-2\pi fk/N).$$

Now we can average these coordinates over a set of N samples, obtaining average coordinates X and Y, by just adding them up and dividing by N. Since N is constant this last division only contributes a fixed overall scaling factor to the final answer, and so we shall ignore it. The results depend on f, so we write them $X(f)$ and $Y(f)$ accordingly.

$$X(f) = \sum_{k=0}^{N-1} x(k)\cos(-2\pi fk/N) - y(k)\sin(-2\pi fk/N)$$

$$Y(f) = \sum_{k=0}^{N-1} x(k)\sin(-2\pi fk/N) + y(k)\cos(-2\pi fk/N).$$

We can also write these expressions using complex numbers. A complex number $z = x + \mathrm{j}y$ can be rotated through an angle θ about the origin in the complex plane by multiplying it by $\mathrm{e}^{\mathrm{j}\theta}$: Exercise 4.2 invites you to check this. The rotated version of $z(k)$ is thus $z'(k) = z(k)\mathrm{e}^{\mathrm{j}\theta}$. Summing over a set of N samples as before, we obtain

$$Z(f) = \sum_{k=0}^{N-1} z(k)\mathrm{e}^{\mathrm{j}\theta},$$

or, substituting for θ,

$$Z(f) = \sum_{k=0}^{N-1} z(k)\mathrm{e}^{-2\pi\mathrm{j}fk/N}. \tag{4.1}$$

Each $Z(f)$ value is called a *Fourier coefficient* (or, slightly loosely, a *frequency component* or *Fourier component* of z) at frequency f and the collection of $Z(f)$ values is called the *Fourier transform* of z. More precisely, it is called the *discrete* Fourier transform, because we have only analysed a sequence of sampled (or 'discrete') values of $z(k)$ at a discrete set of frequencies f; there is also a continuous version of the Fourier transform that uses integrals rather than sums, but we shall not need it in this book. The process of calculating the Fourier coefficients is called *Fourier analysis.*

In the example of Figure 4.5, $Z(1) = +0.7$, $Z(7) = -0.3$, and $Z(0)$, $Z(2)$ and all the other Fourier coefficients are zero. The Fourier transform of the signal is thus $(0, +0.7, 0, 0, 0, 0, 0, -0.3, 0, 0, \ldots)$.

4.3 Interpreting the Fourier transform

In the simple case where we are just dealing with motion in a circle, the Fourier coefficients $Z(f)$ will be zero except where f is the actual frequency of the rotation. In that case, after untwisting, all the sample points end up in the same place, and that place is the location of the point at the time of sample number zero. In terms of complex numbers, all the $z'(k)$ are equal to $z(0)$ and so $Z(f) = Nz(0)$.

Since $Z(f)$ is the sum of contributions from N points, the distance of that point from the origin (or, in terms of complex numbers, the *magnitude* of $Z(f)$) is just N times the radius of the circle in which the point moves. The magnitude of $Z(f)$ is called the *amplitude* of that Fourier coefficient. The angle between the line joining the point to the origin and the real axis (the *argument* of $Z(f)$) measures the time delay between this motion and another motion of the same speed where sample number zero is on the positive real axis, i.e., at the three o'clock position; this angle is called the *phase* of the Fourier coefficient.

In the more complicated case where we have a compound motion more than one of the Fourier coefficients will be non-zero. We still talk about the amplitudes and phases of the various Fourier coefficients. The Fourier transform expresses any signal as a sum of circular motions.

When $f = 0$ all the terms $\mathrm{e}^{-2\pi \mathrm{j} fk/N}$ are equal to 1, and so $Z(0)$ is just the sum of the samples in one period of the signal, or N times the average value of the signal. From an analogy with electrical circuits, $Z(0)$ is called the *DC component*, 'DC' standing for 'direct current'.

4.4 How many Fourier coefficients are there?

We get a Fourier coefficient $Z(f)$ corresponding to each rate $f = 0, 1, 2, \ldots$ at which we try to untwist the signal. In the expression for $Z(f)$ the twist $\mathrm{e}^{\mathrm{j}\theta}$ is only evaluated at the points where we have samples of the original signal; in other words, the twist function itself is sampled. It will therefore exhibit aliasing, as described in Chapter 2. Since there are N samples in one period the sampled twist function at rate f will be identical to that at rate $f+N$, $f+2N$, and so on. If the sampled twist function is the same then the Fourier coefficient will be the same, and so $Z(f) = Z(f+N) = Z(f+2N) = \ldots$. The coefficients from $Z(0)$ to $Z(N-1)$ give us all the information there is to be had about the Fourier transform, and the Fourier transform is periodic with period N, just like the signal.

The Fourier transform thus takes a set of N samples of a complex signal and converts it into N complex-valued Fourier coefficients. Furthermore, the transform does not lose any information: as Exercise 4.14 invites you to show, different sampled signals have different transforms. The Fourier transform represents the same information as the samples of the signal, but in a different way.

4.5 Reconstructing a signal from its Fourier transform

The Fourier transform analyses a signal into a sum of Fourier components, each of which corresponds to a series of samples of a point moving in a circle centred on the origin in the complex plane. If the circular motion had radius 1 and started with sample 0 on the positive real axis, its samples would be given by $u(f, k) = \mathrm{e}^{2\pi \mathrm{j} fk/N}$, where f is the index of the Fourier component and k is the sample number. We call $u(f, k)$ a *Fourier basis function*.

As we said above, the magnitude of $Z(f)$ is N times the radius of the circle in which the point moves and its argument tells us the position of the point at sample 0. Therefore

$$v(f, k) = \frac{1}{N} Z(f) u(f, k) = \frac{1}{N} Z(f) \mathrm{e}^{2\pi \mathrm{j} fk/N}$$

is the contribution to sample k of the signal from one Fourier component $Z(f)$. The signal as a whole is given by adding up the contributions from all the Fourier components:

$$z(k) = \frac{1}{N} \sum_{f=0}^{N-1} Z(f) \mathrm{e}^{2\pi \mathrm{j} fk/N}. \tag{4.2}$$

This expression is called the *inverse Fourier transform*. Note the similarity between this expression and that for the forwards Fourier transform, Equation (4.1).

4.6 Real signals

We have so far only considered taking the Fourier transform of a complex-valued signal. The transform can, of course, equally well be applied to a real signal by treating it as a complex signal with the imaginary part of each sample equal to zero.

As we saw in Chapter 2, the (real-valued) samples of a sine wave of frequency f taken at sample rate f_{s} are indistinguishable from samples of a sine wave of frequency $f_{\mathrm{s}} - s$. We might therefore expect to find a relationship between the Fourier coefficients $Z(f)$ and $Z(N - f)$ when the signal f is real-valued, and indeed this is the case. As Exercise 4.3 shows, $Z(N - f)$ is in fact the *complex conjugate* of $Z(f)$: in other words, it has the same real part, but the imaginary part is negated. We write $Z(N - f) = \overline{Z(f)}$.

Figure 4.6 shows an example real signal and the real and imaginary parts of its Fourier transform: the symmetry in the result can be clearly seen.

When the signal is real-valued the coefficients $Z(f)$ and $Z(N - f)$ always come in complex conjugate pairs, and so it makes sense to look at the sum of these two components. The component at frequency f is

$$v(f, k) = \frac{1}{N} Z(f) \mathrm{e}^{2\pi \mathrm{j} fk/N}.$$

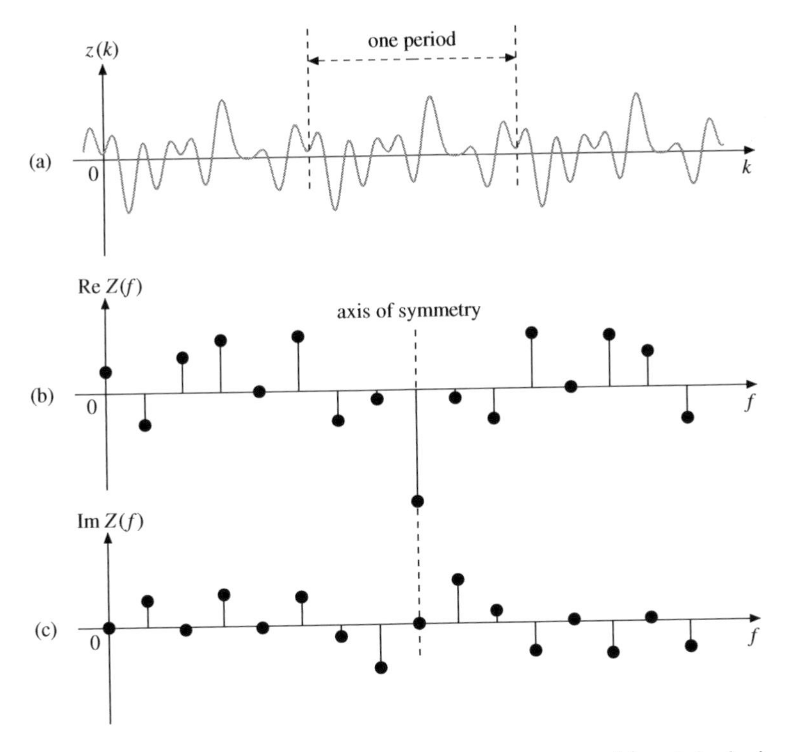

Figure 4.6 A real signal and its frequency components: (a) original signal; (b) real part of its Fourier transform; (c) imaginary part of its Fourier transform.

Splitting into real and imaginary parts,

$$v(f, k) = \frac{1}{N} Z(f)(\cos 2\pi fk/N + \mathrm{j} \sin 2\pi fk/N).$$

The component at frequency $N - f$, which, as the transform is periodic with period N, is the same as the component at frequency $-f$, is

$$\begin{aligned}
v(N-f, k) &= v(-f, k) \\
&= \frac{1}{N} Z(N-f) \mathrm{e}^{-2\pi \mathrm{j} fk/N} \\
&= \frac{1}{N} Z(N-f)\big(\cos(-2\pi fk/N) + \mathrm{j} \sin(-2\pi fk/N)\big) \\
&= \frac{1}{N} Z(N-f)(\cos 2\pi fk/N - \mathrm{j} \sin 2\pi fk/N).
\end{aligned}$$

We can now write down the sum of these two components:

$$\begin{aligned}
v(f,k)+v(N-f,k) &= \frac{1}{N}Z(f)(\cos 2\pi fk/N + \mathrm{j}\sin 2\pi fk/N) \\
&\quad + \frac{1}{N}Z(N-f)(\cos 2\pi fk/N - \mathrm{j}\sin 2\pi fk/N) \\
&= \frac{1}{N}\big(Z(f)+Z(N-f)\big)\cos 2\pi fk/N \\
&\quad + \mathrm{j}\frac{1}{N}\big(Z(f)-Z(N-f)\big)\sin 2\pi fk/N.
\end{aligned}$$

Since $Z(N-f)$ is the complex conjugate of $Z(f)$, the imaginary parts cancel out in $Z(f)+Z(N-f)$ and we are left with just twice the real part of $Z(f)$. Similarly, the real parts cancel out in $Z(f)-Z(N-f)$ and we are left with just twice the imaginary part of $Z(f)$: $Z(f)-Z(N-f)$ is pure imaginary and so $\mathrm{j}\big(Z(f)-Z(N-f)\big)$ is real. The coefficients of both $\cos 2\pi fk/N$ and $\sin 2\pi fk/N$ above are thus real numbers; they are both real functions, and so $v(f,k)+v(N-f,k)$ is also real.

Combining the Fourier components in pairs in this way has given us a way to write a sampled real periodic signal as a sum of sampled sine and cosine waves.

When a real function is expressed like this, the cosine parts (which arose from the real parts of $Z(f)$ and $Z(N-f)$) are called the *in-phase*, or *I*, components and the sine parts (which arose from the imaginary parts of $Z(f)$ and $Z(N-f)$) are called the *quadrature*, or *Q*, components. It is an unfortunate potential source of confusion that both 'in-phase' and 'imaginary' start with the letter 'I'.

Alternatively, we can write $Z(f)$ in terms of its magnitude $A(f)$ and phase $\phi(f)$, which are both real numbers: $Z(f)=A(f)\mathrm{e}^{\mathrm{j}\phi(f)}$. Now, $\overline{\mathrm{e}^{\mathrm{j}\theta}}$, the complex conjugate of $\mathrm{e}^{\mathrm{j}\theta}$, is equal to $\mathrm{e}^{-\mathrm{j}\theta}$: you can check this by writing the exponential out in terms of sin and cos. Thus

$$\begin{aligned}
Z(N-f) &= \overline{Z(f)} \\
&= A(f)\overline{\mathrm{e}^{\mathrm{j}\phi(f)}} \\
&= A(f)\mathrm{e}^{-\mathrm{j}\phi(f)}.
\end{aligned}$$

We now have

$$\begin{aligned}
v(f,k) &= \frac{1}{N}Z(f)\mathrm{e}^{2\pi\mathrm{j}fk/N} \\
&= \frac{1}{N}A(f)\mathrm{e}^{\mathrm{j}\phi(f)}\mathrm{e}^{2\pi\mathrm{j}fk/N} \\
&= \frac{1}{N}A(f)\mathrm{e}^{\mathrm{j}\phi(f)+2\pi\mathrm{j}fk/N} \\
&= \frac{1}{N}A(f)\big(\cos\big(\phi(f)+2\pi fk/N\big)+\mathrm{j}\sin\big(\phi(f)+2\pi fk/N\big)\big)
\end{aligned}$$

and

$$
\begin{aligned}
v(N-f,k) &= \frac{1}{N} Z(N-f) \mathrm{e}^{-2\pi \mathrm{j} fk/N} \\
&= \frac{1}{N} A(f) \mathrm{e}^{-\mathrm{j}\phi(f)} \mathrm{e}^{-2\pi \mathrm{j} fk/N} \\
&= \frac{1}{N} A(f) \mathrm{e}^{-\mathrm{j}\phi(f) - 2\pi \mathrm{j} fk/N} \\
&= \frac{1}{N} A(f) \big(\cos\big(-\phi(f) - 2\pi fk/N\big) + \mathrm{j}\sin\big(-\phi(f) - 2\pi fk/N\big)\big) \\
&= \frac{1}{N} A(f) \big(\cos\big(\phi(f) + 2\pi fk/N\big) - \mathrm{j}\sin\big(\phi(f) + 2\pi fk/N\big)\big) \\
&= \overline{v(f,k)}.
\end{aligned}
$$

Finally, we arrive at

$$
v(f,k) + v(N-f,k) = \frac{2}{N} A(f) \cos\big(\phi(f) + 2\pi fk/N\big).
$$

Again, we see that $v(f,k) + v(N-f,k)$ is real. This time, however, we have expressed it as a single cosine wave with amplitude $\frac{2}{N}A(f)$ and phase offset $\phi(f)$.

We therefore now have a way to write a sampled real periodic signal as a sum of sampled cosine waves with various phase offsets.

Examples

Figure 4.7 shows some examples of real signals and their Fourier transforms. The figure only shows the magnitude of the Fourier coefficients. Since a complex number has the same magnitude as its conjugate, the pattern of magnitudes is exactly mirror-symmetrical in each case. The magnitudes are plotted on a logarithmic axis to make the smaller ones easier to see. Several properties of Fourier transforms become apparent looking at these examples:

(i) The magnitudes of the Fourier coefficients do not depend on where we start taking the samples to be transformed, as long as they represent exactly one period of the signal (compare Figure 4.7(a) and (b));
(ii) A slight misjudgement as to the period of the signal can potentially mean getting different results for many if not all of the Fourier coefficients (compare Figure 4.7(a) and (c));
(iii) A smoothly changing waveform will tend to have a small number of Fourier coefficients that are large in magnitude; a spikier or discontinuous waveform will tend to have a large number of Fourier coefficients that are similar in magnitude (compare Figure 4.7(d) and (i));

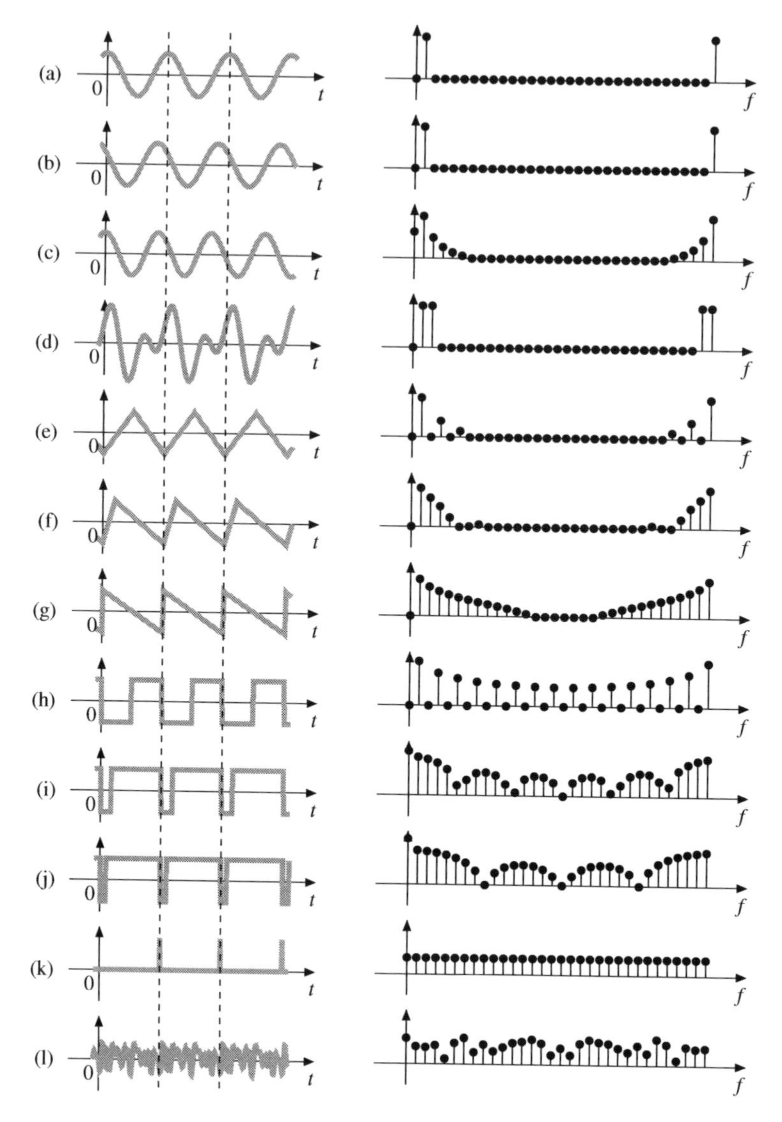

Figure 4.7 Magnitude of the Fourier transform of some example real signals: (a)–(c) sine waves; (d) the sum of two sine waves; (e)–(g) triangular and sawtooth waves; (h)–(j) square waves; (k) an impulse; (l) white noise.

(iv) The Fourier coefficients for a signal with discontinuities will tail off as f increases towards $N/2$ more slowly than a signal that is continuous but which has sudden changes of direction; and the Fourier coefficients for a signal that is continuous but has sudden changes of direction will in turn tail off more slowly than a signal that is smooth (compare Figure 4.7(d), (f) and (i), and Figure 4.7(e), (f) and (g));

(v) If a signal looks the same apart from a horizontal shift of half its period when turned upside-down (i.e., is negated), then every even-numbered Fourier coefficient is zero (compare Figure 4.7(e) and (f), and Figure 4.7(h) and (i));
(vi) The Fourier transform of an impulse is a constant (see Figure 4.7(k));
(vii) The Fourier transform of random noise looks like (different) random noise (see Figure 4.7(l)).

It is possible to formulate precise versions of all these statements and prove them, although we shall not do so here. A couple of the easier ones appear in the exercises; the proofs of some of the others are quite complicated.

4.7 Dealing with non-periodic signals

As we saw in Figure 4.7(c), taking a Fourier transform over an interval that is not a period of the signal gives untidy results. The transform assumes that it is operating on a complete period: it is as if the original signal was like the one shown in Figure 4.8. The discontinuities in this signal explain the form of the result.

In real applications we will often not know in advance what the period of the signal is. Indeed, the purpose of taking the Fourier transform may be to try to determine the period!

In other applications the signal may be gradually changing over time and hence not strictly periodic, but we are prepared to extract a small segment over which we can assume the signal is periodic. For example, Figure 4.9 shows the waveform of a male English speaker saying the 'i' vowel in 'bite'. If you say this word very slowly you will find that the vowel is not pure: it slides from an 'ah' sound to an 'ee' sound. As the figure shows, the waveform looks approximately periodic when a small section is examined with only small variations from cycle to cycle, but is certainly not periodic as a whole.

Even if the signal is known to be periodic, the sampling rate may have been fixed for us, for example because of hardware limitations, and we might only be

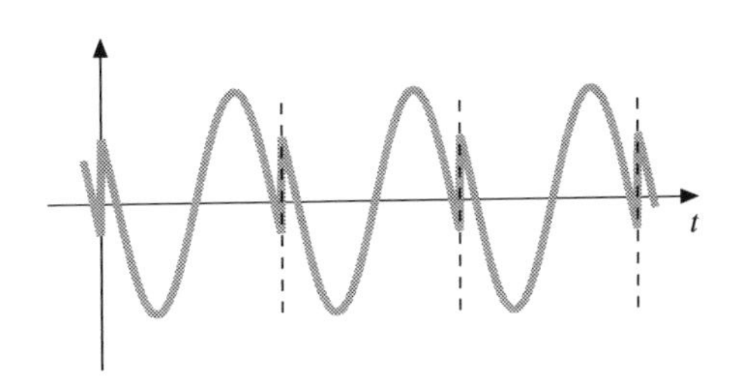

Figure 4.8 Signal that is effectively transformed when applying a Fourier transform to a waveform segment that is not a whole number of periods.

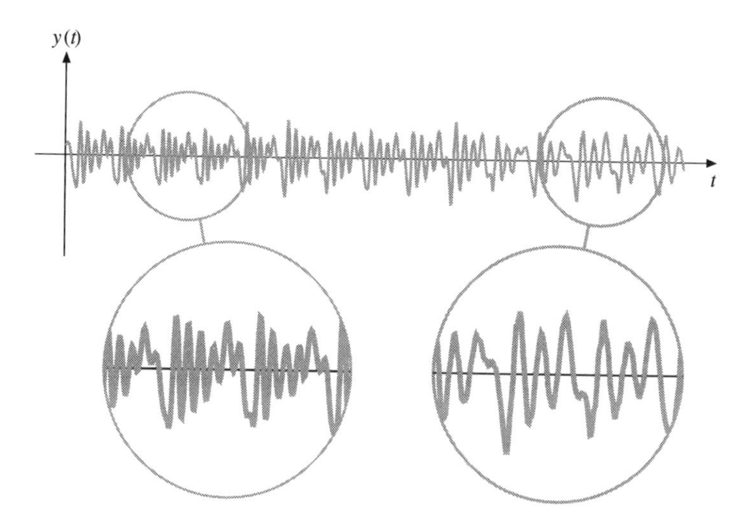

Figure 4.9 A speech waveform might appear locally periodic but in fact changes slowly over time.

prepared to compute Fourier transforms of certain given lengths N (we will see below why this might be the case). Together these restrictions might prevent us from working with a whole number of periods of our signal.

What can we do in such cases? There are three main approaches, which we shall illustrate by considering the problem of estimating the frequency of a sine wave signal by taking a Fourier transform and reporting which coefficient has greatest magnitude. We will extract a fixed number of samples N from the sine wave signal, which will not necessarily include a whole number of periods.

Ignore the problem If the application is not critical we can get away with ignoring the problem. Figure 4.10 shows that this algorithm can work adequately on our test problem. The discontinuities in the signal being transformed (see Figure 4.8) lead to a number of undesired frequency components that only fall off slowly in magnitude (property (iv) in Section 4.6 above). If the original signal were noisy it would be easy for one of these other frequency components to become the one with greatest magnitude, and our algorithm would fail.

Add a reflection to the signal The idea here is to extract a segment of length N from the original signal and then construct a new segment of length $2N$ whose first half is the same as the original segment, and whose second half consists of the samples of the original segment reversed. Figure 4.11 illustrates this idea. Considering the new segment as one period of a periodic signal, we see that at least we have avoided any discontinuities. The signal still has sharp turns at the joins between the copies of the original signal, and so we still have a number of

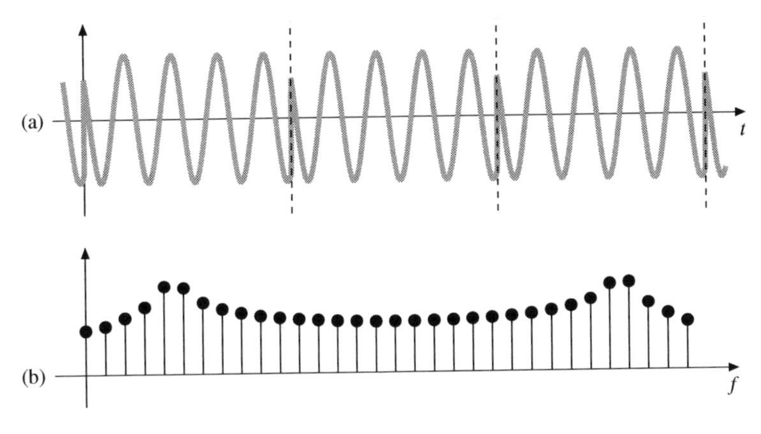

Figure 4.10 Effect of directly taking a Fourier transform of a non-periodic signal.

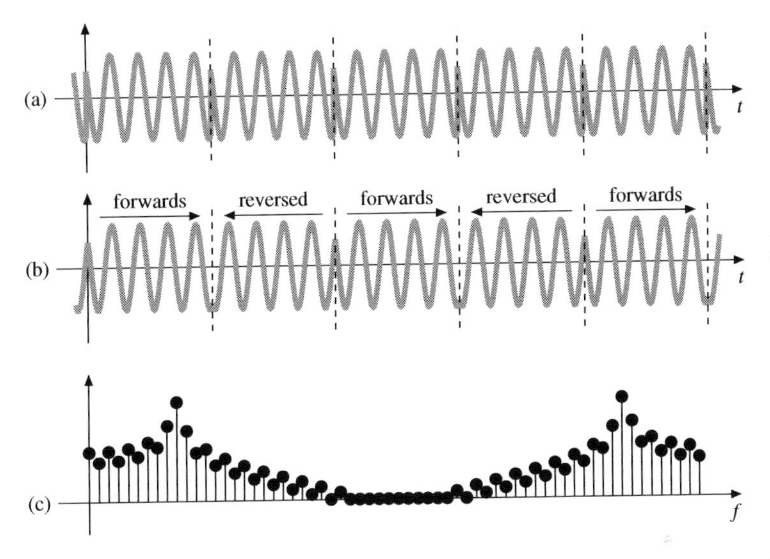

Figure 4.11 Creating a periodic signal: (a) original signal; (b) alternating forwards and reversed copies of a segment of the signal; (c) Fourier transform of (b).

undesired frequency components. These fall off in magnitude more quickly than with the method above, and so the method is more robust.

A variation of this method, called the *discrete cosine transform*, or *DCT*, is widely used in image processing applications. We shall look at it in more detail in Chapter 9.

Use a window function This gives the best results of the three approaches. We extract a segment of length N from the signal and then taper the amplitude of the samples down to zero towards either end of the segment. If the tapering is done smoothly no discontinuities are introduced within the segment, and since the signal is zero at both ends, no discontinuity is introduced when we join the ends together to make a periodic signal. The tapering is done by multiplying the signal point-by-point with a so-called *window function* which is N samples long, the same as the segment. The window function will typically take the value 1 in the middle of the segment so that the full amplitude of the signal is preserved there, and zero, or nearly zero, at either end. Many different window functions have been proposed. A popular choice which works well for most practical applications is the *Hamming window*, given by

$$H(k) = 0.54 - 0.46 \cos \frac{2\pi k}{N-1} \tag{4.3}$$

where k runs from 0 to $N-1$. This function does not quite reach zero at either end: this is a compromise between minimising the amount of discontinuity when the ends of the windowed signal are joined together, and taking at least some account of all the samples in the segment. Figure 4.12 shows the Hamming window and how it is applied to a segment of a signal. As you can see from the Fourier

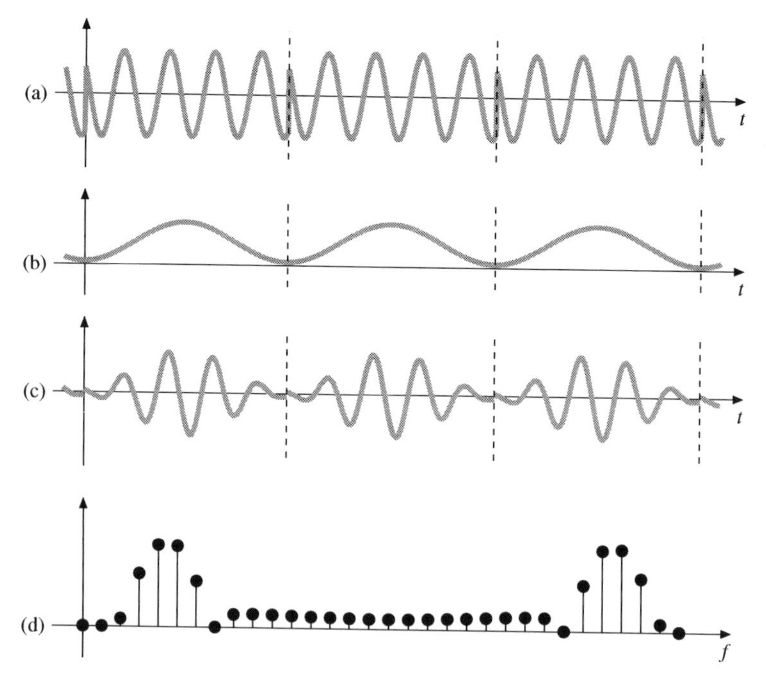

Figure 4.12 Applying a Hamming window: (a) original signal; (b) Hamming window function repeated at the desired period; (c) signal (a) multiplied by signal (b); (d) Fourier transform of signal (c).

transform in the figure, the undesired frequency components are all at a very low level, making this the most robust of the three methods.

4.8 The fast Fourier transform

The *fast Fourier transform*, or *FFT*, is a family of particularly efficient algorithms for computing the Fourier transform of a signal. People sometimes loosely refer to any Fourier transform as an 'FFT', irrespective of what algorithm was actually used to compute it.

Although fast algorithms have been developed to calculate Fourier transforms of signals for an arbitrary number of samples N in one period, the commonest and simplest type of FFT algorithm requires that N is a power of 2. One implementation works as follows.

A common approach in algorithm design is *divide-and-conquer*: expressing a given problem in terms of simpler instances of the same problem, which are then solved and combined. For example, in the 'merge sort' algorithm, a list of numbers is sorted by dividing it into two halves, sorting each half and then merging the sorted halves back together. This is a way of expressing the problem of sorting a list in terms of sorting two smaller lists. We can recursively take the same approach to sorting each of these smaller lists, resulting in still smaller lists; eventually our lists will only have one element, in which case they are already (trivially) sorted.

The fast Fourier transform works in a similar way. Given our vector of N complex samples $z(k)$, we divide it into two parts each of length $N/2$ and calculate the Fourier transform of each part. We then combine the results to obtain the final answer. We shall assume N is even.

The natural division of $z(k)$ into the first $N/2$ samples, from $k = 0$ to $k = N/2-1$ inclusive, and the second $N/2$ samples, from $k = N/2$ to $k = N-1$ inclusive, does not appear to be a promising start. If we apply a Fourier transform to the first $N/2$ samples, it will treat them as if they are one period of a periodic signal – in other words, as if sample 0 could naturally follow on from sample $N/2-1$. In general, however, sample 0 will be very different from sample $N/2-1$, and a discontinuity will be introduced. Using the example in Figure 4.1, it would be as if the point jumped from its position in frame 7 to its position in frame 0.

A better approach is to divide $z(k)$ into the even-numbered ($k = 0, 2, 4, \ldots, N-2$) and odd-numbered ($k = 1, 3, 5, \ldots, N-1$) samples. Let us call the vector of even-numbered samples $a(k)$ and the vector of odd-numbered samples $b(k)$, in both cases with k running from 0 to $N/2-1$. Creating a new signal by discarding samples like this is called *decimation*: here the decimation is by a factor of 2. Each of the decimated vectors of $N/2$ samples could reasonably be considered to be one period of a periodic signal: $a(0) = z(0)$ is a natural successor of $a(N/2-1) = z(N-2)$ in $a(k)$, and $b(0) = z(1)$ is a natural successor of $b(N/2-1) = z(N-1)$ in $b(k)$. It is therefore reasonable to take Fourier transforms of these vectors.

Each of the vectors $a(k)$ and $b(k)$ samples the same signal as $z(k)$ but at half the sample rate. As we saw in Chapter 2, reducing the sample rate generally results in aliasing. In this case, halving the sample rate from N samples per period to $N/2$ samples per period will mean that a frequency f will be confused with a frequency $f + N/2$. The idea behind the FFT is to transform both $a(k)$ and $b(k)$ and use the results to separate out the frequency component at f from the one at $f + N/2$.

First, for simplicity, let us assume that the signal $z(k)$ has no frequency components above half its sample rate – i.e., that there is no aliasing when we split it into the vectors $a(k)$ and $b(k)$. Write $A(f)$ for the Fourier transform of $a(k)$ and $B(f)$ for the Fourier transform of $b(k)$. Since $a(k)$ and $b(k)$ sample the same signal and no aliasing has occurred, the amplitudes of the frequency components reported in $A(f)$ and $B(f)$ will be the same. The phases, however, will in general be different: $A(f)$ tells us the phase of the component relative to $a(0)$, which is the same as $z(0)$, but $B(f)$ reports phase relative to $b(0)$, which is the same as $z(1)$. To correct for this and make the phases in $B(f)$ line up with those in $A(f)$ we need to subtract an amount from the phase of each frequency component corresponding to the time difference between $z(1)$ and $z(0)$. This amount is $\phi = 2\pi f/N$: note that the correction increases with f. We apply this correction by rotating the elements of $B(f)$ backwards by ϕ to create a new vector $C(f)$:

$$\begin{aligned} C(f) &= B(f)\mathrm{e}^{-\mathrm{j}\phi} \\ &= B(f)\mathrm{e}^{-2\pi \mathrm{j} f/N}. \end{aligned}$$

The term $\mathrm{e}^{-2\pi \mathrm{j} f/N}$ is called a *twiddle factor*. Under the assumption that no aliasing occurred when we split $z(k)$ into $a(k)$ and $b(k)$, $C(f)$ should now agree exactly with $A(f)$.

Now suppose that $z(k)$ *only* contains frequency components above or equal to half its sample rate and so aliasing does occur when we split it into $a(k)$ and $b(k)$. Again, the amplitudes of the (aliased) frequency components reported in $A(f)$ and $B(f)$ will be the same, but the phases will differ. To get $B(f)$ to agree in phase with $A(f)$ we again need to apply a correction: since the component at f arose as an alias of a frequency $f + N/2$ in $z(k)$, the correction in this case is $\phi' = 2\pi(f + N/2)/N = \pi + 2\pi f/N$. The corrected values of $B(f)$ are thus

$$C'(f) = B(f)\mathrm{e}^{-\mathrm{j}\phi'}.$$

Under our assumption that aliasing does occur when we split $z(k)$, $C'(f)$ should now agree exactly with $A(f)$. So in this case

$$\begin{aligned} A(f) = C'(f) &= B(f)\mathrm{e}^{-\mathrm{j}\phi'} \\ &= B(f)\mathrm{e}^{-(\mathrm{j}\pi + 2\pi\mathrm{j}f/N)} \\ &= B(f)\mathrm{e}^{-\mathrm{j}\pi}\mathrm{e}^{-2\pi\mathrm{j}f/N} \\ &= -B(f)\mathrm{e}^{-2\pi\mathrm{j}f/N} \\ &= -C(f). \end{aligned}$$

To recap, the effect of the aliasing introduced by the process of decimation is that $A(f)$ includes positive contributions from both $Z(f)$ and $Z(f+N/2)$:

$$A(f) = \frac{1}{2}\big(Z(f) + Z(f+N/2)\big)$$

with the factor of 1/2 arising from the fact that $A(f)$ is a transform of only half the length of $Z(f)$. On the other hand, $C(f)$, the phase-corrected version of $B(f)$ for frequency components below $N/2$, includes a positive contribution from $Z(f)$ but a negative contribution from $Z(f+N/2)$:

$$C(f) = \frac{1}{2}\big(Z(f) - Z(f+N/2)\big),$$

the factor of 1/2 arising as before. Solving these two simultaneous equations we have

$$\begin{aligned} Z(f) &= A(f) + C(f) \\ Z(f+N/2) &= A(f) - C(f). \end{aligned}$$

We can express the above decomposition using a flow diagram as shown in Figure 4.13.

Each of the two $N/2$-point transforms in the diagram can of course be decomposed into two $N/4$-point transforms, $N/4$ complex multiplications for the phase correction, and $N/4$ complex additions and $N/4$ complex subtractions to combine the results. The $N/4$-point transforms can then be expressed in terms of $N/8$-point transforms, and so on.

If N is a power of 2, say $N = 2^n$, this process can be repeated n times until we have expressed the entire N-point transform in terms of multiplications, additions, subtractions, and one-point transforms. The Fourier transform of a single sample is just equal to that sample: consider Equation (4.1) with N set equal to 1. (Some people prefer to think of this subdivision process as stopping when the original is

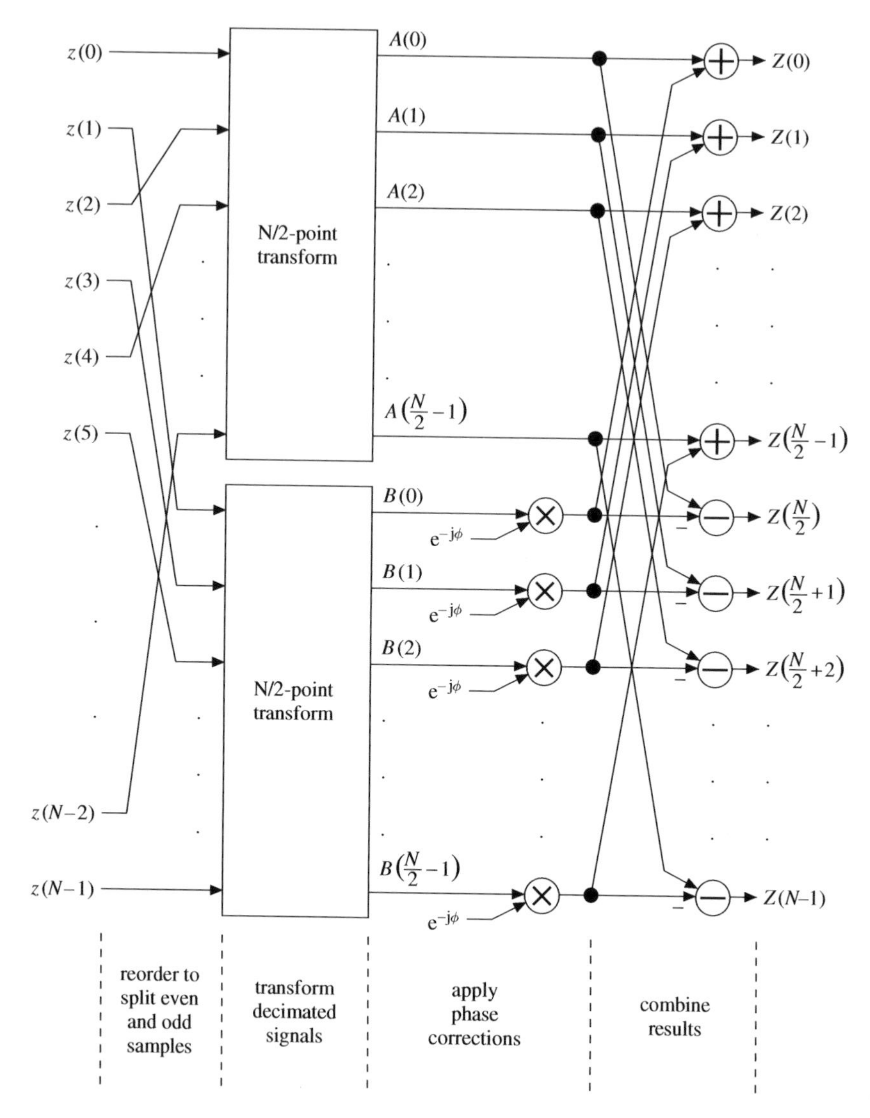

Figure 4.13 The fast Fourier transform algorithm expresses an N-point transform in terms of two $N/2$-point transforms.

represented in terms of 2-point transforms, which they then call *butterflies* from the shape of the signal flow diagram.)

Figure 4.14 shows the expansion of Figure 4.13 in the case $N = 8 = 2^3$ with one-point transforms shown as small crosses within the two-point transforms.

Exercise 4.11 asks you to compare the total number of operations involved in calculating a Fourier transform using the fast algorithm with a direct implementation of Equation (4.1).

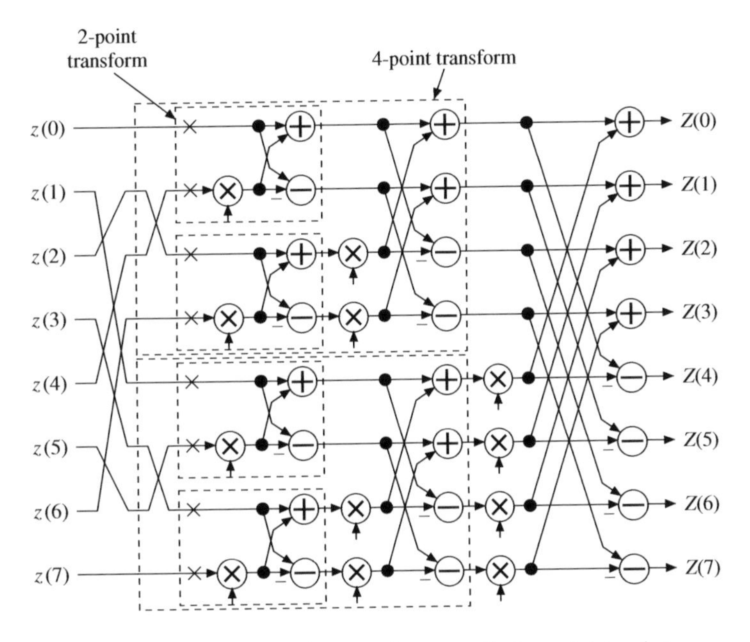

Figure 4.14 Fast Fourier transform applied to eight points.

FFT of real signals

The FFT algorithm does not directly lend itself to efficiently evaluating the transform of a vector of real-valued samples. If you try to take a simple divide-and-conquer approach as above you soon end up with complex intermediate results arising from the phase corrections, and you gain very little over doing a full complex transform. There are techniques for using a complex FFT of size $N/2$ to evaluate the transform of a real vector of length N, but they are usually not necessary. Often you will find that you either only have one transform to do and speed is not critical, or that you have a large number of transforms to do. In the first case the waste involved in using a complex transform, with the imaginary parts of the input vector set to zero, can usually be tolerated. In the second case, which occurs frequently in image processing applications, a technique such as that described in Exercise 4.9 can be used to do two real N-point transforms at once using an N-point complex transform.

Exercises

4.1 We suggested a naïve algorithm for determining the speed of rotation of the point by counting the number of times the point goes upwards past the x axis in one second. What answer does this algorithm give for the motion of Figure 4.4? What answer would we have got if we had chosen to count crossings – say from right to left – of the y axis instead?

4.2 Use the formula

$$e^{j\theta} = \cos\theta + j\sin\theta$$

and the formulae given in the chapter for rotating points in the plane to show that multiplying a complex number z by $e^{j\theta}$ rotates it through an angle θ about the origin in the complex plane.

4.3 (a) Write down versions of the equations for $X(f)$ and $Y(f)$ given in Section 4.2 for the case where $z(k)$ is real, i.e., where $y(k)=0$. Now write down an expression for $X(N-f)$ and show it is equal to $X(f)$, and write down an expression for $Y(N-f)$ and show it is equal to $-Y(f)$. Thus for a real signal, the Fourier coefficients from $Z(N/2)$ to $Z(N)$ are the same as the Fourier coefficients from $Z(0)$ to $Z(N/2)$ but in the reverse order and with the imaginary parts negated.

(b) Show that if $z(k)$ is a real signal then the DC component $Z(0)$ has zero imaginary part. If N is even show that $Z(N/2)$ also has zero imaginary part.

4.4 Show that if $z(k)$ is a complex periodic signal then the Fourier transform of the signal $\bar{z}(-k)$ (i.e., the complex conjugate of $z(k)$ reversed in time) is the complex conjugate of the Fourier transform of $z(k)$. Hence show that if a real periodic signal $z(k)$ satisfies $z(k)=z(-k)$ (i.e., it is mirror-symmetric about $k=0$), its Fourier transform $Z(f)$ is real.

4.5 (a) Prove property (i) in the list in Section 4.6, that the magnitudes of the Fourier coefficients do not depend on where we start taking the samples to be transformed, as long as they represent exactly one period of the signal.

(b) Prove property (v) in the list in Section 4.6, that if a signal looks the same apart from a horizontal shift of half its period when negated, then every even-numbered Fourier coefficient is zero.

(c) Prove that if the even-numbered samples of a signal are all zero, then the second half of the Fourier transform is the same as the first half, but negated.

4.6 If $z(k)$ is a complex signal then it is described by N independent complex numbers, or, equivalently, $2N$ independent real numbers. $Z(f)$ is also described by N independent complex numbers. If $z(k)$ is a real signal then it is described by N independent real numbers: show that in this case $Z(f)$ is also described by N independent real numbers whether N is even or odd. (**Hint:** Consider the even and odd cases separately.)

4.7 (a) Write a program which generates a vector of N random complex numbers and then computes the Fourier transform of this vector. (If your programming language has no built-in Fourier transform operation, you may need to find a suitable library function to call.) Arrange things so that you can time the operation, either using built-in timing functions or by repeatedly calling the

Fourier transform function in a loop with enough iterations that you can measure the time manually. Make sure that you do not include the time to generate the random vector in your measurement.

Plot the time taken to execute a transform of length N against N for $N = 32$, $N = 64$, $N = 128$, $N = 256$, $N = 512$, $N = 1024$, and $N = 2048$, using a logarithmic scale for both axes. If you are using the looped method for timing, you may need to experiment with the iteration count in some cases to get reasonable overall times. Is the machine using a fast Fourier transform algorithm?

(b) If your Fourier transform routine allows, repeat the experiment for $N = 31$, $N = 63$, $N = 127$, $N = 255$, $N = 511$, $N = 1023$, and $N = 2047$. (The larger cases may get rather slow.) Plot the results using logarithmic scales and determine whether the machine is using a fast algorithm for these transform sizes.

4.8 Using a timing procedure as in the previous exercise and a transform length of at least $N = 1024$, determine whether your Fourier transform function detects that its input vector consists only of real-valued samples and uses a faster-executing special case to deal with it.

4.9 Suppose you have two real vectors, $a(k)$ and $b(k)$, whose Fourier transforms you wish to calculate. Suppose we construct the complex vector $c(k) = a(k) + \mathrm{j}b(k)$ and calculate its Fourier transform $C(f)$. Write down $C(f)$ in terms of $A(f)$ and $B(f)$. Using the properties of the Fourier transforms of real vectors, show how to obtain the Fourier transforms $A(f)$ and $B(f)$ from $C(f)$. Assume the lengths of the vectors are even.

Write a program that implements your method, simultaneously transforming two real vectors with a single call to a complex Fourier transform function. Design and implement a way of testing your program.

4.10 Write a program that implements the first algorithm described in Section 4.7, estimating the frequency of a sine wave signal by taking a Fourier transform and reporting which coefficient has greatest magnitude, without using a windowing function. Test it by creating vectors of length $N = 256$ containing (a) 4.0; (b) 4.2; and (c) 4.5 periods of a sine wave. Add random noise at various amplitudes to the vector to determine approximately the point at which the algorithm starts to fail (count a result of either '4' or '5' as correct in case (c)).

Repeat the experiment using the Hamming window. You may find that you have a built-in function for calculating the window function; if not, calculate it yourself using Equation (4.3).

Does using the Hamming window improve the robustness of the algorithm?

Make or obtain a recording of a musical instrument, ideally a flute, but a piano or (undistorted) guitar will do, playing a sustained single note. It is best if the note comes from the middle of the range of the instrument: low notes in particular can work badly. Extract a segment from the middle of the note – the signal will not be steady at the beginning or the end – of length $N = 1024$ samples and run your program on it, using the Hamming window. Calculate the frequency corresponding to the result (you will need to know the sample rate of the recording). What is the smallest difference in frequency that your program can detect?

4.11 Assuming N is even, write down an expression for $u(N)$, the number of complex operations (count additions, subtractions and multiplications as one operation each) involved in calculating an N-point Fourier transform using the fast algorithm in the text, in terms of $u(N/2)$. Assume that the twiddle factor values needed to perform the phase corrections are precalculated and stored in a table.

Since the one-point transform is trivial, $u(1) = 0$. Work out the values of $u(N)$ for $N = 2, 4, 8, 16, \ldots, 2048$. How do these numbers compare with your timing results from Exercise 4.7(a)?

Write down an expression for the number of complex operations $v(N)$ involved in calculating an N-point Fourier transform using Equation (4.1) directly. Again, assume that the constants $e^{j\theta}$ have been precalculated and stored in a table.

What is $v(1024)$? How many times faster would you expect the fast Fourier transform algorithm to be than the direct implementation?

4.12 Expand the product $(a+jb)(c+jd)$ and collect real and imaginary terms. If we use this expression, how many real additions and multiplications are needed to perform one complex multiplication? How many real additions and subtractions are needed to perform one complex addition or subtraction?

How many real operations are involved in calculating a 4-point Fourier transform using the fast algorithm described in the text?

Put $N = 4$ in Equation (4.1) and expand the sum so as to write out expressions for $Z(0), Z(1), Z(2)$ and $Z(3)$ in terms of $z(0), z(1), z(2)$ and $z(3)$, evaluating the $e^{j\theta}$ terms explicitly.

How many real additions, subtractions and multiplications are really needed to work out a 4-point transform?

4.13 Given that if all the outputs of a $N/2$-point transform are zero then all its inputs must be zero, prove that the same is true for an N-point transform. (**Hint:** Work backwards through the the flow diagram in Figure 4.13.) Since the statement is true for a 1-point transform, this implies it is true for any

Table 4.1 *Input ordering for fast Fourier transform.*

Normal binary counting order	Input order
000	000
001	100
010	010
011	110
100	001
101	101
110	011
111	111

transform whose length is a power of 2. (It is in fact true for a transform of any length.)

4.14 If $a(k)$ and $b(k)$ are two complex-valued signals, show that the Fourier transform of $a(k) - b(k)$ is equal to $A(f) - B(f)$, where $A(f)$ is the Fourier transform of $a(k)$ and $B(f)$ is the Fourier transform of $b(k)$.

Given the result of the previous exercise, show that if $a(k)$ and $b(k)$ are different, then $A(f)$ and $B(f)$ are different.

Show more generally that if λ and μ are complex numbers, the Fourier transform of a linear combination of signals in the time domain $\lambda a(k) + \mu b(k)$ is equal to $\lambda A(f) + \mu B(f)$, the same linear combination in the frequency domain. We say that the Fourier transform operation is *linear*.

4.15 Redraw Figure 4.14 so as to remove the reordering stages at the input by permuting the inputs. Check that your inputs, from top to bottom, now read $z(0)$, $z(4)$, $z(2)$, $z(6)$, $z(1)$, $z(5)$, $z(3)$, $z(7)$. Compare this sequence of numbers, written in binary, with the normal binary counting sequence: see Table 4.1.

What pattern do you observe in the binary numbers? If the pattern is not clear, try making a rough sketch of a version of Figure 4.14 with $N = 16$, and then remove the reordering stages at the input by permuting the inputs. You should find that your inputs are now in the order $z(0)$, $z(8)$, $z(4)$, $z(12)$, $z(2)$, $z(10)$, $z(6)$, $z(14)$, $z(1)$, $z(9)$, $z(5)$, $z(13)$, $z(3)$, $z(11)$, $z(7)$, $z(15)$. Make a table of these numbers in binary as above and see if the pattern becomes apparent.

4.16 Embedded systems often have very little spare memory. It would be desirable in some applications to be able to calculate a Fourier transform 'in-place': that is, with the N outputs replacing the N inputs in memory, using as little extra temporary storage as possible. It is clear from the signal flow diagram in Figure 4.14 that the phase adjustment step can be carried out in-place: the unadjusted value is not needed again and so the result of each phase adjustment can replace it in memory. Likewise the combination step can work on one pair of values at a time, replacing them in memory with their sum and difference. Suggest a way of doing the reordering steps in-place. (**Hint:** Use the result of the previous exercise.)

4.17 (a) Show that for any complex numbers a and b, the product of their complex conjugates is the complex conjugate of their product.
(b) Suppose you have a library which includes a forwards Fourier transform routine, but no inverse transform. Using the similarity of Equation (4.1) and Equation (4.2) and the above results, show how to write the inverse Fourier transform in terms of the forwards transform. Implement your procedure and (assuming that your library does include an inverse transform function!) test it on some random complex vectors.

5

Filters

A *filter* is a process which changes the shape of a signal. Many filters can most simply be thought of in terms of amplifying or attenuating the frequency components of the signal. For example, many items of audio equipment include bass and treble tone controls which are connected to a filter inside: the bass control adjusts the amplification of the lower frequency components and the treble control adjusts the amplification of the higher frequency components. More sophisticated items of audio equipment might feature a 'graphic equaliser': a row of tone controls each of which adjusts the amplification of a relatively narrow band of frequencies.

A *low-pass* filter, or *LPF*, is one which preserves (or *passes*) low-frequency components (the components below its *cutoff frequency*) and attenuates (or *blocks*) high-frequency components. Similarly, a *high-pass* filter, or *HPF*, preserves higher frequencies (the ones above its cutoff frequency) and blocks lower ones; and a *band-pass* filter, or *BPF*, blocks all but a given range of frequency components around its *centre frequency*. More complicated combinations of these types are also possible.

One way to build a filter would be to take a Fourier transform of the signal, adjust the frequency components as needed, and then take an inverse transform to return to the time domain. This is indeed the most powerful and general method available, but in many applications there are much less computationally intensive ways to achieve the same result.

5.1 Smoothing a signal

In this section we will look at three filters whose action is to smooth signals.

Filter A Suppose that we wish to smooth a real-valued signal $x(t)$, $t = 0, 1, 2, \dots,$

for example to try to remove noise from it. One way to do this is to average each sample with the previous one to produce an output $y(t)$:

$$y(0) = \frac{1}{2}x(0) + \frac{1}{2}x(-1),$$
$$y(1) = \frac{1}{2}x(1) + \frac{1}{2}x(0),$$
$$y(2) = \frac{1}{2}x(2) + \frac{1}{2}x(1),$$

and in general

$$y(t) = \frac{1}{2}x(t) + \frac{1}{2}x(t-1), \tag{5.1}$$

as shown in Figure 5.1. This is called a *two-point moving-average filter*, and the multiplying factors, here 1/2 and 1/2, are called the *coefficients* or *taps* of the filter. This filter can be thought of as linearly interpolating between pairs of adjacent input samples.

If $x(t)$ is periodic with period N then

$$\begin{aligned} y(N+t) &= \frac{1}{2}x(N+t) + \frac{1}{2}x(N+t-1) \\ &= \frac{1}{2}x(t) + \frac{1}{2}x(t-1) \\ &= y(t). \end{aligned}$$

Thus $y(t)$ is also periodic with period N.

Filter B The output of Filter A at time $t = 1$ is the average of its inputs at times $t = 0$ and $t = 1$. Accordingly it would be natural to call this output sample $y(1/2)$ rather

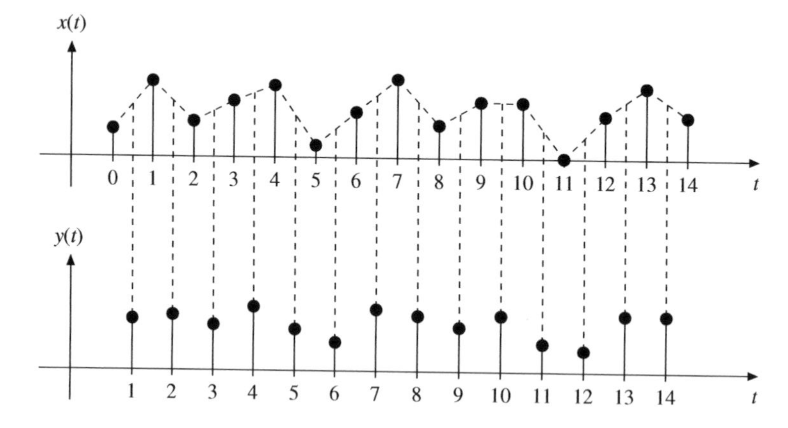

Figure 5.1 A two-point moving-average filter.

than $y(1)$, but unfortunately, this does not fit in with our sampling régime. We can remedy this by averaging the three most recent samples (a three-point moving average): $y(0) = \frac{1}{3}x(0) + \frac{1}{3}x(-1) + \frac{1}{3}x(-2)$, $y(1) = \frac{1}{3}x(1) + \frac{1}{3}x(0) + \frac{1}{3}x(-1)$, and in general

$$y(t) = \frac{1}{3}x(t) + \frac{1}{3}x(t-1) + \frac{1}{3}x(t-2), \tag{5.2}$$

as shown in Figure 5.2. Our y samples now naturally line up in time with the x samples, although offset by one whole sample time. The coefficients of this filter are 1/3, 1/3 and 1/3.

By a similar argument to the one above, if $x(t)$ is periodic with period N then $y(t)$ will be too.

You might perhaps think that it would be more natural if we had written $y(t) = \frac{1}{3}x(t+1) + \frac{1}{3}x(t) + \frac{1}{3}x(t-1)$ to make the expression pleasingly symmetric and make the y samples line up with the x samples without any time offset. However, that would mean that $y(t)$ depended on $x(t+1)$, which is a value of x from the future. A filter which can implicitly 'see into the future' like this is called *non-causal*. It is impossible actually to build a non-causal filter that operates in real time, but it is sometimes a useful theoretical construction. Non-causal filters like this example can be turned into causal ones by adding an overall delay.

Filter C There is no particular reason why all the coefficients should be equal in the averaging filters above, giving equal weight to the x samples. We could, for example, choose coefficients 1/4, 1/2 and 1/4, and write

$$y(t) = \frac{1}{4}x(t) + \frac{1}{2}x(t-1) + \frac{1}{4}x(t-2). \tag{5.3}$$

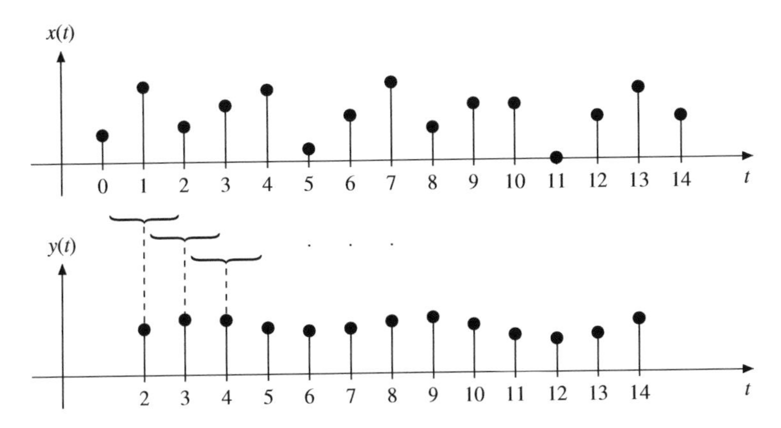

Figure 5.2 Three-point moving-average filter.

This filter might be easier to implement than Filter B, especially in hardware, as divisions by powers of two (such as 2 and 4 here) are generally more efficient than divisions by other numbers. The action of this filter is illustrated in Figure 5.3.

5.2 Analysing a filter

The results obtained in Figure 5.1 to Figure 5.3 differ slightly from one another. In each case it is clear that we have succeeded in smoothing the signal, but is any one of the results 'better' than the others? Is there an objective way to compare the results?

Time domain analysis

One way to understand what a filter does is to see its effect on simple signals. The first thing to try is to give it a constant (or DC) signal: say $x(t) = a$ for all t. For filter A we get

$$y(t) = \frac{1}{2}a + \frac{1}{2}a = a.$$

The filter thus produces a constant, or DC, output of the same magnitude as its input: we say its *DC gain*, the output value divided by the input value when the input is constant, is 1. The same goes for filters B and C, as you can verify for yourself. The reason is that in each case the coefficients of the filter add to 1. For example, in the case of Filter C, we have $1/4 + 1/2 + 1/4 = 1$.

Another simple signal we might try is a *unit impulse*. This is a signal which is always zero except for one sample, which has the value 1: see Figure 5.4. The output of the filter in this case is called its *impulse response*. The results obtained

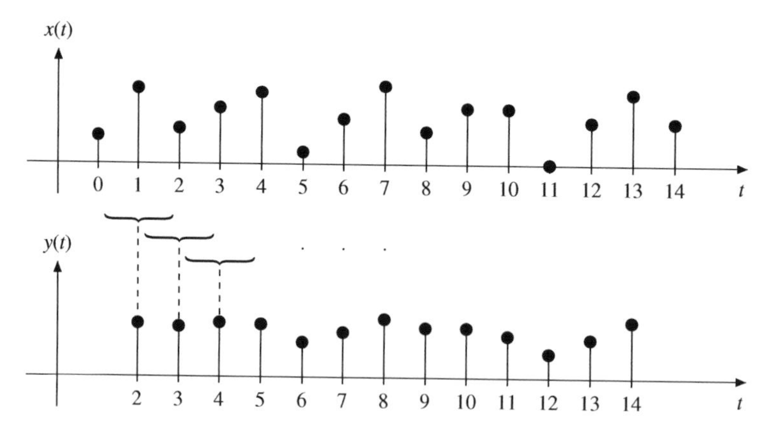

Figure 5.3 A three-point filter with coefficients (1/4, 1/2, 1/4).

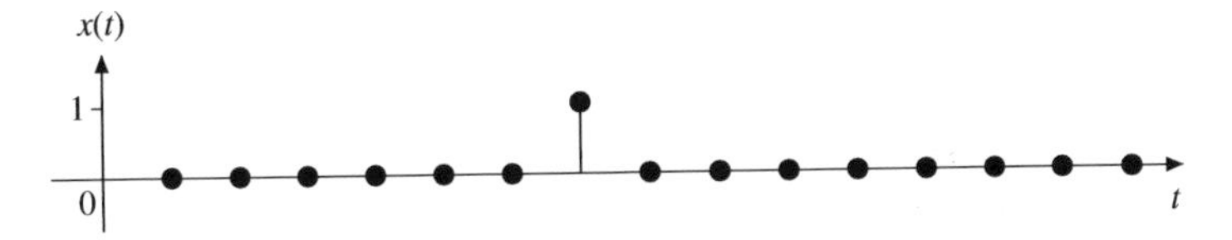

Figure 5.4 A unit impulse.

from applying our three example filters above to a unit impulse are shown in Figure 5.5. As you can verify by looking at the equations of the filters, the output in each case is just the sequence of coefficients of the filter.

The output will be zero before the impulse occurs, and, once all the coefficients have been output, will return to zero and stay there. The impulse response thus has finite duration, giving rise to the name *finite impulse response filter* or *FIR filter* for short.

If we construct a signal which consists of two impulses separated in time and feed it to our filter, the output will be two copies of the impulse response: see Figure 5.6. Sending in a series of impulses of different amplitudes will produce a number of copies of the impulse response, each scaled according to the amplitude of the impulse which produced it: see Figure 5.7.

Things get more complicated when the impulses are close enough together that the responses to them overlap. In this case, the output of the filter is the sum of the overlapped impulse responses, as shown in Figure 5.8. We can find the response of a filter to any given input signal by writing that signal as a sum of weighted unit impulses.

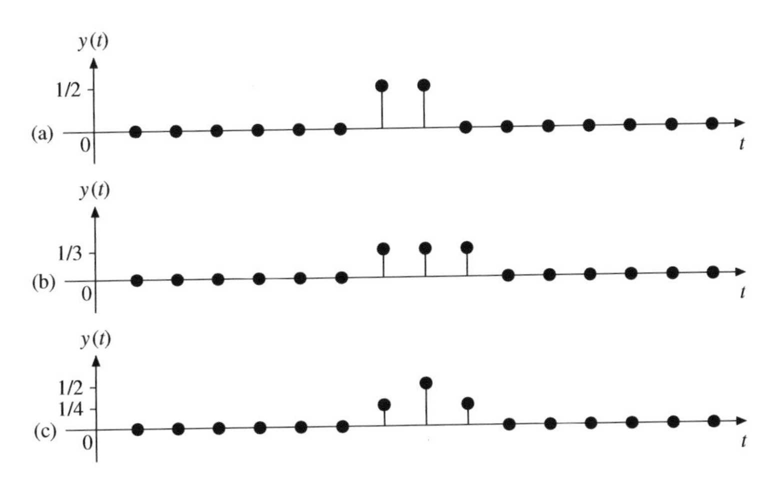

Figure 5.5 Response to a unit impulse of (a) filter A; (b) filter B; (c) filter C.

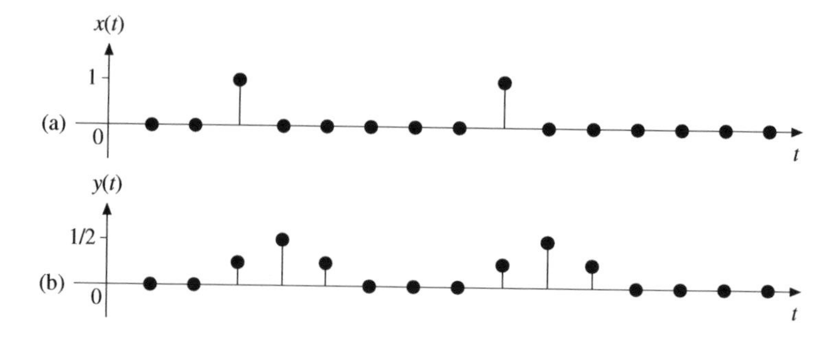

Figure 5.6 (a) Two unit impulses separated in time; (b) output of filter C.

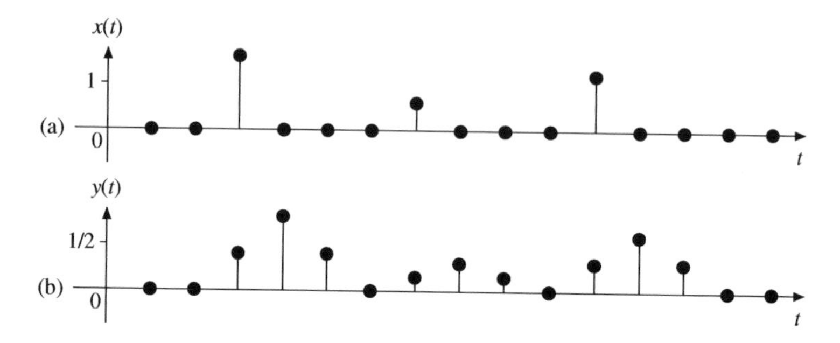

Figure 5.7 (a) Three impulses of differing magnitudes separated in time; (b) output of filter C.

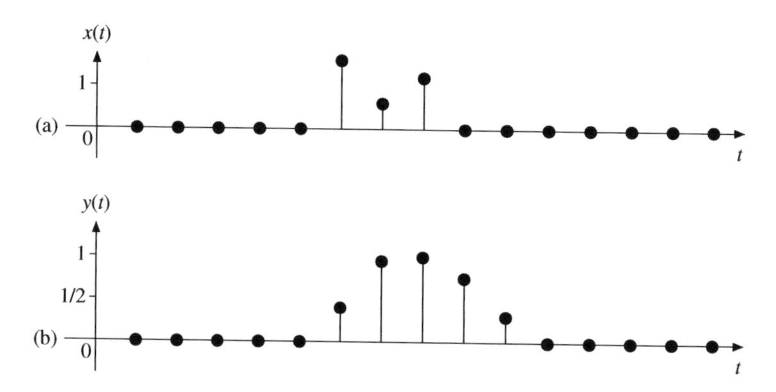

Figure 5.8 (a) Three impulses of differing magnitudes adjacent in time; (b) the output of filter C is the sum of its responses to each.

This process, where a sequence of copies of one signal are overlapped and added together to form a new signal, with the spacing and scaling of the copies being determined by the samples of a second signal, is called *convolution*. We say that the output of an FIR filter is the convolution of its input with its impulse response, and that the filter *convolves* its input with its impulse response. Sometimes the impulse response is referred to as the *kernel* of the convolution. If we convolve a signal $x(t)$ with an impulse response $h(t)$, we write the result as $h * x$. Since this is a function of time, we sometimes have to resort to the clumsy notation $(h * x)(t)$.

Figure 5.9 shows another way of looking at the process of convolution, using the same data as we used in Figure 5.8. The individual samples of the input, $x(t)$, are displayed across the top of the table, while the samples of the impulse response, $h(t)$, are displayed down the left-hand side. Each cell in the table shows the contribution made by a pair of samples, $x(t_0)$ from the input signal and $h(t_1)$ from the impulse response, to the final result: this contribution is the product of the sample values (the sample of the impulse response scaled by the value of the input sample), placed at position $t_0 + t_1$.

If we add the contributions down each column, we obtain the three overlapping scaled copies of the impulse response shown in the bottom row. Adding the signals across this row gives us the result of the convolution in the bottom right-hand corner, as expected.

Alternatively, we can add the contributions across each row, giving the results in the right-hand column. Now we get copies of the *input signal*, in each case shifted in time and scaled according to the samples of the *impulse response*. Adding up the right-hand column again gives us the result of the convolution in the bottom

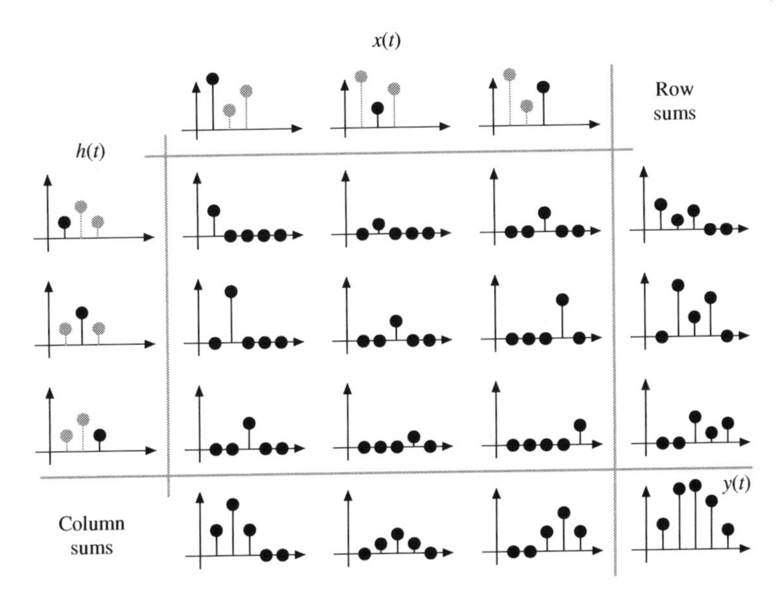

Figure 5.9 The output of an FIR filter (bottom right) can be seen as the sum of its responses to the impulses in its input signal (adding up the column sums in the grid) or as a sum of time-shifted copies of its input signal scaled according to its coefficients (adding up the row sums in the grid).

right-hand corner: we have simply added up the same set of contributions in two different ways. When implementing a convolution in a digital signal processing system we can use whichever of these two approaches is the more convenient.

The above analysis tells us that convolution is a *commutative* operation. In other words, convolving signal A with signal B gives the same result as convolving signal B with signal A: $A * B = B * A$.

We said that each possible pairing of samples, $x(t_0)$ and $h(t_1)$, gives rise to a contribution $h(t_1) \cdot x(t_0)$ to the value of the convolution at position $t_0 + t_1$. Writing $t = t_0 + t_1$ (and thus $t_0 = t - t_1$) we can add up these contributions as follows:

$$(h * x)(t) = \sum_{t_1=0}^{N-1} h(t_1) \cdot x(t - t_1) \qquad (5.4)$$

and you can see that this is indeed a general version of Equations (5.1), (5.2) and (5.3) with $h(t)$ being the coefficients of the filter in each case. Note that the range of the sum must be set so that all non-zero summands are included.

Thinking of the filter as calculating each output sample as a weighted sum of its input samples, we see that the contribution of an input sample to an output sample t sample times later is given by $h(t)$.

Figure 5.10 and Figure 5.11 show two ways of drawing a block diagram of an FIR filter with four coefficients. The first is called the *direct* structure, and the second is called the *transposed* structure. You can check that the two structures compute the same function of their input signal.

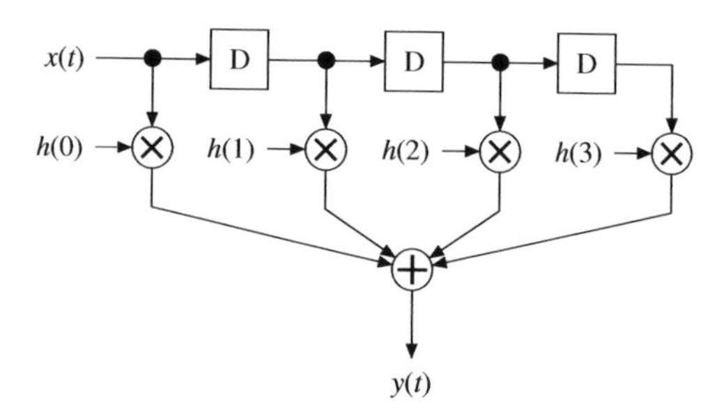

Figure 5.10 Direct form FIR filter structure.

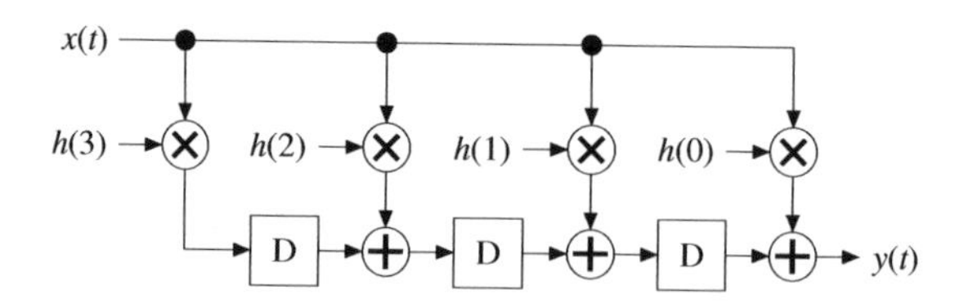

Figure 5.11 Transposed form FIR filter structure.

In this section we have looked at the action of the filter in terms of how the individual samples of the input signal and the impulse response interact: this is called a *time domain analysis*. In the next section we will use Fourier transforms to describe the behaviour of a filter, a process called *frequency domain analysis*.

Frequency domain analysis

Let us return to our first example above, Filter A, which averages pairs of adjacent samples, and analyse it again in a slightly different way. We will assume that the input $x(t)$ is an arbitrary periodic signal with period N samples. As we said above, this means that $y(t)$ is also periodic with period N.

Let us create a new signal $x_1(t)$ which is just the same as $x(t)$ but delayed by one sample time:

$$x_1(t) = x(t-1).$$

We can now write $y(t)$ as

$$y(t) = \frac{1}{2}x(t) + \frac{1}{2}x_1(t)$$

as shown on the left in Figure 5.12. Notice that sample number $N-1$ in $x(t)$ has become sample number 0 in $x_1(t)$ because we have assumed that $x(t)$ is periodic.

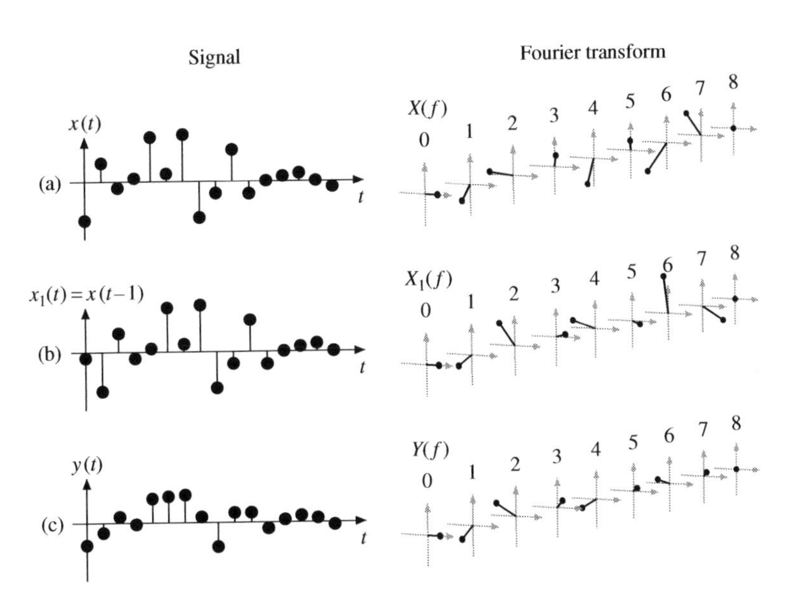

Figure 5.12 The two-point moving-average filter in the time and frequency domains: (a) original input; (b) input delayed by one sample; (c) sum of (a) and (b).

The Fourier transforms of $x(t)$ and $x_1(t)$, $X(f)$ and $X_1(f)$ respectively, are shown on the right in Figure 5.12. Don't forget that even though our signals are real-valued, their transforms are complex-valued. As we saw in Exercise 4.14, if a signal is a linear combination of other signals (as y is of $x(t)$ and $x_1(t)$ here) then its transform is the same linear combination of the transforms of those signals. Hence

$$Y(f) = \frac{1}{2}X(f) + \frac{1}{2}X_1(f).$$

We calculated the Fourier transform in Chapter 4 by 'untwisting' one period of the signal and then adding up the results over that one period. As far as calculating the transform is concerned, delaying the signal by one sample time is equivalent to advancing the untwisting function by one sample time, since the relative movement between the signal and the untwisting function is the same. Advancing the untwisting function by one sample is in turn equivalent to adding an extra rotation of $-2\pi f/N$ to it, which is the same as multiplying it by $e^{-2\pi jf/N}$. The Fourier components $X_1(f)$ are thus given by

$$X_1(f) = X(f)e^{-2\pi jf/N}.$$

In general, if we delay $x(t)$ by k samples, the Fourier transform of the delayed version $x_k(t)$ will be given by

$$X_k(f) = X(f)e^{-2\pi jfk/N}.$$

If you look closely at Figure 5.12 you should be able to see that the Fourier coefficients $X_1(f)$ are the same as those of $X(f)$, but rotated clockwise by an amount that increases with frequency. At the bottom right of Figure 5.12 we see the result of adding $\frac{1}{2}X(f)$ to $\frac{1}{2}X_1(f)$ to obtain $Y(f)$.

Now,

$$\begin{aligned} Y(f) &= \frac{1}{2}X(f) + \frac{1}{2}X_1(f) \\ &= \frac{1}{2}X(f) + \frac{1}{2}X(f)e^{-2\pi jf/N} \\ &= (\frac{1}{2} + \frac{1}{2}e^{-2\pi jf/N})X(f) \end{aligned}$$

and so we can express the effect of Filter A in terms of multiplying its Fourier transform by a sequence of complex numbers $H_A(f)$ thus:

$$Y(f) = H_A(f)X(f),$$

where

$$H_A(f) = \frac{1}{2} + \frac{1}{2}e^{-2\pi jf/N}.$$

$H_A(f)$ is called the *frequency response* of the filter. The argument of $H_A(f)$ is called the *phase response* of the filter. (Sometimes, especially in applications where the phase response of the filter is not important, the term 'frequency response' is used to refer just to the magnitude of $H_A(f)$.)

The magnitude and phase of $H_A(f)$ are plotted in Figure 5.13. As you can see, the magnitude drops off smoothly with increasing frequency, reaching zero at frequency $N/2$; from frequency $N/2$ to $N-1$ it then increases again, forming a mirror image of the first half of the response.

We now revisit Filter C, given by Equation (5.3). Following the same procedure as above, we can write $Y(f)$, the Fourier transform of $y(t)$, in terms of the Fourier transforms of delayed versions of $x(t)$ thus:

$$Y(f) = \frac{1}{4}X(f) + \frac{1}{2}X_1(f) + \frac{1}{4}X_2(f).$$

Continuing as above, we obtain

$$Y(f) = (\frac{1}{4} + \frac{1}{2}e^{-2\pi jf/N} + \frac{1}{4}e^{-4\pi jf/N})X(f)$$

and our frequency response $H_C(f)$ is thus

$$H_C(f) = \frac{1}{4} + \frac{1}{2}e^{-2\pi jf/N} + \frac{1}{4}e^{-4\pi jf/N}.$$

The magnitude and phase of $H_C(f)$ in this case are plotted in Figure 5.14.

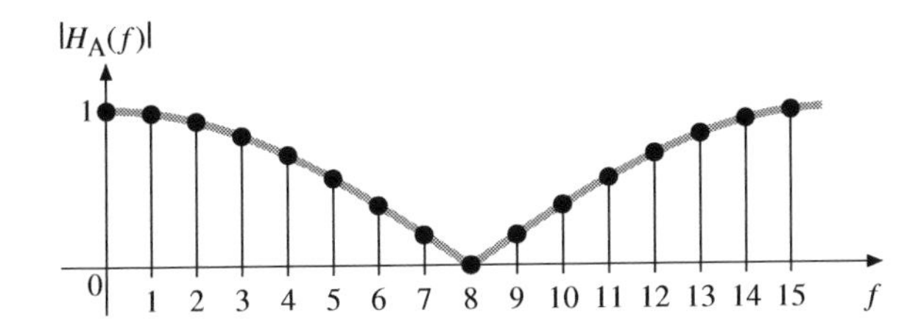

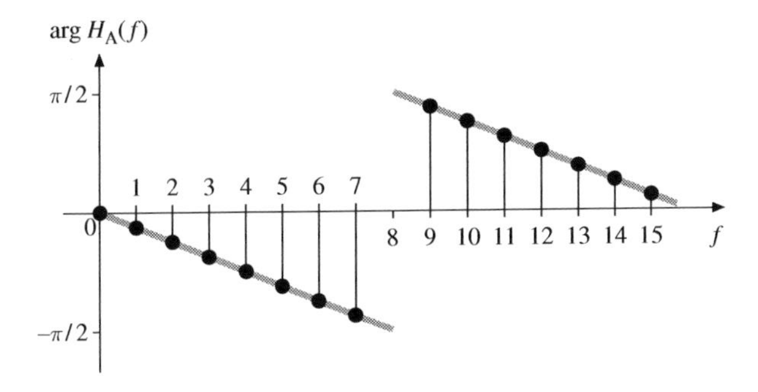

Figure 5.13 Magnitude and phase response of the two-point moving-average filter.

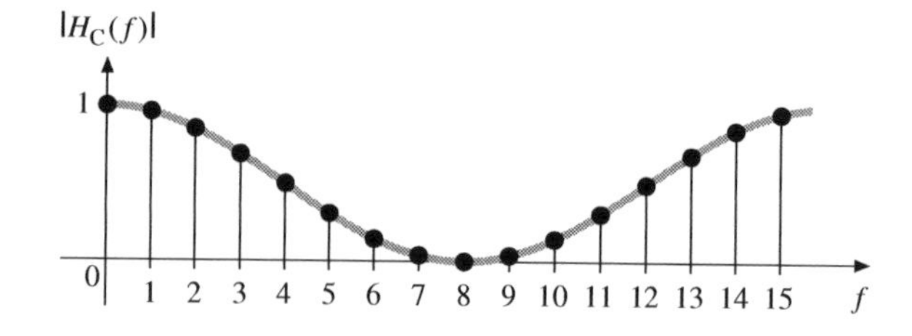

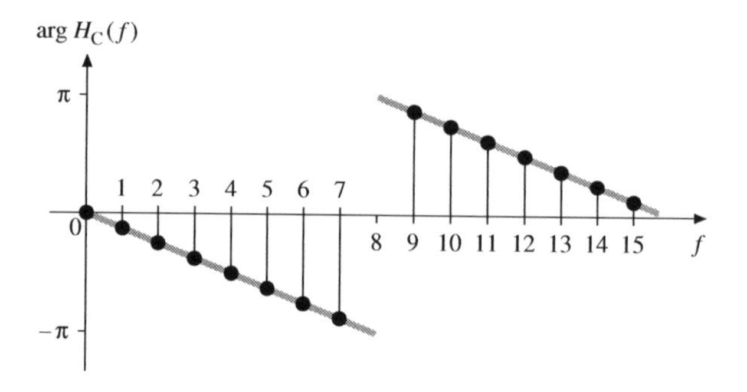

Figure 5.14 Magnitude and phase response of filter C.

You may have noticed a similarity between the Fourier transform equations in Chapter 4 and the expressions we are obtaining for the frequency responses of our filters. Indeed, the expression for $H_C(f)$ above is just the Fourier transform of the sequence $(\frac{1}{4}, \frac{1}{2}, \frac{1}{4}, 0, 0, 0, \ldots, 0)$, as you can see by putting these numbers in Equation (4.1).

5.3 Convolution in the frequency domain

The general form of the equation for an FIR filter along the lines of Equation (5.3) is

$$y(t) = h(0)x(t) + h(1)x(t-1) + h(2)x(t-2) + \cdots$$

or more compactly

$$y(t) = \sum_{k=0}^{N-1} h(k)x(t-k)$$

or, still more compactly,

$$y = h * x$$

where $h(k)$ is the impulse response of the filter, which is the same as the sequence of values of the filter coefficients.

The frequency response of the filter is the Fourier transform of the impulse response: for periodic signals, the transform of the output of the filter is equal to the transform of the input multiplied by the frequency response.

This is illustrated in Figure 5.15 which shows signals and impulse responses on the left (in the time domain), and magnitudes of frequency components and responses on the right (in the frequency domain) for the two-point moving-average filter. To get from a signal in the time domain on the left to its frequency domain counterpart on the right, we take a Fourier transform; to get back, we take an inverse Fourier transform. On the left we do a time-domain convolution to calculate the output of a filter; on the right we do a point-by-point multiplication of complex numbers in the frequency domain.

Calculating a convolution in this way, when we employ a fast Fourier transform to move between time and frequency domains, is called *fast convolution*.

When we use a Fourier transform to calculate convolutions it treats its input as periodic, and so there is an interaction between the samples at the beginning of the input sequence and those at the end. For example, $(h * x)(0)$ in Figure 5.15 is the average of $x(0)$ and $x(N-1)$, where N is the transform size. It is also necessary that the two sequences being convolved have the same length. We say that the Fourier transform method calculates the *circular convolution* of $x(t)$ and $h(t)$. If this

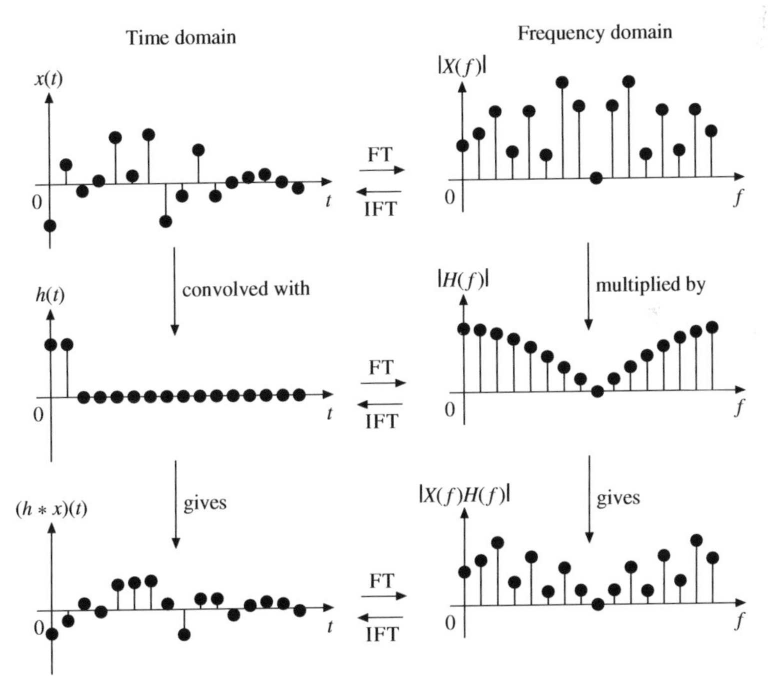

Figure 5.15 Convolution in the time domain (left) corresponds to pointwise multiplication in the frequency domain (right).

interaction is not wanted, the input must be extended with zero samples and a larger transform size used. The transform size chosen must then be large enough to contain the full result of the convolution: Exercise 5.1 asks you to work out how big this is.

5.4 Correlation

Related to the operation of convolution is that of *correlation*, also called *cross-correlation*. The correlation of two real sequences $h(t)$ and $x(t)$ is written $h \star x$ (note the subtle difference between the star symbol used to denote correlation and the asterisk symbol used to denote convolution), and is defined by

$$h(t) \star x(t) = h(-t) * x(t).$$

In other words, the correlation of $h(t)$ and $x(t)$ is the same as the convolution of the reverse of $h(t)$ with $x(t)$. As a special case of Exercise 4.4, when $h(t)$ is real the Fourier transform of $h(-t)$ is the complex conjugate of the Fourier transform of $h(t)$. In the frequency domain, therefore, the correlation of two real sequences involves point-by-point multiplication of the complex conjugate of the Fourier transform of $h(t)$ with the Fourier transform of $x(t)$.

If the Fourier transforms of $h(t)$ and $x(t)$ are similar then each of these point-wise multiplications involves multiplying a complex number by its approximate complex conjugate. Now, if $z = re^{\mathrm{j}\theta}$ then $\bar{z} = re^{-\mathrm{j}\theta}$ and $z\bar{z} = re^{\mathrm{j}\theta}re^{-\mathrm{j}\theta} = r^2$, which is a non-negative real number. As you can verify from Equation (4.2), a signal which has mostly large positive real frequency components will tend to have a large positive sample at time 0, and we can use this as a way to measure the similarity of two signals: simply calculate $(h \star x)(0)$.

By extension of this, $(h \star x)(1)$ measures the similarity of the signals $h(t)$ and $x(t)$ when offset by one sample relative to one another, and in general $(h \star x)(k)$ measures their similarity when offset by k samples. Correlation can therefore be used to find copies of one signal $h(t)$ in another, $x(t)$: peaks in the correlation function correspond to offset copies of $h(t)$ hiding in $x(t)$. Section 11.4 shows how to use this technique to construct a communication scheme that is relatively immune to interference. Since the correlation with $h(t)$ is a convolution with the reverse of $h(t)$, we can think of it in terms of applying a filter to a signal: in such contexts, $h(-t)$ is called a *matched filter* for $h(t)$.

The correlation of a signal $x(t)$ with itself is called its *autocorrelation function*. The Fourier transform of the autocorrelation function, which, from the remarks above, is the square of the magnitude of the Fourier transform of $x(t)$, is called the *power spectrum* of $x(t)$. Some definitions of 'power spectrum' include an extra constant scaling factor; and the term 'spectrum', or 'frequency spectrum', is also used to refer simply to the Fourier transform of $x(t)$.

Correlations of long sequences can be calculated using the fast Fourier transform, exactly analogously to its use in calculating fast convolutions described above. Such implementations are called *fast correlation* algorithms.

5.5 Designing FIR filters

Designing an FIR filter usually involves producing a set of coefficients (or equivalently, an impulse response) $h(t)$ such that the frequency response $H(f)$ meets a given set of specifications. These specifications can be expressed in many different ways. For example, the complex values of $H(f)$ may be given in their entirety, usually with a certain allowable error or tolerance; or the magnitudes may be specified at all frequencies but the phases left unspecified; or the magnitudes may be specified with strict tolerances at some frequencies and less strict tolerances at others. The objective may be to match the given response as well as possible with a given number of filter coefficients, or to do as well as possible using arithmetic operations that work to a given accuracy.

There are algorithms available to help design FIR filters under all these different types of constraint. Below we will give one general method that gives reasonable results for most practical problems, and a quick way to calculate suitable coefficient values directly for some simple kinds of filter; if you have a particularly demanding application, such as a very cost-sensitive filter to be implemented in hardware or a very speed-critical filter to be implemented in software, you would be wise to investigate the huge range of alternative methods and tools available.

The windowing method

The idea behind the windowing method is simple. We will assume that $x(t)$, the input to our filter, is a real signal and that we want the output $y(t)$ to be real as well. Looking at Figure 5.15, we see that if we are given a filter response $H(f)$, where f runs from 0 to $N-1$, we can simply calculate the corresponding time-domain impulse response $h(t)$ by taking its inverse Fourier transform. Unfortunately, there are a couple of snags.

The first snag is that we ought to make sure that our impulse response is real. Otherwise if (for example) $x(t)$ happens to be an impulse, the output of our filter will be complex. As we saw in Chapter 4, this puts the following constraints on $H(f)$. We will assume that N is even.

(i) $H(0)$ and $H(N/2)$ are both real;
(ii) $H(N-f)$ is the complex conjugate of $H(f)$: in other words, the second half of $H(f)$, from $H(N/2+1)$ to $H(N-1)$, must look like the first half reflected but with the imaginary parts negated.

These requirements mean that we are only allowed to specify $H(f)$ from $f=0$ to $f=N/2$ inclusive. The phases of $H(0)$ and $H(N/2)$ must both be zero (i.e., $H(0)$ and $H(N/2)$ must both be real), and we are free to set $H(1)$ to $H(N/2-1)$ as we wish. The remainder of $H(f)$, from $H(N/2+1)$ to $H(N-1)$, is then determined for us by the symmetry constraint.

Often the phase parts of the frequency response are specified as zero, or are left unspecified. In these cases we just make $H(f)$ real and ensure it is mirror-symmetric about $H(N/2)$: this satisfies both the constraints above.

Let us try a practical example. Suppose we have a signal sampled at 32 kHz and we wish to remove all frequency components above 4 kHz: as we shall see later in this chapter, we would require just such a filter if we wished to convert the sample rate of the signal from 32 kHz to 8 kHz.

This is an example of a low-pass filter, and the frequency response we would ideally like is shown in Figure 5.16(a): the response is 1 over the range of frequencies we wish to preserve, and zero over the range we wish to remove. We will fix the phase response of the filter to zero over the entire frequency range for simplicity.

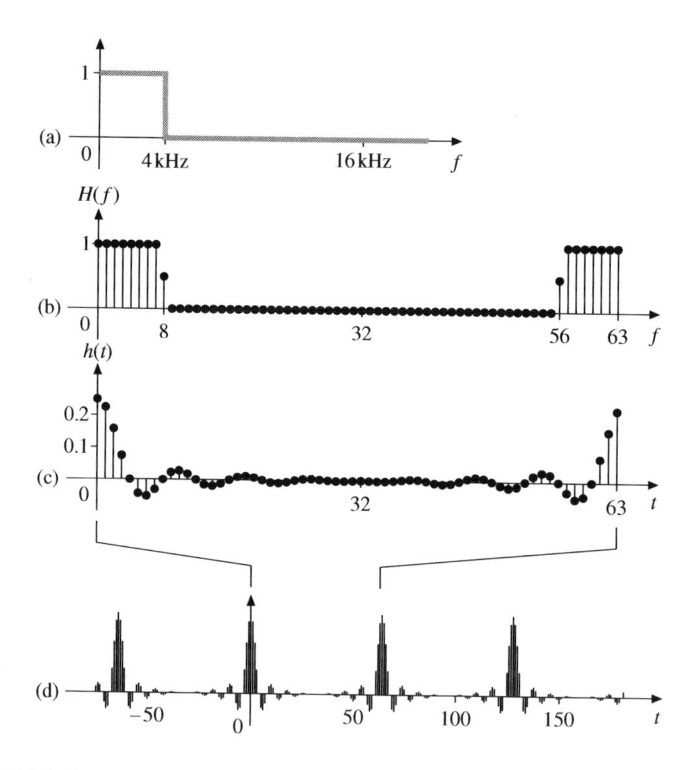

Figure 5.16 Attempt to design a low-pass filter: (a) ideal response; (b) sampled response; (c) inverse Fourier transform of (b); (d) signal (c) viewed as one period of a periodic waveform.

We represent the response in sampled form as $H(f)$ in Figure 5.16(b). For this example we have chosen $N = 64$: we will discuss the choice of N below. The 64 points in $H(f)$ represent frequencies from 0 Hz up to our input sample rate, 32 kHz. Thus the value at $H(f)$ represents the response to a frequency $(f/64) \cdot 32$ kHz, or $500f$ Hz. We therefore set $H(0)$ to $H(7)$, representing frequencies from 0 Hz to 3500 Hz, to 1; $H(9)$ to $H(32)$, representing frequencies from 4500 Hz to 16 000 Hz, to zero; and $H(8)$, representing the transition frequency of 4000 Hz exactly, we set somewhat arbitrarily to 1/2. The second half of $H(f)$, from $H(33)$ up to $H(63)$, is a mirror image of $H(1)$ to $H(31)$ as discussed above.

We can now take an inverse Fourier transform of $H(f)$ and obtain $h(t)$, the impulse response of our filter, as shown in Figure 5.16(c). And here we see the second snag.

Because of the way $h(t)$ was obtained, there is an underlying assumption that it is a periodic signal with period N samples, as shown in Figure 5.16(d). To be precise, $h(t)$ is not the impulse response of the filter, but its response to a series of impulses, one every N samples. It is therefore important to ensure that N is big enough that there is no significant overlap between the response to one impulse and the response to the next: in other words, the impulse response must decay within N samples to a level that is negligible as far as our application is concerned.

In Chapter 4 we discussed taking the Fourier transform of a non-periodic signal. We applied a window function to the signal to extract a segment that we could then safely treat as a single period. Here we face the reverse problem: we have a periodic signal from which we wish to extract a segment which we can then treat as a non-periodic signal. We can proceed in a similar way, applying a windowing function to the calculated impulse response of the filter. Figure 5.17 shows how a Hamming window (Equation (4.3)), shifted so that it is centred about $t = 0$, can be applied in our example. We now have a segment of waveform to use as the impulse response $h(t)$ of our filter.

Note that the segment of waveform we have extracted includes some coefficients with negative indices, which means that the filter is non-causal. This is a consequence of our setting the phases of all the frequency response samples to zero. We could go back and work out what phases would be required to make the resulting windowed filter causal, but it is simpler just to calculate the filter coefficients as we have described and then shift the result in time.

It may come as a surprise that this low-pass filter, whose effect is essentially to produce a smoothed version of its input signal, has some negative coefficients. Intuitively we might expect that a smoothing filter would have to take the form of some positive weighted average of a range of input samples, like the three examples we looked at near the start of this chapter; but this intuition is wrong.

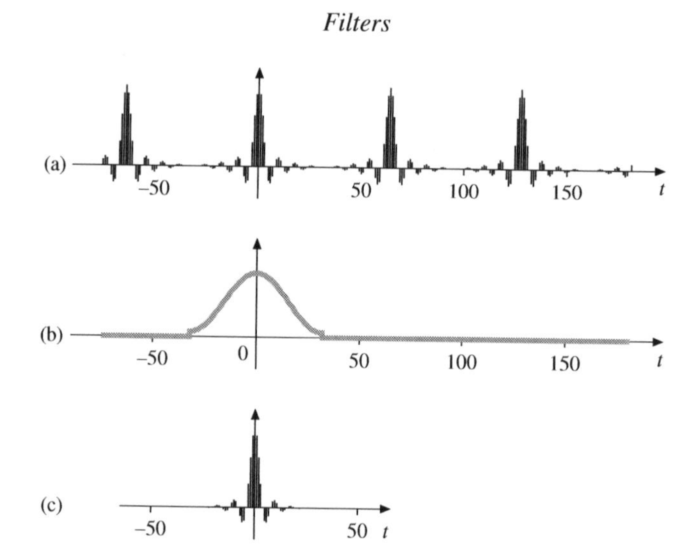

Figure 5.17 Applying a windowing function to the impulse response of a filter: (a) original (periodic) impulse response; (b) window function; (c) signal (a) multiplied by signal (b).

How closely does our filter design match our original specifications in Figure 5.16(a)? Figure 5.18 shows the magnitude of the Fourier transform of our final design compared to the original specification.

Direct calculation of filter coefficients

It is possible to calculate the coefficients required for some simple kinds of filter directly. We shall not prove the results in this section.

First we define the *sinc function*

$$\text{sinc } x = \frac{\sin \pi x}{\pi x}$$

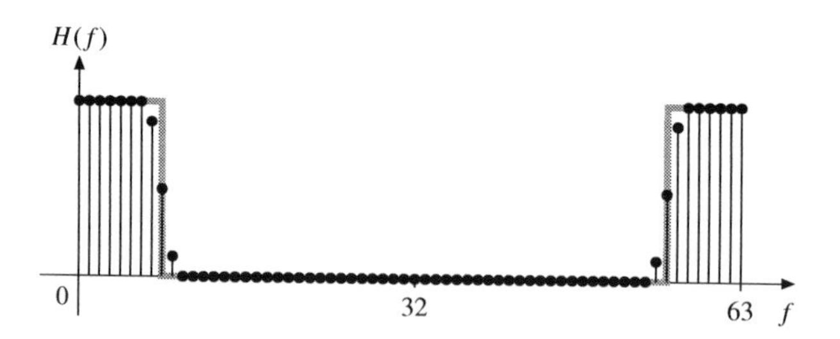

Figure 5.18 Actual magnitude response of windowed filter (black samples) superimposed on ideal response (grey).

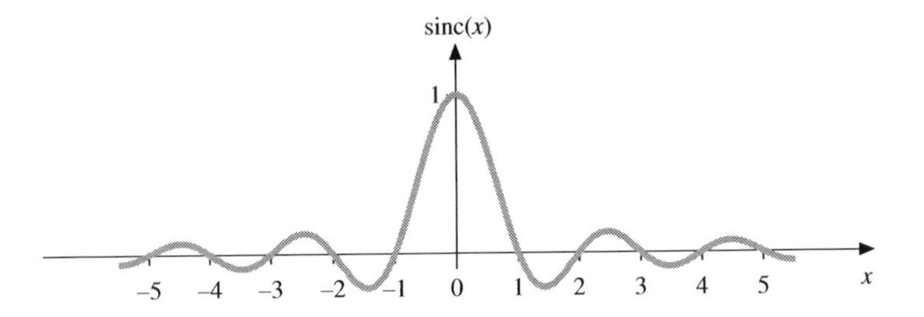

Figure 5.19 The sinc function.

where sinc (0) is defined to be 1: see Figure 5.19. Note that $\operatorname{sinc} x = \operatorname{sinc}(-x)$; and that for all integer values of x other than zero, $\operatorname{sinc} x = 0$.

Let $w(t)$ be a window function centred on $t = 0$ like the one shown in Figure 5.17(b). Then

$$h(t) = w(t) \operatorname{sinc} kt \tag{5.5}$$

with $k \leq 1$ is the impulse response of a (non-causal) low-pass filter with cutoff frequency equal to $kf_s/2$, where f_s is the sample rate of the signal being filtered. When $k = 1$ the cutoff frequency is equal to $f_s/2$, the Nyquist limit for f_s. The impulse response $h(t)$ is then the sinc function evaluated at integer points, which, as we remarked above, is simply an impulse.

The exact DC gain of this low-pass filter depends on the window function chosen; for reasonable choices of window function it is approximately proportional to $1/k$. Because of the symmetry of the sinc function, the frequency response of a low-pass filter constructed in this way is entirely real: i.e., the phase response is zero.

From Exercise 4.14 we know that taking linear combinations of signals in the time domain is equivalent to taking the same linear combinations in the frequency domain. We can use this to construct more complicated filters based on the low-pass design above. For example, consider the filter whose impulse response is

$$g(t) = w(t)(\lambda \operatorname{sinc} k_0 t - \mu \operatorname{sinc} k_1 t)$$

where λ and μ are real constants. This is the difference of two filters whose cutoff frequencies are $k_0 f_s/2$ and $k_1 f_s/2$. Suppose that $k_0 < k_1$. With suitable choice of λ and μ we can arrange for the responses of the two filters to be equal, and thus cancel out, over the range of frequencies from zero up to $k_0 f_s/2$. We are thus left with a filter which only passes frequencies between $k_0 f_s/2$ and $k_1 f_s/2$: we have constructed a band-pass filter. If we set $k_1 = 1$, we have a high-pass filter. More complicated responses can be constructed by combining several low-pass filters in this way.

Design compromises

The compromises in the design of the FIR filter stem from the fact that, for most simple filter specifications, the ideal impulse response is infinite rather than finite. To make a finite impulse response we inevitably have to chop off parts of this ideal infinite signal, which we do using the windowing function. Roughly speaking, the smaller the magnitude of the signal in the parts we discard, the better the response of our filter will match the ideal response.

It is therefore important to ensure that the ideal impulse response samples that fall outside the window are small: this constrains the choice of the size of the window, and, if we are using the windowing design method, it also constrains the choice of N.

It is essential to check before applying the window that the body of the window function encompasses all the samples in the impulse response of significant magnitude. (What 'significant' means will depend on your application.) If the maximum length of your impulse response is fixed, for example by hardware or software performance constraints, you may have to consider making the frequency response specification less demanding.

We mentioned in Chapter 4 that if a signal contains discontinuities the magnitudes of its Fourier components will only become smaller in magnitude slowly with increasing frequency. A similar thing happens in the case of the inverse transform: discontinuities in the frequency domain translate into slow decay of signals in the time domain. The smoother the frequency response we aim for, the more rapidly the impulse response will tail away. Figure 5.20 shows some smoother versions of the example frequency response above along with the corresponding impulse responses, produced by reducing the size of the window. As you can see, the smoother the frequency response, the shorter the impulse response.

If we want evaluation of Equation (5.4) to be efficient we need the impulse response to be as short as possible. The compromise is thus as follows: you can have a short FIR filter – one with only a small number of non-zero coefficients – that is efficient to evaluate with an undemanding frequency response; or you can have a long FIR filter that is slow to evaluate with a very sharp frequency response.

Windowing can affect the whole of the frequency response of the filter. In particular, the DC gain of the resulting filter can differ from the desired value: in Figure 5.20(c) the DC gain of the filter was considerably reduced when we used a narrow window. In an image processing application, for example, it may be important to ensure that the DC gain of a filter is 1, so that the filtering process does not change the overall brightness of the image. In such circumstances it is possible to scale all the filter coefficients by the same factor so as to correct the

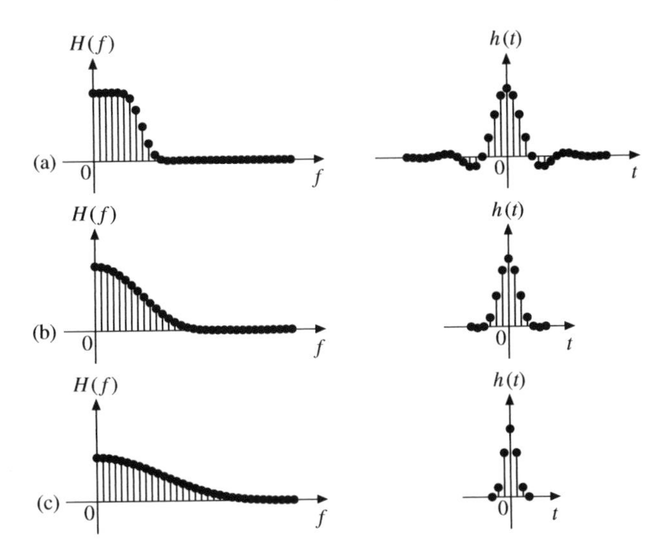

Figure 5.20 Effect of window size on the frequency response of a filter: (a) wide window; (b) medium window; (c) narrow window.

DC gain, although this may increase the deviation of the actual response from the ideal at other frequencies.

5.6 Interpolation

A common problem in digital signal processing is *interpolation*, which we touched on in Chapter 2: given a signal sampled at certain times, generate some new sample points at times in between the original samples. To take a simple example, if we generated one new sample between each pair of original samples we could interleave the interpolated samples and the original ones to create a new signal at double the original sample rate: see Figure 5.21. This would be useful if we wanted to play an audio signal sampled at, for example, 16 kHz using hardware only capable of playing samples at 32 kHz, or as part of a system to scale an image from one resolution to a higher one.

In the filter design example above we initially produced a non-causal filter (one where the impulse response $h(t)$ was non-zero for some negative values of t) and then modified it to make a causal filter. The modification involved delaying the windowed, and therefore finite in extent, impulse response so that $h(t)$ was zero for all negative t. We noted that this delay corresponds to modifying the phase response of the filter: indeed, as we saw in the discussion of the fast Fourier transform in Chapter 4, a delay of one sample in the time domain corresponds to a multiplication by $e^{-2\pi j f/N}$ in the frequency domain.

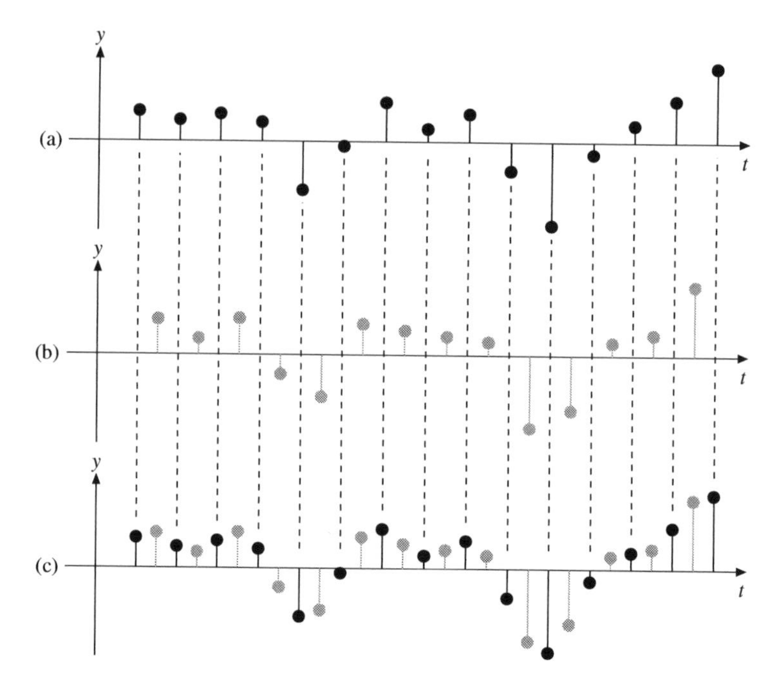

Figure 5.21 Upsampling by a factor of 2: (a) original signal samples; (b) interpolated samples; (c) signals (a) and (b) interleaved produce the final result.

Delaying a signal by k samples in the time domain is therefore equivalent to k-fold multiplication in the frequency domain by $e^{-2\pi jf/N}$, or, in other words, a multiplication by $(e^{-2\pi jf/N})^k = e^{-2\pi jfk/N}$ in the frequency domain. Conversely, advancing a signal by k samples in the time domain is equivalent to multiplication in the frequency domain by $e^{2\pi jfk/N}$.

There is no reason why k has to be an integer in the above expressions. If we let $k = 1/2$ we get the result that advancing a signal in the time domain by half a sample is equivalent to a multiplication in the frequency domain by $e^{2\pi jf/2N}$.

Let us now use this idea to make an interpolator. We start with a filter that does nothing: its frequency response has magnitude 1 at all frequencies and its phase response is zero at all frequencies. Since this filter does not have any effect on a signal, its impulse response is simply an impulse: see Figure 5.22(a).

Now we advance the impulse response by half a sample. We do this by multiplying the frequency response by $e^{2\pi jf/2N}$. Since the frequency response was originally 1, the new response is just $e^{2\pi jf/2N}$. Figure 5.22(b) shows this response and its time-domain counterpart, calculated using an inverse Fourier transform. To make the impulse response real, we need to ensure that the second half of the Fourier transform is the complex conjugate of the first half in reverse order: see Section 4.6. For the purposes of Figure 5.22 we used a 16-point transform; we

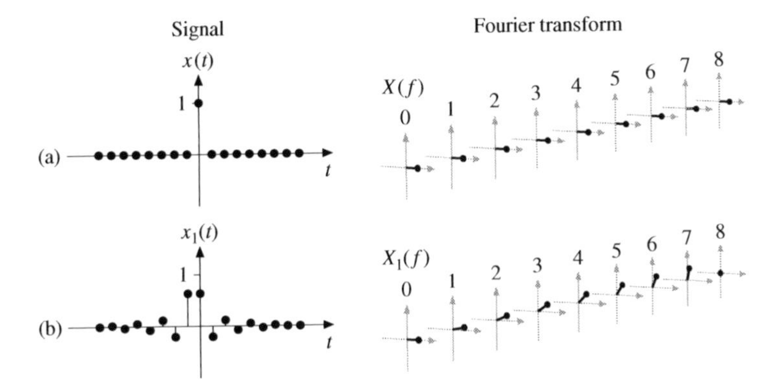

Figure 5.22 A unit impulse advanced by half a sample: (a) unit impulse in the time domain (left) and frequency domain (right); (b) impulse advanced by half a sample.

could calculate the response to greater accuracy using a longer transform if we wished.

So Figure 5.22(b) is the response of a filter which advances its input signal by half a sample in the time domain. Its output samples are therefore samples of a signal which is identical to its input signal but which has been advanced by half a sample time; looking at it the other way around, we can say that they are samples of the original signal at times falling exactly mid-way between the original samples.

Note that – as you might expect given that it advances a signal in time – the filter as we have shown it is not causal. This can be remedied in a practical interpolator by adding an overall delay to the system.

The main part of the ideal impulse response of a filter which advances its input signal by half a sample is (−0.058, 0.071, −0.091, 0.127, −0.212, 0.637, 0.637, −0.212, 0.127, −0.091, 0.071, −0.058). Compare this impulse response with the two-point moving-average filter we considered at the start of this chapter, which does a similar job and whose impulse response is (0, 0, ..., 0, 0.5, 0.5, 0, ..., 0, 0): our new filter clearly requires more computing power to evaluate, but on the other hand we obtain a uniform, or 'flat', frequency response, rather than the one shown in Figure 5.13.

We will give an expression below that will let you calculate the ideal impulse response of a filter like this directly rather than using a Fourier transform.

Sample rate conversion

Sample rate conversion, or *resampling*, is the process of turning a set of samples of a signal taken at one sample rate into a set of samples representing the same signal at a different sample rate. For example, we may have some audio from a CD sampled at 44.1 kHz and want to play it using hardware that can only operate at a sample rate of 32 kHz. Alternatively, we may have a scanned image at a resolution of 600 pixels by 450 pixels that we wish to scale up to use as a screen background at a resolution of 1600 pixels by 1200 pixels. The first of these examples, going from a higher sample rate to a lower one, is called *downsampling*; the second, going from a lower rate to a higher one, is called *upsampling*.

There are of course simple approaches to this problem, involving repeating samples (in the case of upsampling) or dropping samples (in the case of downsampling) as necessary to get the desired output sample rate, but the results are rarely satisfactory. Doing the job properly requires some slightly more sophisticated signal processing.

First let us consider upsampling by an integer ratio. The filter described above can be used to double the sample rate of a signal as shown in Figure 5.21. The idea can be extended to increasing the sample rate by other ratios: for example, if we want to triple the sample rate of a signal, we simply construct two filters, one corresponding to an advance in the time domain of 1/3 sample, the other to an advance of 2/3 sample. The first of these produces output samples that naturally fall 1/3 sample later than those of the original signal, while the second produces samples that naturally fall 2/3 sample later than those of the original signal. Taking samples from the original signal and the two interpolated versions in turn will yield the desired output: see Figure 5.23. As you can see, the underlying signals in Figure 5.23(a), (b) and (c) are the same, but signal (b) is advanced in time (i.e., shifted to the left) relative to signal (a) by 1/3 sample time and signal (c) is advanced in time relative to signal (a) by 2/3 sample. We have offset plots (b) and (c) horizontally to show how the interpolated samples fit between the original samples.

Downsampling requires more care. Our original signal will contain frequency components up to half its sample rate (the Nyquist limit, discussed in Chapter 2). The output signal we produce will only be able to represent frequencies up to half the new sample rate: since the output sample rate is lower than the input sample rate, there must inevitably be some information loss in the downsampling process. Let us suppose that we wish to convert a signal from a sample rate of 32 kHz to 8 kHz. As we saw in Chapter 2, simply dividing the signal into groups of four consecutive samples and discarding all but the first of each group (decimating by a factor of 4) will introduce undesirable aliasing, mixing together frequency components that we will not be able to separate again. The first step in the downsampling process is thus to filter the input signal to ensure that it contains no frequencies above half the target sample rate (4 kHz in this example). We can then discard three out of every four samples without introducing aliasing.

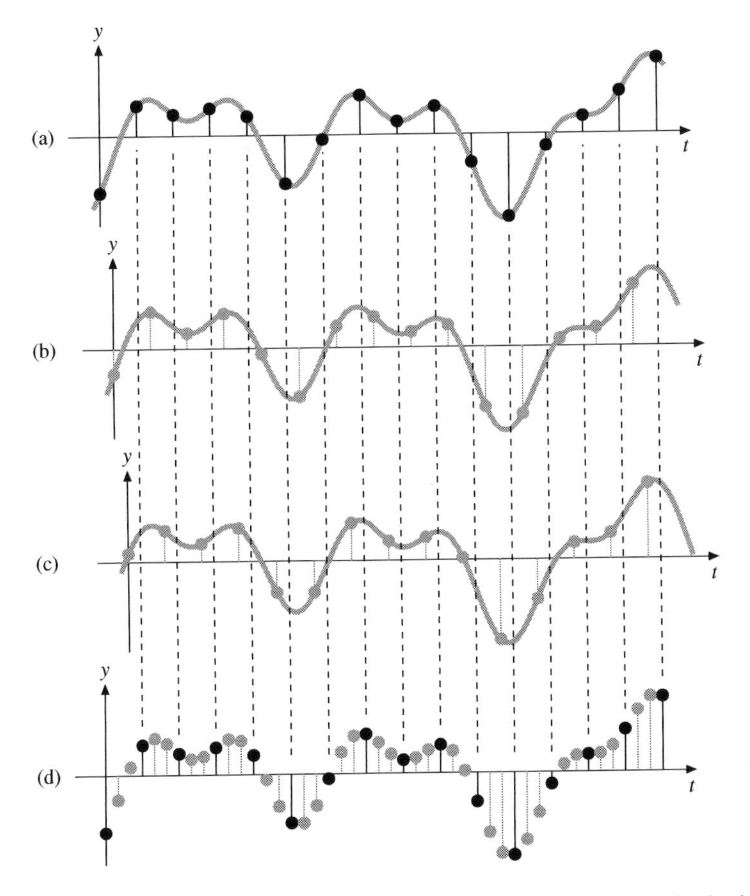

Figure 5.23 Upsampling by a factor of 3: (a) samples of original signal; (b) reconstructed samples of original signal delayed by one-third of a sample time; (c) reconstructed samples of original signal delayed by two-thirds of a sample time; (d) samples (a), (b) and (c) interleaved produce the final result.

The filter we need, called an *anti-aliasing filter*, is in this case the one we designed as an example of the windowing method in Section 5.5. The principle of the downsampling process is illustrated in Figure 5.24.

In practice we can make an optimisation by observing that there is clearly no point in having our low-pass filter calculate samples that we are immediately going to discard. We therefore only evaluate the filter output for the points we are going to keep, combining the two elements of Figure 5.24 into a single process. When the ratio between the input and output sample rates is large, this gives a huge saving in the number of operations required.

Note that in this example there are four different downsampled results that we could obtain depending on the relative alignment of the groups of four samples

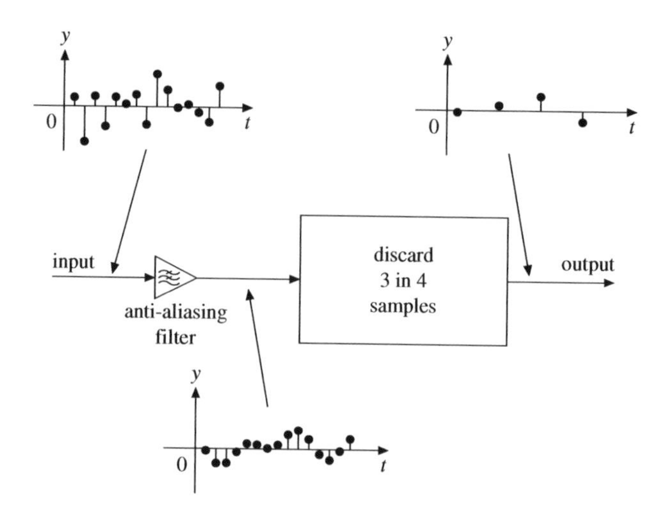

Figure 5.24 Downsampling.

and the input signal. These results are all sampled versions of the same signal, but with the samples taken at different time offsets.

Non-integer ratios

If resampling is required with a non-integer ratio between input and output rates, the task is more complicated.

If the ratio is a fraction whose numerator and denominator are small integers then it is possible to proceed in two stages, first upsampling to a suitable higher sample rate and then downsampling to the desired final result. For example, suppose we wish to convert an audio signal from a 48 kHz sample rate to a 32 kHz sample rate, a ratio of 3 : 2. We first upsample by a factor of 2 to 96 kHz, and then downsample by a factor of 3 to 32 kHz. The process is illustrated in Figure 5.25(a), where the sample rates used at each point in the process are shown alongside the signals.

The graphs across the top of the figure show the frequency components present in the signal at each stage, with f_N being the Nyquist limit in each case. Note that an anti-aliasing filter with a cutoff frequency of 16 kHz (the Nyquist limit for the output sample rate of 32 kHz) is needed in front of the decimation stage. Again, we could combine these last two processes to reduce the amount of computation required, but there is an even more efficient way to proceed.

When we designed an interpolator in Section 5.6 we started with a filter having a flat frequency response and then modified its phase to move its time-domain response by a fraction of a sample. There is no reason why we have to start with a flat frequency response: if we wished to combine a filtering function with interpolation, we could do so in a single filter by simply starting from a different initial frequency response.

In this case we design an interpolator with an input sample rate of 48 kHz and an output sample rate of 96 kHz. Rather than starting the design with a flat frequency response, however, we start with a low-pass frequency response with a cutoff frequency of 16 kHz. This is effectively the same as putting the anti-aliasing filter before the interpolator rather than after it, as shown in Figure 5.25(b). We then immediately follow the interpolator with a three-to-one decimator to reduce the sample rate down to 32 kHz. Since the signal leaving the interpolator will not contain any frequency components above half the final output sample rate, 16 kHz, the decimator will not introduce any aliases.

The final step is to modify the interpolator in the same way as before so that it does not compute any output samples that will be immediately discarded. The system is now a particular case of the more general sample-rate conversion method that we will describe next.

A polyphase interpolation filter

If the ratio of the sample rates is not a simple fraction we need a more general approach. On the one hand, we prefer in general to do as little computation as

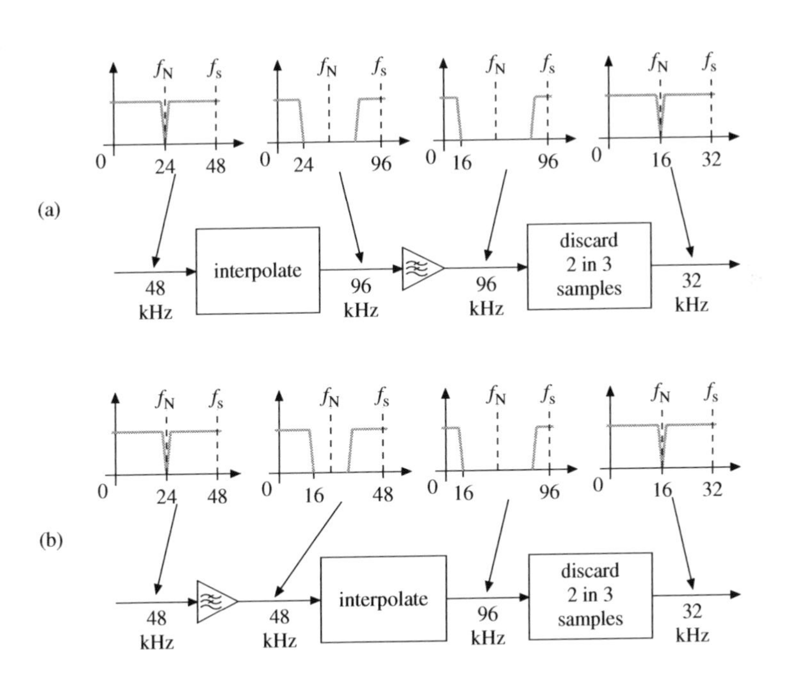

Figure 5.25 Downsampling by a simple fractional ratio: (a) simple strategy; (b) alternative strategy that can be implemented more efficiently.

possible; on the other hand, it seems that we cannot avoid doing at least one filtering calculation for each output sample we generate.

As an example, we will consider converting from the CD audio sample rate of $f_i = 44\,100\,\text{Hz}$ to the sample rate of $f_o = 32\,000\,\text{Hz}$ used by many other items of audio equipment. The ratio between these frequencies is not a particularly simple one.

If we only had to generate a single sample of output data, we would proceed as follows. First, we would calculate the position of the desired output sample in terms of the input sample indices. For example, suppose we want the sixth output sample: we will say that the output offset $t_o = 6$. This sample occurs $t_o/f_o = 6/32\,000 = 187.5 \times 10^{-6}\,\text{s}$ after the sample with index zero. In terms of the input samples, this is $187.5 \times 10^{-6} \times f_i = 187.5 \times 10^{-6} \times 44\,100 = 8.26875$ samples after the sample with index zero: see Figure 5.26 (where the possibility of aliasing has been ignored). We shall call this number the input offset t_i. Observe that the input offset is equal to the output offset multiplied by the ratio of the input sample rate to the output sample rate: $t_i = t_o f_i / f_o = 8.26875$.

Next we observe that the output sample rate is lower than the input sample rate, and so there is the possibility of aliasing during the conversion. An anti-aliasing filter is therefore needed. Figure 5.27(a) shows its ideal frequency response: the cutoff frequency is $\frac{1}{2} \cdot 32\,000 = 16\,000\,\text{Hz}$. Converting this to sampled form suitable for operation at the input sample rate of 44 100 Hz gives us Figure 5.27(b) (we have chosen $N = 32$).

Now we advance the impulse response by $t_i = 8.26875$ input samples by multiplying the frequency response by $e^{2\pi \cdot 8.26875 f/N}$: Figure 5.27(c) shows a few

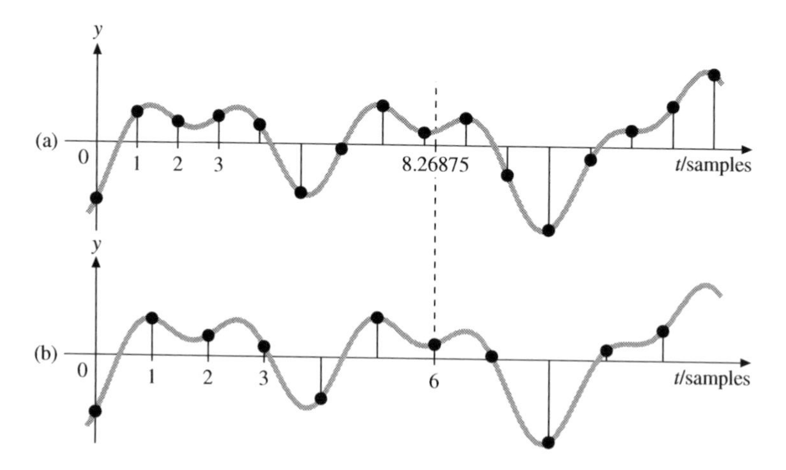

Figure 5.26 Polyphase interpolation: (a) original signal sampled at 44.1 kHz; (b) signal resampled at 32 kHz.

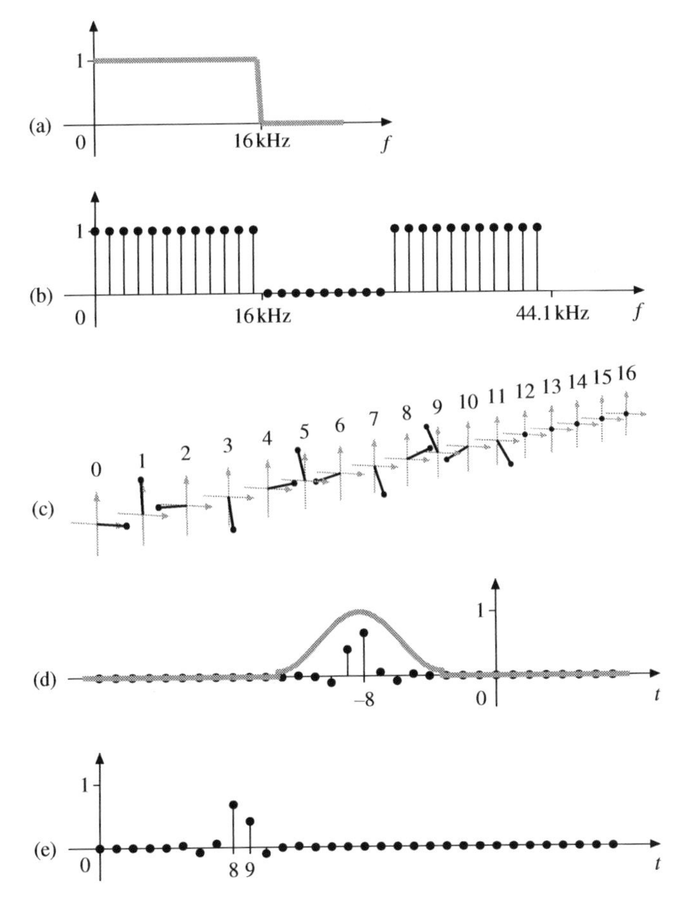

Figure 5.27 Design of an interpolation filter: (a) ideal magnitude response of low-pass filter; (b) sampled magnitude response of low-pass filter; (c) frequency components multiplied by a complex exponential to offset response by 8.26875 samples in time; (d) inverse Fourier transform of (c) gives filter impulse response (window function superimposed in grey); (e) implied weighting of input samples.

samples of the (now complex) result. Returning the response to the time domain by taking an inverse Fourier transform yields Figure 5.27(d). This is the impulse response of a filter which, at time $t = 0$, will produce our desired interpolated sample. We have applied a window to the filter coefficients as shown, centred at time $t = -8.26875$. This impulse response is real.

Figure 5.27(e) shows the impulse response of the filter reversed in time. As you can verify using Equation (5.4), this illustrates the contribution from each input sample $x(t)$ to the interpolated output sample. As we would hope, the largest contributions are from the eighth and ninth samples of the input.

We could in principle do the above sequence of calculations for each output sample we wish to generate, but that would be extremely time-consuming. We could precalculate and store the filter coefficients for each possible output sample position, but this idea is only feasible if we want to work with short output sequences.

In the example above we calculated the coefficients of a filter to generate an output sample positioned at an offset of $t_i = 8.26875$ from the beginning of the input. Suppose we wanted to produce an output sample positioned at an offset of $t_i + 1 = 9.26875$ samples in the input: then we could simply use the same filter coefficients, but shifted along in time by exactly one sample. In other words, the values of the filter coefficients depend only on the fractional part of the offset, their position being determined by the integer part of the offset.

This suggests the following idea. First we precalculate sets of filter coefficients for various fractional input offsets from 0 to 1. Then, when we need an output sample at offset t_o, we calculate the corresponding input offset $t_i = t_o f_o / f_i$. We use the fractional part of t_i as an index into a precalculated table of filter coefficients, and we then shift these coefficients along in time by a number of samples given by the integer part of t_i. We apply the filter and obtain an output sample.

Figure 5.28 shows the impulse responses of a set of filters designed for the input and output sample rates above corresponding to fractional input offsets ϕ from 0 to 0.9 in steps of 0.1. As you can see, the filter for offset 0 is dominated by the coefficient at time $t = 0$. As we increase the fractional offset, the coefficient at time $t = -1$ increases until at offset 0.5 the coefficients at times $t = 0$ and $t = -1$ are equal. By offset 0.9 the coefficient at $t = -1$ dominates. As we would expect, the shape of the coefficient values is similar to that for offset 0, but shifted by one sample.

Thinking in terms of the contributions of the input samples to the interpolated output sample, we can say that the majority of the contribution moves gradually from input sample 0 to input sample 1 as the offset increases from 0 to 1.

The set of filters in Figure 5.28 is sufficient to let us produce output samples positioned to the nearest tenth of a sample in the input. If we wanted a sample at time $t_i = 8.26875$, we could round off the fractional part 0.26875 and use the filter for offset 0.3: as you can see, the impulse response shown in Figure 5.28 for a phase offset of 0.3 samples is rather similar to that shown in Figure 5.27(d). The rounding introduces a small error in the result, analogous to the phenomenon of 'jitter' that we discussed in Chapter 3. As usual, there is a compromise to be made, here between the accuracy of the results and the amount of storage required to hold the coefficients of the precalculated filters.

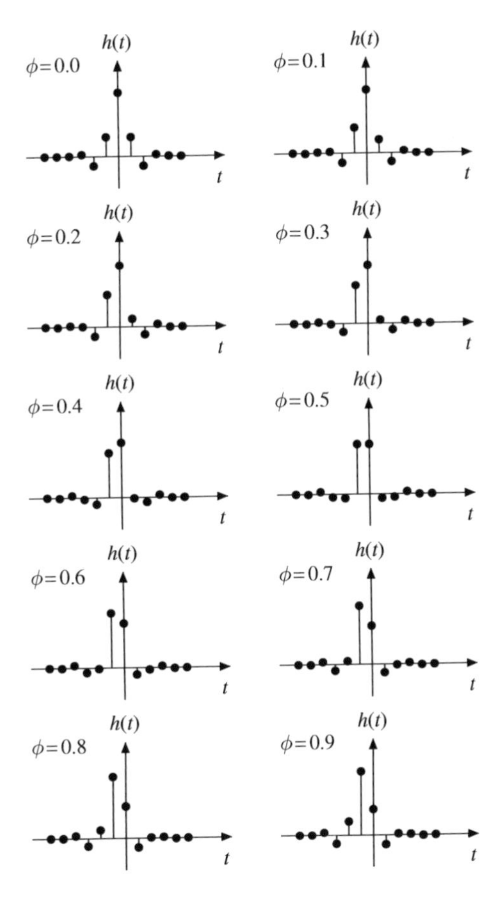

Figure 5.28 Impulse responses of a set of polyphase interpolation filters with phase offsets from 0 to 0.9 samples in steps of 0.1.

An economical way to get a better answer with a small number of filters is to add a linear interpolation step. In the above example, we would calculate output samples at offsets $t_l = 8.2$ samples and $t_r = 8.3$ samples. Note that t_l lies to the left of the desired sample, and t_r lies to its right. We then linearly interpolate between the results for t_l and t_r to produce a result for t_i.

The system of filters we have described above is called a *polyphase interpolation filter*. The word 'polyphase' is used because the set of filters we use have different phase responses (although they all fit within the same overall magnitude response envelope).

In the example above we started from a low-pass filter response because we wanted to avoid aliasing in the result. It is of course possible to start from any desired filter response, thus combining sample rate conversion and arbitrary filtering into a single operation.

Direct calculation of interpolation filter coefficients

It is possible to calculate the coefficients required for an interpolation directly using an extension of the technique we used above to construct a low-pass filter. Again we shall simply state the most important result without proof.

The coefficients of a filter whose output at time 0 corresponds to the interpolated value of its input signal at time ϕ are given by

$$h(t) = w(t+\phi)\,\mathrm{sinc}\,k(t+\phi)$$

where w is a windowing function as before.

Notice that this is just a rewriting of Equation (5.5) with $t+\phi$ replacing t on the right-hand side; in other words, it is the set of coefficients for a low-pass filter shifted in time by ϕ. In a downsampling application low-pass filtering is required to prevent aliasing and k is set equal to the ratio of the two sample rates involved, which will be less than 1. In an upsampling application, where there is no risk of aliasing, k is set equal to 1.

Multi-rate systems

An FIR filter has to evaluate Equation (5.4) once for each output sample. The number of arithmetic operations per second involved is therefore proportional to the length of the filter (the number of summands in Equation (5.4)) and to the output sample rate of the filter. It therefore makes sense to try to design a system so that each filter runs at as low a sample rate as possible. As long as we take care to avoid aliasing we can process signals in a given frequency band at a lower sample rate than the input. This can lead to a significant saving in total operation count, even when taking into account the operations required to convert from the higher sample rate to the lower one. We shall return to this idea in Chapter 8 when we discuss audio compression schemes such as MP3 where the signal is split into a number of frequency bands which are processed separately.

A system that uses a variety of different internal sample rates to process signals is called a *multi-rate system*.

5.7 Infinite impulse response filters

We now turn to the second main class of filter, the *infinite impulse response*, or *IIR* filter.

As the name implies, an IIR filter has an impulse response that goes on (at least in theory) for ever. It achieves this by recirculating the output samples it generates back into the filter, each output sample being dependent on the previous output sample.

A simple example of an IIR filter is a *decaying-average filter*, where each output sample is a weighted average of the previous output sample and the current input sample:

$$y(t) = ax(t) + by(t-1) \tag{5.6}$$

where $a + b = 1$. Figure 5.29 shows the result of applying this filter to the same signal as the one we used to illustrate simple FIR filters in Figure 5.1 to Figure 5.3: here we have set $a = 1/10$ and $b = 9/10$.

Figure 5.30 shows the impulse response of the filter, from which the reason for the name 'decaying-average filter' should be apparent. The time taken for the output to decay to a level $1/\mathrm{e} \approx 0.36788$ times its initial peak value (where e is the base of natural logarithms) is called the *time constant* of the filter, often written using the symbol τ. In this example, each sample is 0.9 times the previous one, and so the time constant is given by

$$0.9^{\tau} = \frac{1}{\mathrm{e}}.$$

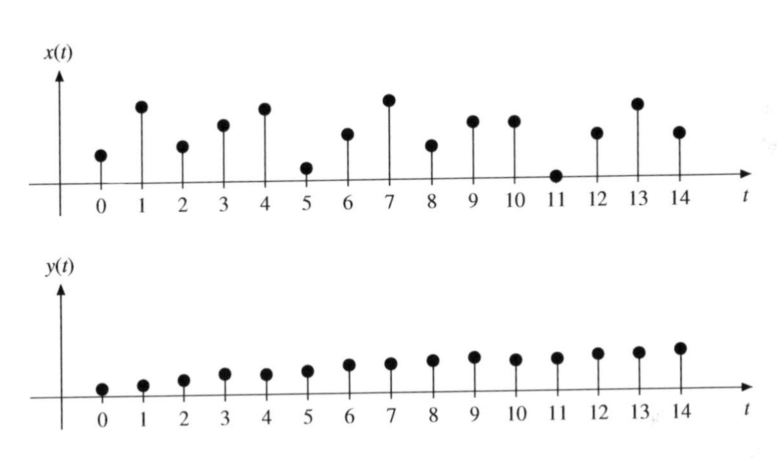

Figure 5.29 Input and output of a decaying-average filter.

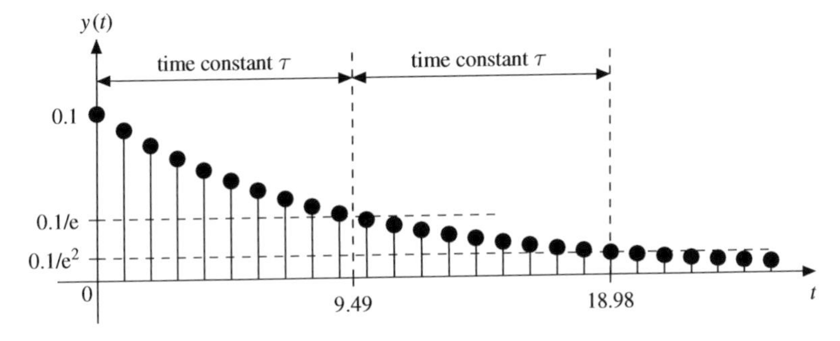

Figure 5.30 Impulse response of a decaying-average filter.

Taking logarithms of both sides, we obtain

$$\tau \log 0.9 = \log \frac{1}{e} = -1$$

and so

$$\tau = \frac{-1}{\log 0.9} = 9.49 \text{ samples.}$$

Although in the case of this filter the output decays gently (the filter is *stable*), it is possible to construct filters like this where the output signal increases in magnitude without limit. Such filters are called *unstable* and are rarely of practical use.

More generally, we can take a weighted average of several previous output samples and combine them with an input sample thus:

$$y(t) = ax(t) + \sum_{t_1=1}^{N} b(t_1)y(t - t_1). \tag{5.7}$$

Comparing this equation with Equation (5.4) we can see that the summation applies an FIR filter with coefficients $b(k)$ to the output signal $y(t)$, with the restriction that $b(0) = 0$. This is represented in the block diagram of Figure 5.31.

Let us work out the frequency response of this IIR filter. We shall start by assuming that the signals involved are periodic. (This is a reasonable assumption if the impulse response of the filter decays gently, as with periodic input the output will eventually settle to a signal which is also approximately periodic.) As before, write $X(f)$ for the Fourier transform of the input signal and $Y(f)$ for the Fourier transform of the output signal. If the response of the FIR filter block is $H(f)$ the Fourier transform of the signal at point A in Figure 5.31 will be $A(f) = Y(f)H(f)$. The Fourier transform of the signal at point B is just $aX(f)$

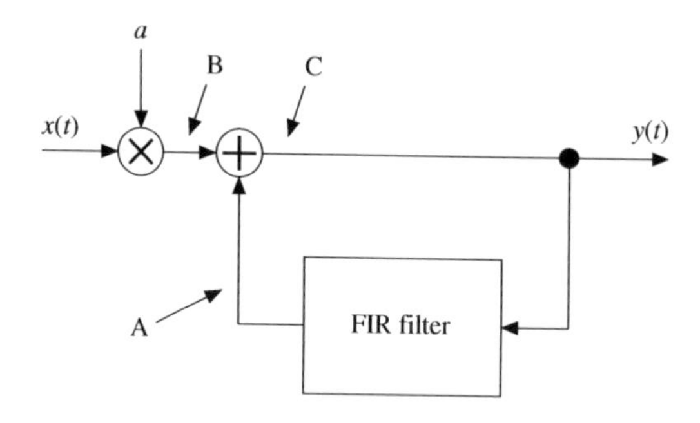

Figure 5.31 Simple infinite impulse response filter.

and so the Fourier transform of the signal at point C is $Y(f)H(f) + aX(f)$. But this is equal to $Y(f)$ and so we have

$$\begin{aligned} Y(f) &= Y(f)H(f) + aX(f) \\ Y(f)\big(1 - H(f)\big) &= aX(f) \\ Y(f) &= \frac{aX(f)}{1 - H(f)} \end{aligned}$$

and the frequency response of the filter is equal to

$$\frac{Y(f)}{X(f)} = \frac{a}{1 - H(f)}. \tag{5.8}$$

This expression is useful for working out the frequency response of a filter which is already known to be stable. It is of less use in designing a filter from scratch to meet a given frequency response specification, since applying this equation naïvely will probably result in an unstable filter.

Indeed, the problem of designing IIR filters to given specifications is a difficult one, and sophisticated software packages are available to help with the task. As in the case of FIR filters, there is a wide range of different strategies available, many of which produce IIR filters with particular structures. It is common for such programs to produce an IIR filter and an FIR filter intended to be used in series as shown in Figure 5.32, or a cascade of such pairs of filters as shown in Figure 5.33. Using this regular structure simplifies the design process. The blocks outlined in the latter figure are called *biquad sections*.

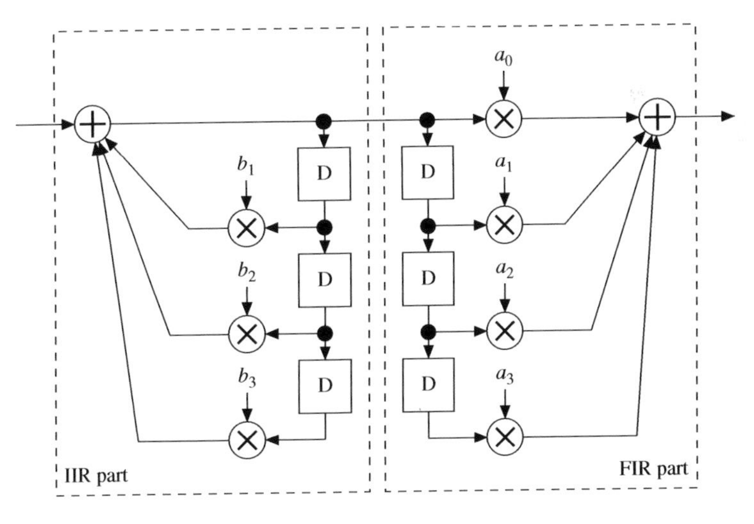

Figure 5.32 General infinite impulse response filter.

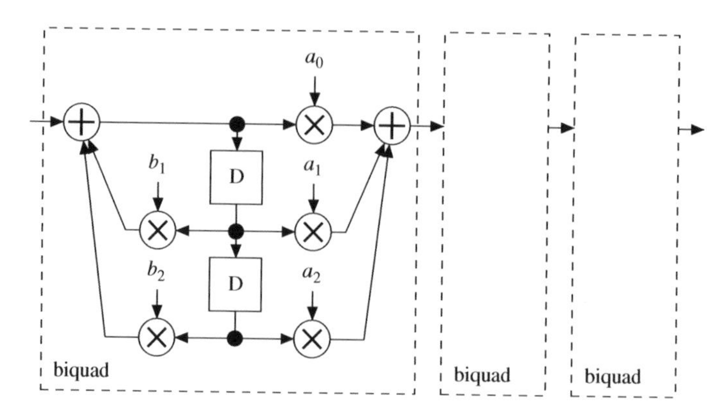

Figure 5.33 Cascade of biquad sections.

Resonators

A useful example of an IIR filter is the *resonator*, of which a simple design is shown in Figure 5.34. This filter has a magnitude response that is sharply peaked around a certain frequency f. The sharpness of the peak is set by the value of r, which must be less than 1 if the filter is to be stable. The impulse response of the filter is a sinusoidal wave that decays slowly. The closer r is to 1, the sharper the peak of the response and the slower the decay of the impulse response. Figure 5.35 shows the frequency response of a few examples of resonators for various values of r and f, calculated using Equation (5.8), and corresponding impulse responses. We have normalised the frequency scale so that $f_s = 1$.

Resonators can be used to detect the presence of specific frequency components in a waveform. For example, DTMF (dual tone multi-frequency) or 'touch tone' telephones produce a combination of two sine waves of different frequencies when a button is pressed, the frequencies used being dependent on the position of the button on the keypad. A robust DTMF decoder can be made using a set of resonators, one tuned to each of the frequencies used. The amplitudes of the outputs of the resonators are compared to determine which button was pressed.

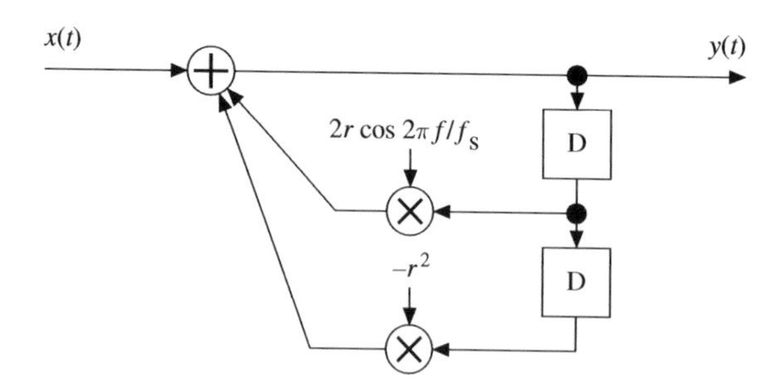

Figure 5.34 Simple resonator.

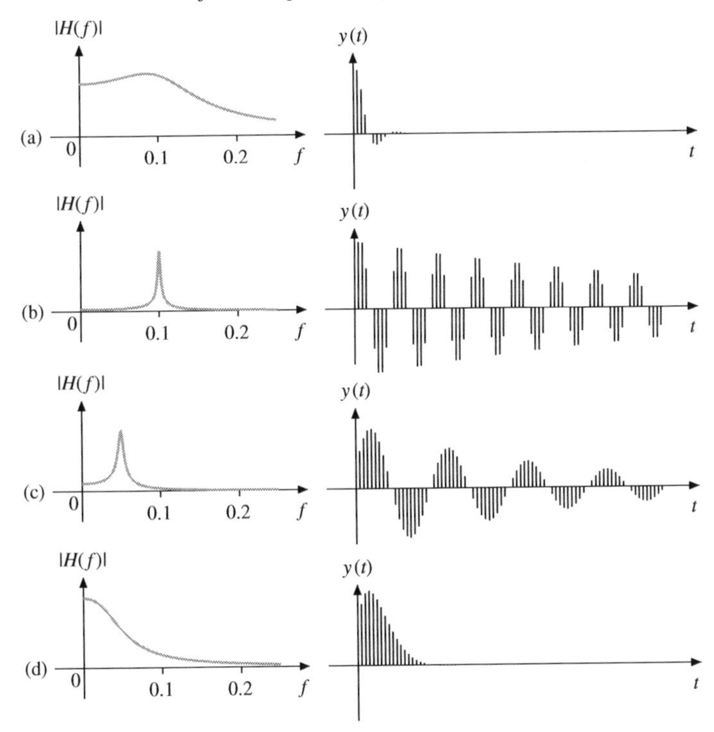

Figure 5.35 Frequency and impulse responses of resonators with various parameters: (a) $r = 0.7,\ f = 0.1$; (b) $r = 0.99,\ f = 0.1$; (c) $r = 0.98,\ f = 0.05$; (d) $r = 0.8,\ f = 0.025$.

Advantages and disadvantages of IIR filters

The main advantage of the IIR filter over its FIR cousin is that, for a given number of arithmetical operations, the IIR filter can usually obtain a closer approximation to a desired frequency response. Alternatively, for a given accuracy of frequency response, it requires fewer operations. It is thus the method of choice where there are tight performance constraints and where the disadvantages listed below can be overcome.

The first disadvantage of the IIR filter is its potential instability. An FIR filter is guaranteed to be stable; an IIR filter needs careful design to ensure that it is stable. Numerical considerations come into play here too: as we shall see in Chapter 7, even a seemingly innocent rounding off of the coefficient values can turn a stable IIR filter into an unstable one.

The second disadvantage of the IIR filter is the complexity of the design methods required. It is especially difficult to create an IIR filter with a specified phase response, which makes them unsuitable for many applications where phase is important.

As we shall see in Chapter 12, a consequence of the fact that each output sample of an IIR filter depends on the previous output sample is that implementations on modern processors with deep pipelines can run much more slowly than expected.

It is worth noting that most analogue filters – in other words, those constructed using analogue electronic components such as capacitors and inductors and operating on continuously varying voltages – have an infinite impulse response. Like their digital cousins, good analogue filters are complicated to design and prone to instability. Just as small changes to coefficients can make a digital IIR filter unstable, small variations in component values can make the difference between a stable analogue filter and an unstable one. And, again as with digital filters, it is particularly difficult to create an analogue filter with a given phase response.

5.8 Filtering complex sequences

We have so far only considered applying our filters to real-valued signals, but there is no reason why we should not use them on complex signals: the equations are the same.

If an FIR filter has real coefficients then we can apply it to a complex signal by operating separately on its real and imaginary parts to generate the real and imaginary parts of the result. The same goes for IIR filters.

For example, we can consider the periodic signal corresponding to samples of a point moving in a circle in the complex plane

$$x(t) = \mathrm{e}^{2\pi \mathrm{j} ft/N} = \cos 2\pi ft/N + \sin 2\pi ft/N$$

where N is the number of samples in one period and f is the number of revolutions in that period.

If we apply a filter with impulse response $h(t)$ to this signal, its output will be given by

$$\begin{aligned} y(t) &= \sum_{t_1=0}^{N-1} h(t_1) \cdot x(t - t_1) \\ &= \sum_{t_1=0}^{N-1} h(t_1) \cdot \mathrm{e}^{2\pi \mathrm{j} f(t-t_1)/N} \\ &= \sum_{t_1=0}^{N-1} h(t_1) \cdot \mathrm{e}^{2\pi \mathrm{j} ft/N} \cdot \mathrm{e}^{-2\pi \mathrm{j} ft_1/N} \\ &= \mathrm{e}^{2\pi \mathrm{j} ft/N} \sum_{t_1=0}^{N-1} h(t_1) \cdot \mathrm{e}^{-2\pi \mathrm{j} ft_1/N} \\ &= x(t) \sum_{t_1=0}^{N-1} h(t_1) \cdot \mathrm{e}^{-2\pi \mathrm{j} ft_1/N}. \end{aligned}$$

The summation in this last expression is just $H(f)$, the Fourier coefficient of $h(t)$ (see Equation (4.1)) at frequency f, and so we have

$$y(t) = x(t)H(f).$$

The output of the filter is just the same as the input, but scaled by $H(f)$ (which may be complex).

This gives us another way of seeing that the frequency response of a filter, as defined by the possibly complex scaling factor it applies to the complex exponential function above, is just the Fourier transform of its impulse response.

FIR filters with complex coefficients

It is possible to design FIR filters with complex coefficients, by and large using the same techniques as in the real case. For example, the windowing method described in this chapter can be used to produce filters with complex coefficients. The main difference is that the mirror-symmetry constraint, which we had to impose on the frequency response to ensure that the impulse response is real, disappears.

The equations in the previous section, showing that the frequency response of a filter is the Fourier transform of its impulse response, also hold for filters with complex coefficients.

Exercises

5.1 How many non-zero samples are there in the output of a five-point moving-average filter supplied with the input sequence (0, 0, 0, 0, 0, 0, 0, 1, 2, 4, 3, 4, 2, 1, 0, 0, 0, 0, 0, 0)? In general, if the coefficients of a filter span a range of m samples ($m = 5$ in this example) and the non-zero samples in the input span a range of n samples ($n = 7$ in this example), what is the maximum number of non-zero output samples you would expect?

5.2 Write down the frequency response of Filter B using the frequency-domain analysis method of Section 5.2. Either calculate the magnitude of this response algebraically and sketch it, or write a program to calculate the magnitude and plot it.

5.3 A smoothing filter is implemented which adds up the samples in a window of 7 consecutive samples in its input, shifting the window by one sample for each output value generated so that successive windows overlap. What is the DC gain of this filter?

5.4 Show that the effect of applying Filter C (Equation (5.3)) to a periodic sequence of samples is the same as that of applying Filter A (Equation (5.1)) to the sequence twice in succession. Explain why we should expect the magnitude of the frequency response of Filter C to be the square of that of Filter A, and verify this claim either numerically or algebraically.

5.5 If a filter is designed so that its output is always equal to its input, its impulse response will simply be an impulse. By thinking about the action of the filter in the frequency domain, show that the Fourier transform of an impulse (Figure 4.7(k)) has all coefficients equal to 1.

5.6 (a) Let the sequence $a(t)$ be (2, 2, 1, 0, 0, 0, 0, 0) and the sequence $b(t)$ be (1, 3, 0, 0, 0, 0, 0, 0), with t running from 0 to 7 in each case. Calculate by hand the convolution of these two sequences.
(b) Using the computer, take the Fourier transforms of these sequences, $A(f)$ and $B(f)$, and multiply the results together point-by-point to obtain $C(f) = A(f)B(f)$. Find $c(t)$, the inverse transform of $C(f)$. Check that $c(t)$ agrees with your result above. (The machine may introduce small numerical errors in its calculation. In particular, you may find that it reports $c(t)$ not as real-valued, but as complex-valued with imaginary parts that are all approximately zero.)
(c) Calculate 122×31 by long multiplication. Compare this process with your hand calculation of the convolution in part (a).
(d) Calculate 103×212 using Fourier transforms. Try calculating 153×292: what goes wrong in this case? Show how to work out the correct product from the result you get.
(e) Estimate the number of elementary operations (additions and multiplications) involved in using long multiplication to calculate the 2048-digit product of two 1024-digit numbers. Using the results from Exercise 4.11, estimate the number of elementary operations required using the Fourier transform method with a transform size of 2048 points (note that the transform size has to be big enough to hold the result).

5.7 You wish to calculate the full convolution (i.e., not the circular convolution) of two signals, $x(t)$ and $h(t)$, using fast Fourier transforms. The sequence $x(t)$ is 256 points long and $h(t)$ is 128 points long. Unfortunately, your fast Fourier transform library only offers forwards and inverse Fourier transforms on sequences of length 256. Show how to calculate the convolution by splitting $x(t)$ into two parts, convolving them separately, and combining the results. What is the total number of forwards and inverse Fourier transforms you need to perform? What would your answer be if the length of $x(t)$ were 257? 258? 259?

5.8 The coefficients of most of the example FIR filters we looked at are symmetric about the middle coefficient (or about a point between the two middle coefficients if the length of the filter is even). Show how to use this symmetry to evaluate Equation (5.4) more quickly on a system where multiplication is much slower than addition, performing only one multiplication for each pair of symmetrically-placed points.

5.9 Design by direct calculation a low-pass filter with a cutoff frequency of 1 kHz suitable for use at a sample rate which your computer's audio output supports, preferably around 32 kHz. Use a Hamming window (Equation (4.3)) 150 samples wide. Plot the impulse response of your filter.

Apply your filter to sine waves of various frequencies and check that its behaviour is as expected. Try a range of frequencies around the cutoff frequency (say from 800 Hz to 1200 Hz) and measure the amplitude of the output. Plot your results.

Repeat the experiment with windows (a) 75 samples wide and (b) 300 samples wide, plotting your results on the same axes.

5.10 Suppose you have laboriously worked out the coefficients for a polyphase interpolation filter with phase offset $\phi = 0.25$ samples. How could you quickly work out a set of coefficients for $\phi = 0.75$ samples?

5.11 Construct a signal $x(t)$ where each sample is greater than or equal to 0 and less than or equal to 1, such that the result of applying the interpolating filter of Figure 5.22 to it includes at least one sample whose value is strictly greater than 1.

5.12 The text gave an example to show how to convert an audio signal from a 48 kHz sample rate to a 32 kHz sample rate by first upsampling by a factor of 2 to 96 kHz and then downsampling by a factor of 3 to 32 kHz. Why is this a better approach than first downsampling by a factor of 3 to 16 kHz and then upsampling by a factor of 2 to 32 kHz?

5.13 In a practical implementation of a polyphase interpolation filter the number of precalculated filters is often chosen to be a power of 2, such as $2^4 = 16$. We then work with input sample offsets in steps of 1/16 from 0/16 to 15/16 inclusive.

If the input sample offset is represented as a fixed-point binary number (see Chapter 7), with five bits after the binary point, give a procedure for determining which set of filter coefficients to use.

5.14 Show that in Equation (5.6) if the filter has a DC gain of 1 then $a + b = 1$.

5.15 Find a pair of values a and b such that the impulse response of the filter described by Equation (5.6) increases in magnitude without limit rather than decaying gradually.

6
Likelihood methods

If you flip a fair coin there is a fifty-fifty chance it will land heads uppermost. We say that the *probability* of this outcome is 0.5.

Philosophers have written reams on what this last statement means, and more generally on the question of assigning a probability to an outcome which is uncertain. An outcome can be uncertain either because it is in the future, or because it is in the past but our information about it is incomplete.

Paraphrasing their positions broadly, some (the 'frequentists') say that the statement in the first paragraph means 'if you repeat the coin flipping experiment many times, in the long run half the results will be heads'. Others (the 'Bayesians') say that the statement means that if you flip a coin once, my strength of belief in the result being heads – either before you flip it or after you flip it, but before you show it to me – can be represented by 0.5 on a scale from 0 ('definitely not the case') to 1 ('definitely the case').

Both frequentists and Bayesians believe in the equations we shall present in the rest of this chapter, though Bayesians lose less sleep worrying about what some of them mean. As engineers we shall ignore the philosophers and use whichever interpretation fits our needs best: it turns out that the Bayesian interpretation of probability is usually the most helpful.

6.1 Probability and conditional probability

We will use capital letters to stand for events. For example, we can write H for the event 'the coin lands heads uppermost' and T for the event 'the coin lands tails uppermost'. We then write, for example, $P(H)$ for the probability of H. The probability of something is expressed as a real number from 0 to 1 inclusive. For a frequentist, $P(H) = 0$ means 'H will never happen however many times you try' and $P(H) = 1$ means 'H will happen every time you try'; for a Bayesian,

$P(H) = 0$ means 'I have unshakeable belief that H is false' and $P(H) = 1$ means 'I have unshakeable belief that H is true'.

If we have a number of events E_i no two of which can happen simultaneously they are said to be *mutually exclusive*.

Furthermore, if exactly one of the set of events *must* happen, we write

$$\sum_i P(E_i) = 1 \tag{6.1}$$

and in general we can simply add up the probabilities of mutually exclusive events to obtain the probability of any one of them happening.

It is Equation (6.1) that determines that the overall scale of probabilities runs from 0 to 1.

If we roll a fair die we have no reason to prefer any of the six possible outcomes over the others, and so they have equal probability. The outcomes are mutually exclusive, and one of them must happen, so the probability of each is $1/6$.

Intersection of events

What is the probability that, if we flip a coin twice, the result is two heads? There are four possible outcomes, which we can write with the obvious notation as HH, HT, TH, TT, and which (in the absence of any sleight of hand) are all equally probable. The answer is therefore $1/4$. We could obtain this result by multiplying the probability of getting a head on the first flip, $1/2$, with the probability of getting a head on the second flip, also $1/2$. The first head restricts the possible outcomes to the first two in the list above, HH and HT, and the second result chooses between these two outcomes.

What is the probability that, if we roll a fair die, the result is an even number? Even and odd are equally probable (there are three numbers of each kind on the faces of the die), and so the answer must be $1/2$.

What is the probability that the result is greater than 3? Again, there are three numbers on the faces of the die greater than 3 and three less than or equal to 3, so the answer is again $1/2$.

What is the probability that the result is both an even number and greater than 3? (This is called the *joint probability* of the two events. The joint probability of two events A and B is written $P(AB)$ or $P(A \cap B)$.) From the result above we might expect to be able to work this out by multiplying the probability the result is even, $1/2$, by the probability that the result is greater than 3, also $1/2$, obtaining $1/4$. But looking at the problem another way, we see that the possible results that are both an even number and greater than 3 are 4 and 6, giving us a total probability of $1/6 + 1/6 = 1/3$. Which is correct?

The difficulty in the last example is that the events 'even result' and 'result greater than 3' are not *independent*. In other words, if I tell you that the result is even, there are now two ways it could be greater than 3 (the result could be 4 or it could be 6) versus only one way it could be less than or equal to 3 (if the result is 2). 'Greater than 3' is now twice as likely as 'less than or equal to 3'. The situation is summarised in Figure 6.1, which shows the four combinations of possible cases in four separate boxes. The situations in the four boxes are mutually exclusive, and so we can add up their probabilities.

Figure 6.1 lets us easily work out the answers to the above questions and more besides. For example:

(i) The probability the result is even is the sum of the probabilities in the left-hand column of the table.
(ii) The probability the result is greater than 3 is the sum of the probabilities in the bottom row of the table.
(iii) The probability the result is both even and greater than 3 is the probability in the bottom left-hand corner.

The fact that the event 'even result' is not independent of the event 'result greater than 3' is shown by the fact that the probabilities read in order across the top row of the table are different from those read in order across the bottom row of the table.

Union of events

Figure 6.1 also shows us how to work out the probability of either of two events (or both) happening if they are not mutually exclusive. The probability that a

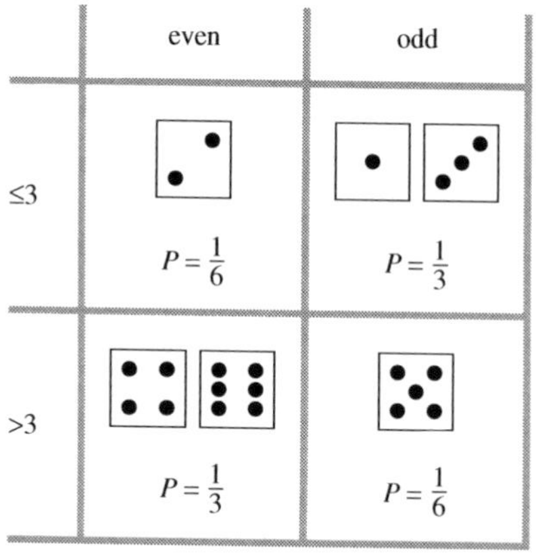

	even	odd
≤3	$P = \frac{1}{6}$	$P = \frac{1}{3}$
>3	$P = \frac{1}{3}$	$P = \frac{1}{6}$

Figure 6.1 Possible outcomes when rolling a die.

roll of a die is either greater than 3 or even or both is obtained by adding up the probabilities in the top left, bottom left and bottom right boxes. We see that simply adding up $P(\text{result even})$ (the sum of the left column) and $P(\text{result} > 3)$ (the sum of the bottom row) is wrong because it double-counts the box in the bottom left corner, the probability that both events occur. The probability of either or both of events A and B occurring is thus given by

$$P(A \cup B) = P(A) + P(B) - P(AB).$$

Note that if the events are in fact mutually exclusive, $P(AB) = 0$ and the expression reduces to $P(A) + P(B)$ as we would expect.

Conditional probability

The *conditional probability* of an event is its probability after some extra piece of information is taken into account. This extra information may take the form of the outcome of another event or it may be some piece of knowledge obtained from elsewhere. The notation used is to write the extra information after a vertical bar symbol, so $P(A \mid B)$ is the probability of event A when piece of information B has been taken into account. It is read aloud as the 'probability of A given B'.

In the example above the probability of the event 'die roll is greater than 3' changed when I told you that the result was even. Using Figure 6.1 again, being given this extra piece of information is equivalent to saying that the combination of events that occurred is definitely in the left-hand column of the table: see Figure 6.2. The probability that the result is greater than 3 given that it is even is the probability in the bottom left-hand corner box expressed as a fraction of the total probability that remains after we have removed the right-hand column, or

$$P(\text{result} > 3 \mid \text{result even}) = \frac{\frac{1}{3}}{\frac{1}{6} + \frac{1}{3}} = \frac{2}{3}.$$

You can verify for yourself that $P(\text{result} \leq 3 \mid \text{result even}) = 1/3$ and thus that $P(\text{result} > 3 \mid \text{result even})$ is indeed twice as big as $P(\text{result} \leq 3 \mid \text{result even})$ as we said above.

In general for two events A and B we can construct a table as in Figure 6.3, where we have used the notation $\overline{A}$ to represent the event 'not A'. (Some people write A^{C}, or 'the complement of A'.) You should be able to convince yourself from the figure that

$$P(A \mid B) = \frac{P(AB)}{P(AB) + P(\overline{A}B)} = \frac{P(AB)}{P(B)}.$$

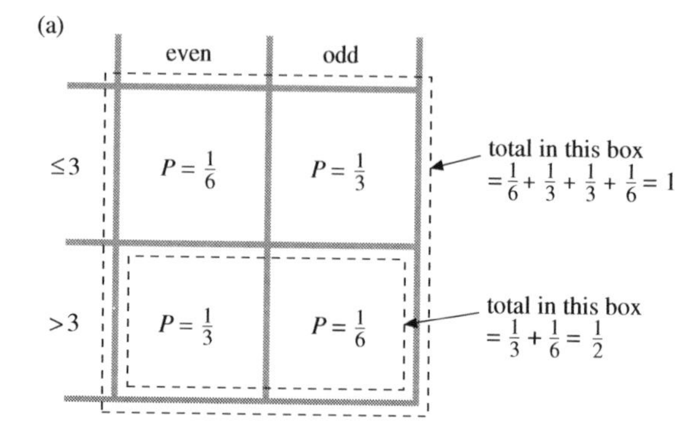

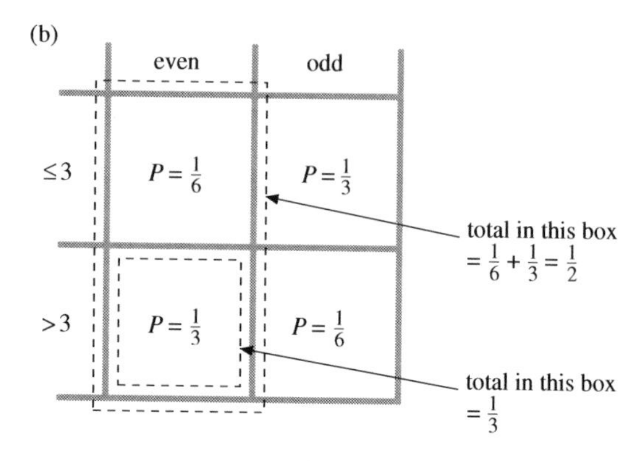

Figure 6.2 (a) Probability of a roll greater than 3 is $\frac{1}{2}/1 = \frac{1}{2}$; (b) probability of a roll greater than 3, given that the roll is even, is $\frac{1}{3}/\frac{1}{2} = \frac{2}{3}$.

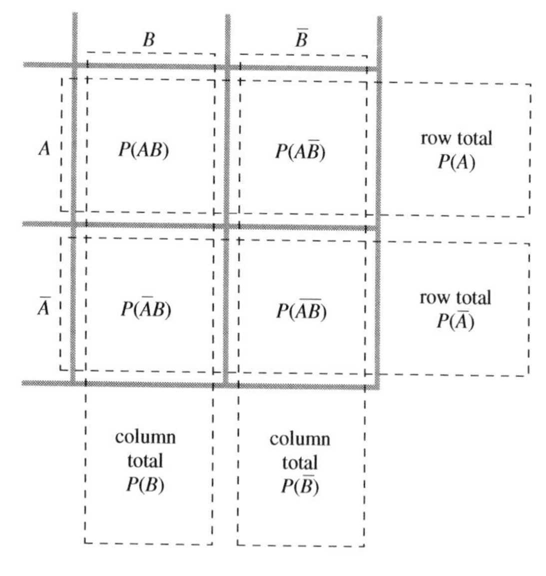

Figure 6.3 Events and their complements: the four combinations for two events.

Similarly, by exchanging the labels A and B, or by transposing the table, we can write

$$P(B \mid A) = \frac{P(AB)}{P(A)}$$

and so

$$P(AB) = P(A \mid B)P(B) = P(B \mid A)P(A)$$

and

$$P(A \mid B) = \frac{P(B \mid A)P(A)}{P(B)}. \tag{6.2}$$

This last equation is called *Bayes' theorem.*

In these equations, $P(A)$ represents what we knew about A before we were aware of piece of information B, and is called the *prior probability*. $P(A \mid B)$ represents what we know about A after piece of information B came along, and is called the *posterior probability*.

In other words, Equation (6.2) is a 'recipe for learning': it tells you how you should update your beliefs about A in the light of new information B. As more information comes to light we can apply the equation iteratively, the calculated posterior probability at each iteration becoming the prior probability for the next iteration.

6.2 Probability and signal processing

How is probability useful in signal processing? Very often we receive a signal that has been corrupted by noise or interference of some kind. For example, suppose we are trying to build a simple speech recognition system that can work in a car to recognise the driver saying one of two words, 'on' and 'off', to turn the headlights on and off. The signal we get from the microphone includes sound from the engine, from other people talking in the car, from the radio, from the wind outside or any number of other sources. The recognition system might receive a segment of waveform W and we need to find a way to decide which spoken word it corresponds to.

One way to phrase this question is in terms of conditional probabilities. We can ask 'what is the probability that the driver said "off" given that the waveform samples are the ones we have?', which we can write as $P_0 = P(\text{'off'} \mid W)$, and, similarly, 'what is the probability that the driver said "on" given that the waveform samples are the ones we have?', which we can write as $P_1 = P(\text{'on'} \mid W)$.

Using conditional probabilities lets us express the other knowledge we have about the situation as contributions to the prior probability. In this example, the system may know the speed of the engine and hence be able to estimate its contribution to the background noise. We can represent such extra pieces of knowledge by K and include them after the vertical bar in our expressions for probability as $P_0 = P(\text{'off'} \mid W, K)$.

More subtly in this example, we could use the current state of the headlights as an extra piece of information: presumably if they are currently off, it is much more likely that the driver will say 'on' than 'off', and vice versa. Indeed, a system that just flips the state of the headlights every time it hears either word might well manage to fool a driver for some time into thinking that it was correctly differentiating between the words 'on' and 'off'.

Of course, there are many other things that the driver could have said besides 'off' or 'on', but those possibilities do not interest us. It does mean, however, that we should not expect P_0 and P_1 to add up to 1. We can still decide between the alternatives by seeing which of P_0 and P_1 is greater; and we can see how confident we are in our decision by looking at the ratio of P_0 to P_1, called a *likelihood ratio*:

$$\Lambda = \frac{P_0}{P_1}.$$

The likelihood ratio compares two different hypotheses (in this case the hypothesis that the word said was 'off' versus the hypothesis that the word said was 'on') as ways of explaining the same data (in this case the waveform W).

If $\Lambda < 1$ we prefer the 'on' hypothesis, and the closer Λ gets to zero, the more we prefer it; if $\Lambda > 1$ we prefer 'off', and the larger Λ gets, the more we prefer it. If $\Lambda = 1$ we have no reason to prefer either hypothesis over the other. Our speech recognition system might prompt the driver to repeat his command if it found $\Lambda \approx 1$.

The advantage of using the likelihood ratio is that we do not need to take into account the other possibilities besides P_0 and P_1: we are just comparing these alternatives against one another. There is also no need to worry if the method we use to arrive at P_0 and P_1 introduces an overall constant scaling factor, since it will cancel out when we divide one by the other.

As we shall see below, it is often easier to work with the logarithms of probabilities. We can then also work with the logarithm of the likelihood ratio

$$\log \Lambda = \log P_0 - \log P_1.$$

On the log-likelihood scale 0 represents the situation where there is no preference between the two alternatives, positive values represent a preference for the event associated with P_0, and negative values prefer the event associated with P_1. The magnitude of the value in each case indicates the strength of the preference.

Of course, any common scaling factor in P_0 and P_1 will again cancel out when we calculate the logarithm of the likelihood ratio.

6.3 Noise

Let us now turn to a more concrete example. Figure 6.4(a) shows a signal $a(t)$ that at any given time takes on one of the values $+1$ or -1. This signal is transmitted, for example over a radio link, and we receive it as $b(t)$ which is equal to $a(t)$ with noise $n(t)$ added as shown in Figure 6.4(b) and 6.4(d): $b(t) = a(t) + n(t)$. The noise in this example is called *additive*. It is also *white*, which means that on average the magnitudes of all the Fourier components of $n(t)$ are the same; and it is *Gaussian*, which means that if you make a histogram of the sample values of $n(t)$ using suitably small bins (the *distribution* of the noise) the result will look like a bell curve (a 'normal distribution'): see Figure 6.4(c).

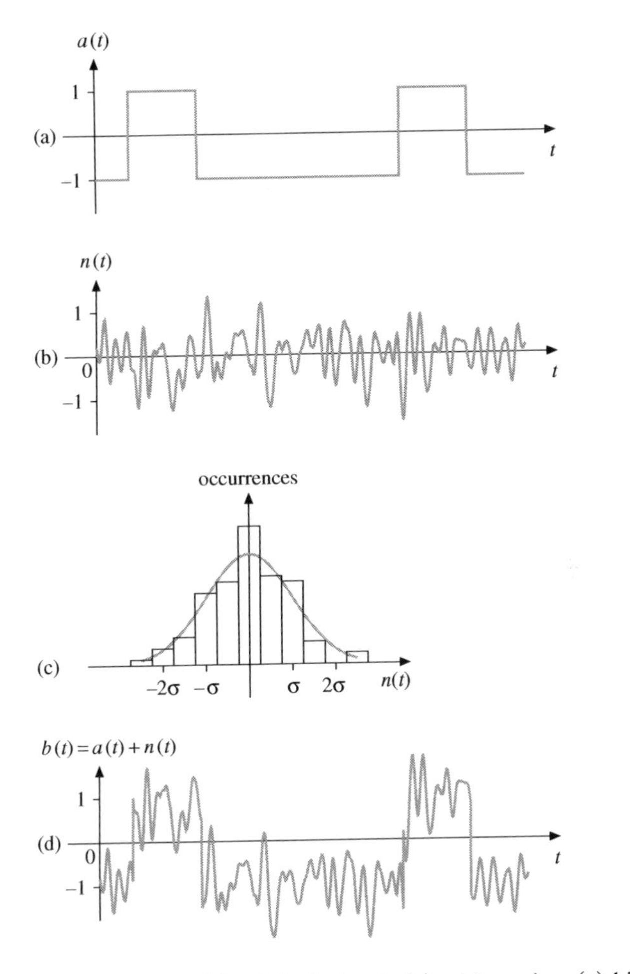

Figure 6.4 A signal in noise: (a) original signal; (b) white noise; (c) histogram shows distribution of noise samples; (d) sum of signal and noise.

The phrase 'additive white Gaussian noise' is sometimes abbreviated to 'AWGN'.

A normal distribution is characterised by its *mean* μ, which is the value about which the distribution is centred, and its *standard deviation* σ, which is a measure of how widely spread out the distribution is. The bigger σ, the wider the distribution. The example in Figure 6.4 has mean $\mu = 0$ and standard deviation $\sigma = 0.5$. Figure 6.5 shows some more examples. As a rule of thumb, approximately 68% of the samples will on average fall within one standard deviation of the mean (i.e., between $\mu - \sigma$ and $\mu + \sigma$); about 95.4% within two standard deviations of the mean (between $\mu - 2\sigma$ and $\mu + 2\sigma$); about 99.73% between $\mu - 3\sigma$ and $\mu + 3\sigma$; about 99.994% between $\mu - 4\sigma$ and $\mu + 4\sigma$; about 99.99994% between $\mu - 5\sigma$ and $\mu + 5\sigma$; and about 99.9999998% between $\mu - 6\sigma$ and $\mu + 6\sigma$. As you can see from these figures, the normal distribution has 'very thin tails': in other words, the probability of a random sample being more than a couple of standard deviations away from the mean is very tiny indeed.

The histograms in Figure 6.4 and Figure 6.5 tell us the probability that a given (quantised) noise sample will have a given value. If we refine our histogram by making the quantisation steps smaller and smaller, the result tends to a smooth curve called the *probability density function*, or *PDF*, of the noise distribution,

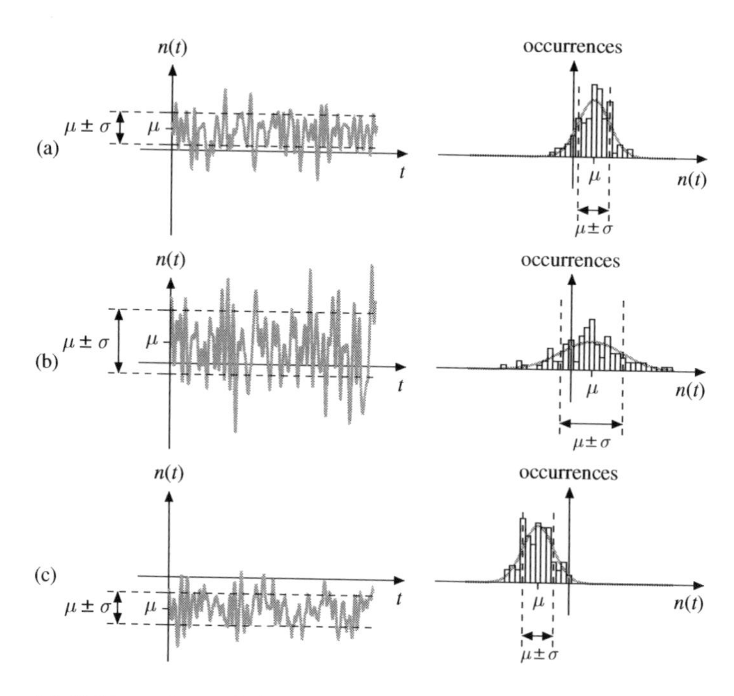

Figure 6.5 White noise and the distribution of its samples: (a) $\mu = 2, \ \sigma = 1.5$; (b) $\mu = 2, \ \sigma = 3$; (c) $\mu = -3, \ \sigma = 1.5$.

shown in grey in Figure 6.4(c) and on the right in Figure 6.5. The full expression for the probability density function of Gaussian noise in terms of μ and σ is

$$P(x) = \frac{1}{\sigma\sqrt{2\pi}} e^{-\frac{1}{2}(x-\mu)^2/\sigma^2}. \tag{6.3}$$

Probability density functions are always scaled (or *normalised*) so that the area under the curve is 1. Although they are functions of continuous variables, they can be treated exactly as if they were probabilities of discrete events.

It is not useful to talk about the probability of a continuous variable taking on a particular value. If we have a distribution like the one above, the chance of a random sample hitting any particular given value exactly is zero: we are looking at an infinitesimally thin slice of the continuous distribution. It does, however, make sense to talk about *quantised* samples taking on given values, since each such sample corresponds to a range of values before quantisation and therefore to a slice of the distribution with positive thickness and to a probability that is in general positive. The normalisation of the probability density function is designed so that we can approximate the distribution of quantised samples by

$$P(n(t) = x) = \Delta \cdot \frac{1}{\sigma\sqrt{2\pi}} e^{-\frac{1}{2}(x-\mu)^2/\sigma^2}$$

where Δ is the width of the quantisation interval. In this example, the approximation is a good one as long as Δ is small compared with σ, or, in other words, as long as the histogram of the quantised samples looks smooth.

If we write $y = (x-\mu)/\sigma$, the number of standard deviations x is away from the mean, and hide the term $\Delta \cdot \frac{1}{\sqrt{2\pi}}$ as a constant of proportionality, we get the more useful form

$$P(n(t) = x) \propto \frac{1}{\sigma} e^{-\frac{1}{2}y^2} \tag{6.4}$$

or the even more useful

$$\log P(n(t) = x) = -\frac{1}{2}y^2 - \log\sigma + k \tag{6.5}$$

where the constant of proportionality is subsumed by k.

In many practical applications σ is the same throughout our calculations, and we can therefore incorporate its value into k. If we only ever compare probabilities by subtracting their logarithms to work out log likelihood ratios, k will cancel out and so can be ignored. We can then proceed as if the log probability were equal to $-\frac{1}{2}y^2$.

How could we try to recover the signal $a(t)$ from the noisy data $b(t)$, armed with the knowledge that $a(t) = \pm 1$? Since we know that each of the noise samples is independent of the others, we can consider each sample individually. Suppose

$b(0) = 0.3$, and assume that the quantisation of $b(t)$ is sufficiently fine that the approximations above are good. Either $a(0) = +1$, in which case $n(0) = b(0) - a(0) = -0.7$, or $a(0) = -1$, in which case $n(0) = b(0) - a(0) = +1.3$. To decide between these two possibilities, given our received sample, we calculate the likelihood ratio

$$\Lambda = \frac{P(a(0) = +1 \mid P(b(0) = 0.3))}{P(a(0) = -1 \mid P(b(0) = 0.3))}.$$

We now apply Bayes' theorem (Equation (6.2)) to top and bottom of the fraction separately:

$$\begin{aligned}\Lambda &= \frac{P(b(0) = 0.3 \mid a(0) = +1)P(a(0) = +1)/P(b(0) = 0.3)}{P(b(0) = 0.3 \mid a(0) = -1)P(a(0) = -1)/P(b(0) = 0.3)} \\ &= \frac{P(b(0) = 0.3 \mid a(0) = +1)P(a(0) = +1)}{P(b(0) = 0.3 \mid a(0) = -1)P(a(0) = -1)}.\end{aligned}$$

At this point we have to decide on a prior distribution over the values of $a(0)$. We may have some reason to believe that one value is more likely than the other before seeing the received sample. For example, the transmission could be using a code where the signal $a(0) = +1$ is used twice as often as $a(0) = -1$, in which case we could reflect this bias by setting $P(a(0) = +1) = 2/3$ and $P(a(0) = -1) = 1/3$. For this example, however, we will assume that the two possibilities are equally likely and so $P(a(0) = +1) = P(a(0) = -1) = 1/2$. We are thus left with

$$\begin{aligned}\Lambda &= \frac{P(b(0) = 0.3 \mid a(0) = +1)}{P(b(0) = 0.3 \mid a(0) = -1)} \\ &= \frac{P(n(0) = -0.7)}{P(n(0) = +1.3)}\end{aligned}$$

Recall that in this example $\mu = 0$ and $\sigma = 0.5$. The value $n(0) = -0.7$ is thus $-0.7/0.5 = -1.4$ standard deviations away from the mean; $n(0) = +1.3$ is $1.3/0.5 = 2.6$ standard deviations away from the mean. We thus have

$$\begin{aligned}\Lambda &= \frac{\frac{1}{\sigma\sqrt{2\pi}} e^{-\frac{1}{2}(-1.4)^2}}{\frac{1}{\sigma\sqrt{2\pi}} e^{-\frac{1}{2}(+2.6)^2}} \\ &= \frac{e^{-0.98}}{e^{-3.38}} \\ &= 11.02.\end{aligned}$$

This means that we prefer the hypothesis $a(0) = +1$ to the hypothesis $a(0) = -1$ by a factor of 11.02.

Is your noise really Gaussian?

You may have found the factor of 11.02 above surprisingly large. After all, $b(0) = 0.3$ is only slightly closer to $a(0) = +1$ than it is to $a(0) = -1$: should the ratio of probabilities really be that big?

The reason the factor is so large is, as we pointed out above, that the tails of the normal distribution are very thin. If the received sample $b(0) = 0.3$ is to arise from a transmitted sample $a(0) = -1$, the noise sample $n(t)$ would have to be 2.6 standard deviations away from its mean, which would be rather unlikely.

Exercise 6.6 asks you to show that if we had used $\sigma = 0.6$ in the example above, the likelihood ratio would be only $e^{1.67} = 5.294$; with $\sigma = 0.4$ the result would be approximately 42.52. As you can see, small changes in the value of σ lead to very large changes in the likelihood ratio.

In a practical application we very often do not know the standard deviation of the noise and are forced to guess it. Alternatively, the noise may be *heteroscedastic* – in other words, its standard deviation may not be constant. If there is even a small error in our estimate of the standard deviation, the magnitudes of the likelihood ratios we calculate become virtually meaningless. We can, however, still safely compare likelihood ratios with 1 to see which of two hypotheses we prefer.

Now let us consider a sampled audio signal received from a (rather poor-quality) analogue radio transmission, with the standard deviation of the background 'hiss' being one tenth of the typical level of the foreground signal. The tails of the normal distribution are so thin that if this noise were truly Gaussian the probability of a noise sample being comparable in amplitude to the foreground signal would be in the region of 10^{-23}. In other words, we would expect to hear a 'click' of this amplitude roughly every 10^{23} samples, or once in about 10^{11} years at a sample rate of 32 kHz: this is clearly not realistic.

This result and the sensitivity of the likelihood ratio to changes in σ can be attributed to the thin tails of the normal distribution. The problem is that the noise present on real-world signals is very rarely Gaussian: the normal distribution is not a good *model* for the noise because it underestimates how often unlikely events occur.

However, a more subtle effect is also at work here. Suppose that we had collected noise samples from a radio over a period of a day and found that the distribution of our samples was indeed well approximated by the Gaussian curve. When we then encounter a 'click', what should we conclude? The obvious explanation is that this is an exceptional event, perhaps due to a nearby thunderstorm or some other source of interference that was not present when we calibrated the model. The lesson here is that the likelihood method is only as good as the noise model used; and if the basic noise model is Gaussian then great care is required when doing calculations on noise samples that are many standard deviations away from the mean. If you wish to do accurate calculations in such situations, you will probably need a more sophisticated noise model. In this example, you might want to add in a small probability of an 'exceptional' sample to account for clicks.

The Laplace distribution

Unfortunately the normal distribution has some convenient mathematical properties that make it very tempting to assume that noise is Gaussian even when it is not.

An alternative, which is also reasonably convenient mathematically, is to model noise using a *Laplace* distribution (also called a *double exponential* distribution). The full expression for this distribution is

$$P(x) = \frac{1}{2b} e^{-|x-\mu|/b}. \tag{6.6}$$

Here μ is again the mean of the distribution, and b plays a similar role to σ in the case of the normal distribution, setting its overall width. To get distributions of comparable width, we need to set $b = \sigma/\sqrt{2}$.

As before we can set $y = (x - \mu)/b$ and obtain

$$\log P(x) = -|y| - \log 2b.$$

Again, if b is constant throughout our calculations it will cancel out when we work out likelihood ratios.

One way to think of the Laplace distribution is that (apart from scaling factors) we can obtain it by replacing $\frac{1}{2}y^2$ by $|y|$ in Equation 6.5. Both $\frac{1}{2}y^2$ and $|y|$ are functions that turn negative numbers into positive ones, and both are suitable ways of measuring the absolute size of an error which can have either sign. However, y^2 grows quadratically rather than linearly, and this is why the tails of the Gaussian are so thin: larger errors are disproportionately more unlikely than smaller ones.

Figure 6.6 shows a plot of the two distributions on the same axes. The plots are scaled so that their bodies appear to have approximately the same width, i.e., with $b = \sigma/\sqrt{2}$. As the enlarged part shows, the tails of the Laplace distribution are much thicker than those of the normal distribution.

There are many other distributions that can be used to model noise. In critical applications, it is important to try to obtain real examples of the signal and noise you will be dealing with so that you can see what model fits the noise best.

Evaluating a noise model

How can we compare two different models of noise to see which fits our data better? Figure 6.7 shows a plot of 30 noise samples $n(t)$. Suppose you knew

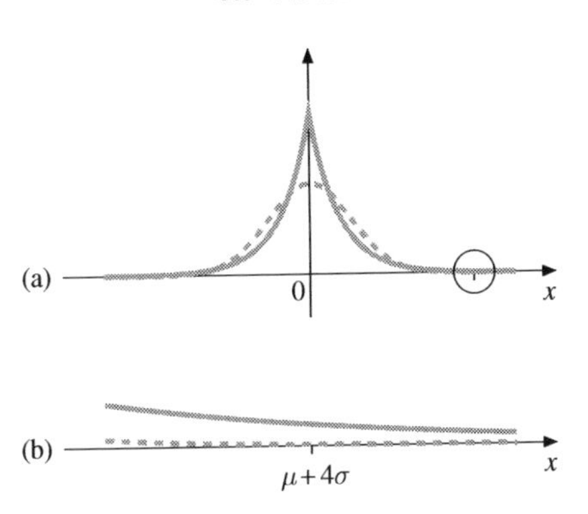

Figure 6.6 The Laplace distribution (solid) and the normal distribution (dotted): (a) main view; (b) close-up of circled part of tail in (a).

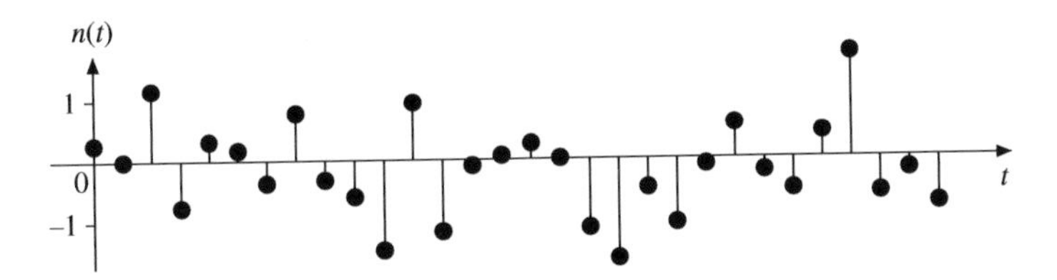

Figure 6.7 Samples from a mystery distribution.

that they were either Gaussian-distributed noise with mean $\mu = 0$ and standard deviation $\sigma = \sqrt{2}$ or Laplace-distributed noise with mean $\mu = 0$ and size $b = 1$. Call these alternatives 'Hypothesis A' and 'Hypothesis B' respectively. Phrased slightly more formally, the question we are trying to answer is 'is the probability of Hypothesis A being true, given the values of $n(t)$ that we have, greater than or less than the probability of Hypothesis B being true, again given those values?'. Put like this, it is clear that we should try to calculate a likelihood ratio:

$$\Lambda = \frac{P(\text{Hypothesis A} \mid n(t))}{P(\text{Hypothesis B} \mid n(t))}.$$

Before you read any further, look again at Figure 6.7 and decide which hypothesis you think is more likely, and by what factor: you may want to refer back to the plots of the distributions in Figure 6.6.

As before, we can apply Bayes' theorem to the top and bottom of the fraction to obtain

$$\begin{aligned}\Lambda &= \frac{P(n(t) \mid \text{Hypothesis A})P(\text{Hypothesis A})/P(n(t))}{P(n(t) \mid \text{Hypothesis B})P(\text{Hypothesis B})/P(n(t))} \\ &= \frac{P(n(t) \mid \text{Hypothesis A})P(\text{Hypothesis A})}{P(n(t) \mid \text{Hypothesis B})P(\text{Hypothesis B})}.\end{aligned}$$

Again we need to decide on our prior probabilities for the two hypotheses. To keep things simple, we shall give them equal probabilities, so that we have

$$\Lambda = \frac{P(n(t) \mid \text{Hypothesis A})}{P(n(t) \mid \text{Hypothesis B})}.$$

We can evaluate the likelihood functions on the top and bottom of this fraction by simply using the expressions for the two distributions given in Equations (6.3) and (6.6). Under the assumption that the noise samples are independent of one another, we can simply multiply the individual probabilities together. We get the slightly fearsome-looking

$$P(n(t) \mid \text{Hypothesis A}) = \prod_t \frac{1}{\sigma\sqrt{2\pi}} \mathrm{e}^{-\frac{1}{2}(n(t)-\mu)^2/\sigma^2} \tag{6.7}$$

and

$$P(n(t) \mid \text{Hypothesis B}) = \prod_t \frac{1}{2b} \mathrm{e}^{-|n(t)-\mu|/b}. \tag{6.8}$$

Carrying out these calculations for the noise samples plotted in Figure 6.7, the results we obtain are

$$P(n(t) \mid \text{Hypothesis A}) = 3.70171 \times 10^{-19}$$

and

$$P(n(t) \mid \text{Hypothesis B}) = 1.20952 \times 10^{-17}.$$

We thus prefer Hypothesis B to Hypothesis A by a factor of about 33.

This example demonstrates a practical benefit of working with logarithms of probabilities and of likelihood ratios. Even in this simple case, where we were only looking at thirty samples in total, the probabilities we calculated were very small. In a practical example we might be dealing with many thousands of values, and the resulting probabilities could easily be too small to represent conveniently on a computer. We will discuss the representation of numbers in more detail in Chapter 7. The reason that the numbers involved are so tiny is that there is a huge number of possible noise sequences, and we are looking at the probability of just one of them.

Working with the logarithms of probabilities throughout keeps the range of numbers involved much more manageable. The products in Equations (6.7) and (6.8) become sums of logarithms, and we can compare logarithms of probabilities to decide between hypotheses as easily as we can compare the probabilities themselves.

Care is needed if a probability is ever zero, since the logarithm of zero is not defined. The simplest approach is to substitute a tiny probability that the machine can represent – perhaps 10^{-30} – should zero ever occur.

Although multiplying simplifies to addition when we represent probabilities by their logarithms, addition of probabilities becomes more complicated. Exercise 6.8 demonstrates what we need to do.

Evaluating a noise model with parameters

In the example above we compared two noise models assuming that we knew their parameters: the mean μ and standard deviation σ in the case of the Gaussian noise, μ and b in the case of the Laplace-distributed noise. In practice we may not know these numbers in advance. We can still compare models, but more care is required when we decide on our prior probabilities.

To illustrate the procedure we shall repeat the above example using the same input data, but without assuming we know in advance the precise values of σ (for the Gaussian noise hypothesis) or b (for the Laplacian noise hypothesis). We shall, however, assume that we know $\mu = 0$ in both cases.

We will again say that we have no prior reason to choose between the two types of distribution, and so we shall set prior probabilities $P(\text{Gaussian}) = 1/2$ and $P(\text{Laplacian}) = 1/2$.

Turning first to the Gaussian case, we have to express what prior knowledge we do have about σ. Let us say that we believe σ is somewhere between 0 and 4, with a two-to-one preference for the middle of this range over the ends. A probability density function that represents this belief (using the Bayesian interpretation of 'probability') is shown in Figure 6.8(a): you should check that the distribution is normalised, i.e., that the area under the distribution is 1. This is our prior distribution on σ. In practice the actual shape of the prior distribution usually does not have much effect on the results, as long as the true value of the variable whose value we are trying to deduce does not receive a prior probability of zero: in this case, for example, we will be in trouble if the true value of σ is greater than 4.

Likewise we can make a prior distribution for b in the Laplacian case, as shown in Figure 6.8(b). This has the same shape as our prior distribution for σ, but scaled

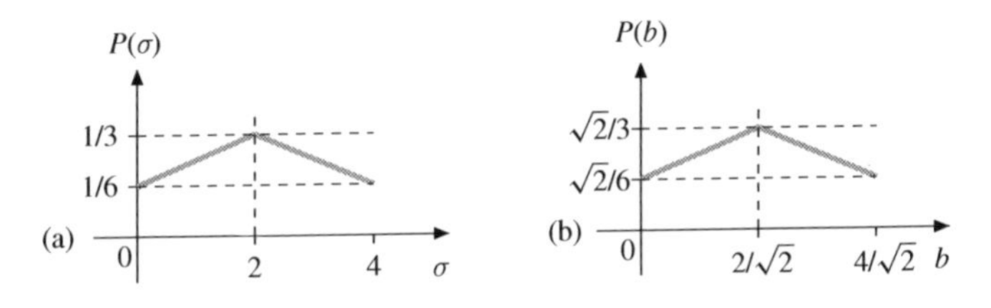

Figure 6.8 Example prior distributions (a) for the standard deviation σ of a normal distribution; (b) for the parameter b of a Laplace distribution.

by a factor of $\sqrt{2}$ as discussed above. The density function has been normalised so that the area underneath it is again 1.

We can now consider a joint hypothesis $G(k)$, 'the noise is Gaussian with parameter $\sigma = k$'. The prior probability of this hypothesis, $P(G(k))$, is the prior probability that the noise is Gaussian (which we set to 1/2) multiplied by the prior probability that $\sigma = k$, which we read off from Figure 6.8(a). Bayes' theorem tells us that

$$P(G(k) \mid n(t)) = \frac{P(n(t) \mid G(k))P(G(k))}{P(n(t))}.$$

We can work out $P(n(t) \mid G(k))$ using the formula in Equation (6.7) substituting k for σ. The resulting likelihood function, ignoring the term $P(n(t))$, is shown by the solid line in Figure 6.9.

Similarly, we consider a joint hypothesis $L(k)$, 'the noise is Laplacian with parameter $b = k$', whose prior probability is the prior probability that the noise is Laplacian (again, 1/2) multiplied by the prior probability that $b = k$, which we get from Figure 6.8(b). We can apply Bayes' theorem again and obtain the likelihood function

$$P(L(k) \mid n(t)) = \frac{P(n(t) \mid L(k))P(L(k))}{P(n(t))}.$$

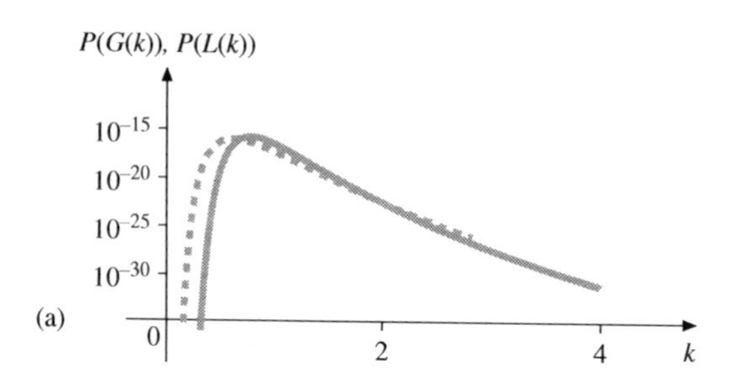

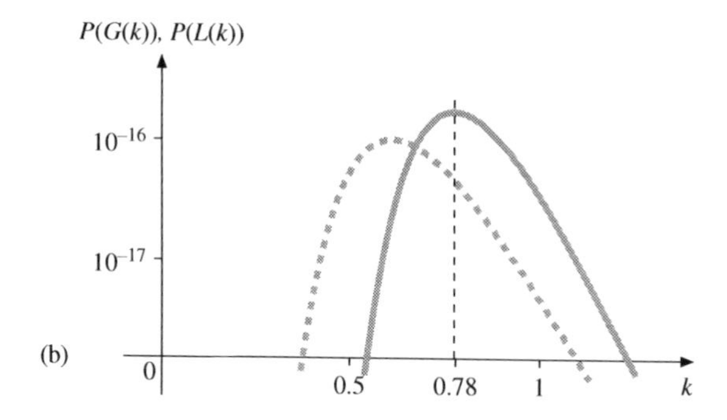

Figure 6.9 Posterior distributions on k, $P(G(k))$ solid and $P(L(k))$ dotted: (a) main view; (b) close-up of the peaks of the curves.

As before, we work this out using Equation (6.8) substituting k for b, and the result is shown by the dotted line in Figure 6.9.

Comparing the close-ups of the two curves in Figure 6.9(b) we see that the most likely hypothesis is $G(0.78)$, that the noise samples $n(t)$ are Gaussian-distributed with σ approximately equal to 0.78. If σ is much less than this the probability of the distribution being Gaussian falls off sharply: underestimating σ puts too many samples into the thin tails of the distribution. This effect is also present, although less pronounced, in the case of the Laplace distribution. (The original samples in Figure 6.7 were in fact generated from a Gaussian distribution with $\sigma = 0.7$.)

It is possible to use the methods described in this section to estimate the values of more than one parameter simultaneously. For example, you might want to estimate both μ and b for a Laplace distribution. The result is a likelihood function of two variables, $P(\mu, b \mid n(t))$ and to find the best model you have to search over the two-dimensional space of values of μ and b. In some cases it is possible to find the most likely parameter values algebraically from the noise samples, but in many cases it is not.

Exercises

6.1 Put on your philosopher's hat and give (a) a frequentist's, and (b) a Bayesian's interpretation of Equation (6.1).

6.2 You wake up with no idea what time it is, and look at your bedside clock. Its seven-segment display shows the time in hours and minutes using the twelve-hour clock, so that the hours digits run from 1 to 12. The digit patterns are like the ones shown in Figure 6.10. The clock is partially obscured by the pile of signal-processing textbooks on your bedside table, as illustrated in Figure 6.11. What is the probability that the time is 2:34? 3:34? 4:34? What is the probability that the units digit of the hours is 2?

Judging by the light streaming through the gap in the curtains, you estimate that the probability that the time is between 11:00 and 11:59 is 1/20; between 12:00 and 12:59, 3/20; between 1:00 and 1:59, 6/20; between 2:00 and 2:59, 6/20; between 3:00 and 3:59, 3/20. between 4:00 and 4:59, 1/20. Given this information, what is the probability that the units digit of the hours is 2?

Figure 6.10 Possible digit patterns displayed by the clock.

Figure 6.11 The bedroom scene.

6.3 Construct a vector of a few hundred thousand independent Gaussian noise samples and play them as audio at a sample rate of 32 kHz or nearby convenient value. Ensure that you have enough samples for several seconds of noise. Apply a twenty-point moving-average filter to the signal (see Chapter 5) and listen to the result. Is the resulting noise (a) Gaussian; (b) white? Explain your answers.

6.4 A popular management consultancy fad is the 'six sigma methodology'. Here 'sigma' measures the standard deviation of some measured output of a manufacturing process, such as the weight of a particular component. The requirements for the component are such that it is rejected if its weight is not within a specified range $w \pm \Delta w$. The aim of the six sigma methodology is to control the process so tightly that $6\sigma < \Delta w$, thus ensuring a low rejection rate.

Proponents claim that their methodology corresponds to a rejection rate of '3.4 defects per million opportunities'. Compare this figure with the numbers given in the text for the proportion of samples falling within six standard deviations of the mean for a normal distribution. What is the most charitable explanation you can think of for the choice of the name 'six sigma'? (Observe that the 'six sigma methodology' implicitly focuses attention on the standard deviation of the results of a process rather than the mean. This is in line with the jargon use of the word 'quality' to mean how precisely repeatable a manufacturing or business process is rather than how useful its results are.)

6.5 Work through the example in Section 6.3 and calculate the likelihood ratio Λ assuming a prior distribution over the values of $a(0)$ given by $P\big(a(0) = +1\big) = 2/3$ and $P\big(a(0) = -1\big) = 1/3$.

6.6 Work through the example in Section 6.3 with $\sigma = 0.6$ instead of $\sigma = 0.5$ and show that the resulting likelihood ratio is approximately 5.294 in favour of the hypothesis $a(0) = +1$.

Repeat with $\sigma = 0.4$ and show that the likelihood ratio in this case is approximately 42.52 in favour of the hypothesis $a(0) = +1$.

6.7 Work through the example in Section 6.3 using a Laplace distribution to model the noise with (a) $b = 0.4/\sqrt{2}$; (b) $b = 0.5/\sqrt{2}$; and (c) $b = 0.6/\sqrt{2}$. Calculate the likelihood ratios in each case. Using the results from the previous exercise, compare the sensitivity of the likelihood ratio to b when noise is modelled using a Laplace distribution with its sensitivity to σ when a normal distribution is used.

6.8 Write and test a function that, given $\log a$ and $\log b$, calculates $\log(a+b)$. Ensure that it works correctly for $-1\,000\,000 < \log a < +1\,000\,000$ and $-1\,000\,000 < \log b < +1\,000\,000$ (at least). (**Hint:** Use the relations

$$\log(a+b) = \log a(1+\frac{b}{a}) = \log a + \log(1+\frac{b}{a})$$

and

$$\frac{b}{a} = \mathrm{e}^{\log b - \log a}$$

and think about what might happen if b is much bigger than a.)

6.9 Record a few seconds of noise samples from an analogue radio not tuned to any station; alternatively, you can record noise from a microphone in a very quiet room, or even disconnect the microphone altogether. Calculate the mean of the noise samples. The mean should be very close to zero. If it is not, subtract the mean from each sample.

Write a program that uses the method in the text to compare the Gaussian and Laplace models for the noise. You will need to look at a histogram of the samples to decide on a suitable range of values for σ and b to examine, and you will need to choose a suitable prior distribution. Calculate the value of the likelihood function at one hundred equally spaced trial values of σ and b and generate a plot like the one in Figure 6.9.

Modify your program so that it only analyses the first n samples of the recording, where you can specify n. Run it with $n = 1, 2, 4, 8, 16$ and so on, until you are using the whole recording. Plot the likelihood function in each case. For small n your plot should look similar to your prior distribution; as n increases the plot should resemble the final result more and more.

Choose a different (but still reasonable) prior distribution and repeat the experiment. How much effect does the choice of the prior have on the final result?

7
Numerical considerations

In this chapter we will look in more detail at how we represent numbers, in particular sample values, in signal processing algorithms. The representations used fall into two main categories: fixed point and floating point.

7.1 Fixed-point representations

In a *fixed-point* system a value is represented by an integer using a suitable scaling factor α. For example, a voltage in the range 0 V to 10 V might be represented by an integer from 0 to 100 using a scaling factor of $\alpha = \frac{1}{10}$ V. We can also write '1 LSB $= \frac{1}{10}$ V', where 'LSB' stands for 'least significant bit': in other words, changing the least significant bit of the binary representation of the integer corresponds to a change of $\frac{1}{10}$ V in the represented value. Figure 7.1(a) shows the representation of the voltage $x = 4.5$ V as a binary number using this scheme. The significance of each bit is also shown. As you can see, the significance of bit b, where the least significant bit is bit 0, is $\alpha 2^b$. If the binary number is viewed as an integer, its value is $n = x/\alpha = 45$. (We shall generally use x to stand for a real number we are trying to represent and n for its representation thought of as an integer.)

As we shall see below, it is sometimes convenient to make the scaling factor a power of 2, such as $\alpha = 1/8 = 2^{-3}$. In this case one of the bits in the fixed-point representation has a significance of exactly 1, and, by analogy with the use of a decimal point, we can imagine that there is a 'binary point' written immediately after that bit. Figure 7.1(b) shows $x = 4.5$ represented in fixed point with a scaling factor of $\alpha = 1/8$, which implies that the binary point is between bits 2 and 3. The representation shown in the figure has five bits before the binary point and three after, and is referred to variously as '5.3', 'Q5.3' or '5Q3' format. We shall use the '5Q3' notation. The bits after the binary point are called the *fraction bits*.

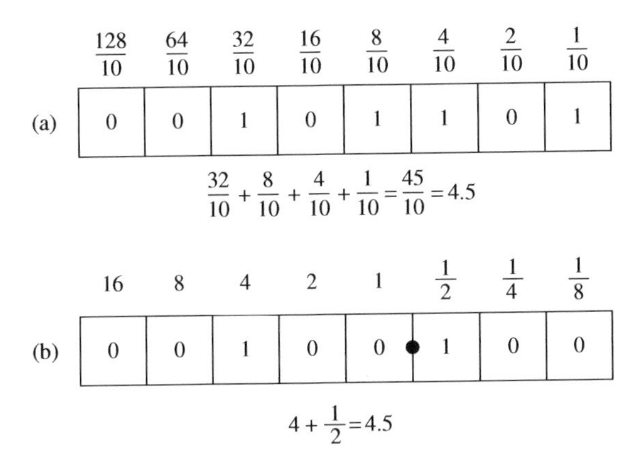

Figure 7.1 The value 4.5 represented in fixed-point notation with (a) $\alpha = 1/10$ and (b) $\alpha = 1/8$.

Some people restrict the use of the term 'fixed point' to refer to schemes where the scaling factor is a power of 2.

Range and precision of fixed-point numbers

If a fixed-point number is expressed in a total of k bits a total of 2^k different representations are possible, from $n = 0$ (all bits zero) to $n = 2^k - 1$ (all bits one) inclusive. The range of possible values is thus $x = 0$ to $x = (2^k - 1)\alpha$ in steps of α.

In uQv format the range of representable values is from 0 to $(2^{u+v} - 1)2^{-v} = 2^u - 2^{-v}$ in steps of 2^{-v}.

Note that we have only considered non-negative or *unsigned* numbers here: negative numbers are discussed in Section 7.2.

Adding and subtracting fixed-point numbers

If two values x_0 and x_1 are represented using a fixed-point scheme with the same scaling factor α then we can simply add them together as integers. Let $n_0 = x_0/\alpha$ and $n_1 = x_1/\alpha$; then $n_0 + n_1 = x_0/\alpha + x_1/\alpha = (x_0 + x_1)/\alpha$ which is the correct answer, also with a scaling factor of α.

If the scaling factor is a power of 2, we can imagine that we are carrying out the addition with the binary points aligned: Figure 7.2 illustrates the addition $4.5 + 1.625$ in fixed-point arithmetic with three bits after the binary point.

If the scaling factors used to represent x_0 and x_1 are not equal, they must be made equal before we can do the addition. We do this by multiplying the

	1	0	0	.	1	0	0	$x_0=4.5$	$n_0=36$
+	0	0	1	.	1	0	1	$x_1=1.625$	$n_1=13$
	1	1	0	.	0	0	1	$x_0+x_1=6.125$	$n_0+n_1=49$

Figure 7.2 Adding fixed-point numbers.

representation of one number or the other by a suitable constant. If x_0 is represented by n_0 with scaling factor α_0, and x_1 is represented by n_1 with scaling factor α_1, then

$$x_1 = \alpha_1 n_1 = \alpha_0 \cdot \frac{\alpha_1}{\alpha_0} n_1$$

and we can make x_1 compatible with x_0 by multiplying n_1 by α_1/α_0. Alternatively, we could have made x_0 compatible with x_1 by multiplying n_0 by α_0/α_1.

If the scaling factors are powers of 2 these multiplications become shifts and can be thought of as 'lining up the binary points' before carrying out the addition. For example, suppose x_0 is represented in 4Q4 format and x_1 is represented in 3Q5 format. To add them together we could shift x_0 up one place, so that it is now in 3Q5 format, and then add it to x_1: the result would also then be in 3Q5 format. Alternatively, we could shift x_1 down one place to convert it to 4Q4 format and then add it to x_0, leaving the result in 4Q4 format. Which route we choose will depend on what format we want our answer in, which in turn depends on the precision and range required of the answer.

The result of the fixed-point addition is in general larger than the numbers being added (the *summands*), and in particular might be too big to be represented in the same format: this is called *overflow*. You might be certain that overflow cannot happen because of what you know about the sizes of the numbers you are adding; otherwise, you need to decide what you are going to do about it.

Processors designed for digital signal processing that use fixed-point arithmetic usually offer a 'clamping' or 'clipping' or 'saturation arithmetic' mode, whereby if the result of an operation is too big to fit in the target representation, it is replaced by the largest representable value. General-purpose processors can emulate this behaviour, of course, but often at a significant cost in speed. Without a mode like this, the result typically 'wraps around' from the maximum representable value to the minimum representable value, in much the same way as the odometer in a car might wrap around from 99 999 to 00 000. The effects of clipping and wrap-around on the output of a program that generates a sine wave are illustrated in Figure 7.3. As you can see, the effect of wrap-around is rather dramatic, resulting in a discontinuous signal; clipping causes less damage to the shape of the waveform.

Another feature offered by some special-purpose processors is a scaling mode, in which the result of an addition is halved. This ensures that the result can be represented in the same format as the summands.

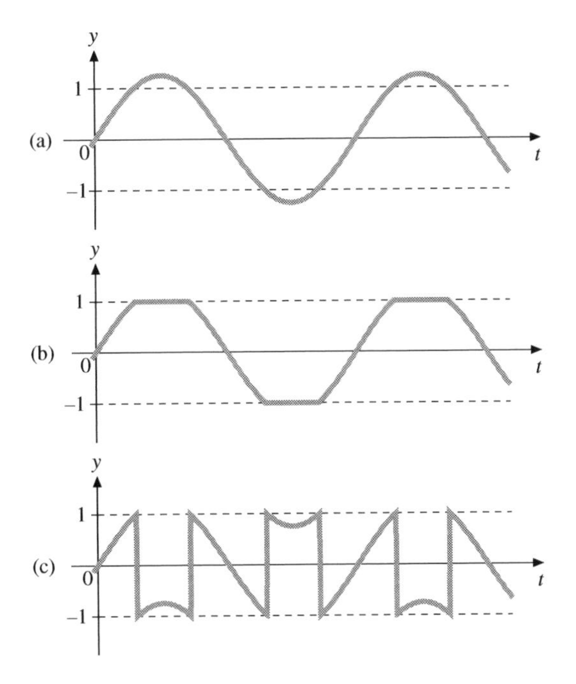

Figure 7.3 (a) Original sine wave; (b) effect of clipping; (c) effect of wrap-around.

Care also needs to be taken to ensure overflow does not occur if a number is shifted up in order to align the binary points.

Subtracting fixed-point numbers involves the same considerations as addition. Section 7.2 discusses the representation of negative numbers.

Multiplying fixed-point numbers

As before, let $n_0 = x_0/\alpha_0$ and $n_1 = x_1/\alpha_1$. Then if we simply multiply the representations n_0 and n_1 as integers, the result is

$$n_0 n_1 = \frac{x_0}{\alpha_0} \cdot \frac{x_1}{\alpha_1} = \frac{x_0 x_1}{\alpha_0 \alpha_1}$$

which is the correct product with scaling factor $\alpha_0 \alpha_1$.

Now suppose the scaling factors are powers of 2, x_0 being represented in $u_0 \mathrm{Q} v_0$ format (a total of $u_0 + v_0$ bits) and x_1 being represented in $u_1 \mathrm{Q} v_1$ format (a total of $u_1 + v_1$ bits). We have $\alpha_0 = 2^{-v_0}$ and $\alpha_1 = 2^{-v_1}$. The product of these two numbers, as integers, will have a total of $u_0 + v_0 + u_1 + v_1$ bits. The scaling factor of the result will be $\alpha_0 \alpha_1 = 2^{-v_0} \cdot 2^{-v_1} = 2^{-(v_0 + v_1)}$, and so the format of the result can be written as $(u_0 + u_1)\mathrm{Q}(v_0 + v_1)$.

1 1 0 . 1	$x_0 = 6.5$	$n_0 = 13$
× 1 . 0 1 0	$x_1 = 1.25$	$n_1 = 10$
1 0 0 0 . 0 0 1 0	$x_0 x_1 = 8.125$	$n_0 n_1 = 130$

Figure 7.4 Multiplying fixed-point numbers.

Figure 7.4 illustrates the multiplication 6.5×1.25 in fixed-point arithmetic with 6.5 (the *multiplicand*) represented in 3Q1 format and 1.25 (the *multiplier*) represented in 1Q3 format. The result is in 4Q4 format.

Observe that the total number of bits in the product is equal to the sum of the numbers of bits in the values we are multiplying together. It would clearly be convenient if we could represent the output of the multiplication using the same format as the inputs, but this is rarely possible. The maximum possible result of the multiplication is the product of the maximum representable inputs, which can be considerably larger than either of them. Only if the format has no bits before the binary point (0Qv format) will the output be in the same range as the inputs.

The usual approach is to represent the result of the multiplication to the same relative accuracy as the inputs, or possibly to slightly better accuracy. If, using the above notation, we assume that $u_0 + v_0 = u_1 + v_1 = k$, we express the result also in a total of k bits using $(u_0 + u_1)\mathrm{Q}(k - u_0 - u_1)$ format. In the example of Figure 7.4 we might choose an output format of 4Q0 or possibly 4Q1.

If the multiplier and multiplicand are both in 4Q4 format, the natural format of the product, as we discussed above, will be 8Q8. If we return this to 4Q4 format, we need to extract an 8-bit segment, or *field*, from the middle of the result: see Figure 7.5. There are two issues here: the first is that the result might overflow the 4Q4 format, and the second is that we are throwing away some of the less significant bits of the result. The latter issue is discussed in the section on rounding below.

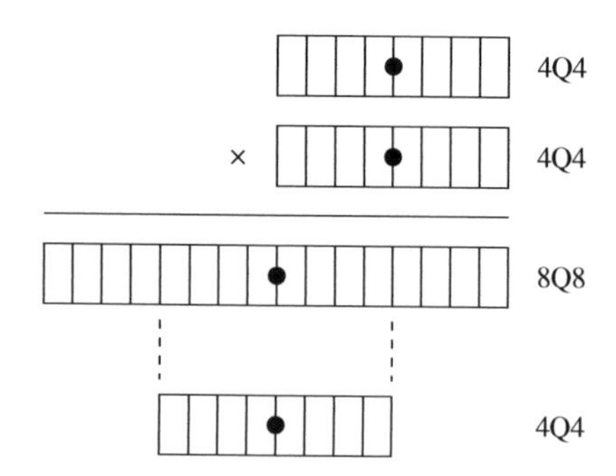

Figure 7.5 Extracting a field from the result of a multiplication.

Processors designed for digital signal processing that use fixed-point arithmetic usually use a representation with only a few bits before the binary point and many bits after. The intention of the manufacturers is that you will arrange your system so that samples, filter coefficients and the like have a maximum magnitude of 1. You can multiply numbers together without risk of overflow, and having a few bits before the binary point means that you can safely add together several of these products, for example if you are implementing an FIR filter.

General-purpose processors are often not designed with efficient fixed-point multiplication in mind. Usually only a simple integer multiplication operation is available, and sometimes only the lower half of the product is calculated. For example, a 32-bit processor might offer a 32-bit by 32-bit multiply instruction. The product of two 32-bit integers can occupy 64 bits, but the processor only generates the bottom 32 of them: in other words, it calculates the product modulo 2^{32}. (One reason for the manufacturers adopting this strategy is that, as we shall see below, this means that the same instruction can be used to multiply signed numbers.) Can you see why this is inconvenient for digital signal processing applications? If we want to represent numbers with a maximum magnitude in the region of 1, we might choose a 1Q31 fixed-point representation. We multiply two such numbers together, yielding a 64-bit result in 2Q62 format. To convert this back to 1Q31 format we need to extract bits 31 to 62 inclusive from it, as shown in Figure 7.6: unfortunately, most of the bits we need are not available from the integer multiplication. Exercise 7.5 investigates how restrictive not having these bits can be. Some more enlightened processor designers include a multiplication instruction that yields the full product (in this example, the 64-bit product of two 32-bit numbers) or an auxiliary instruction that returns the upper half of the product.

A similar problem can arise when trying to implement fixed-point arithmetic using the integer operations available in a general-purpose programming language.

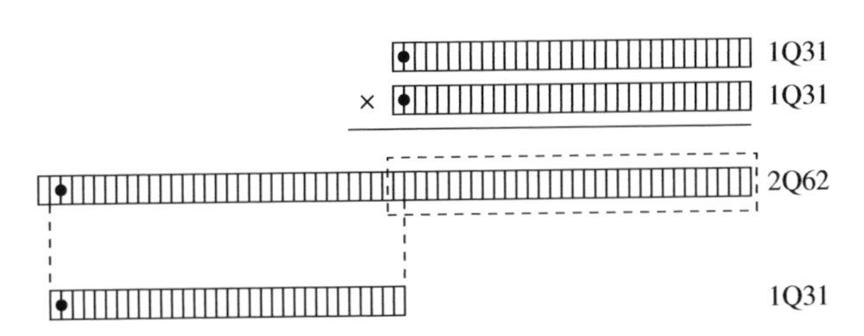

Figure 7.6 Multiplication of two 1Q31 fixed-point values: an ordinary 32-bit integer product of the operands unfortunately produces only the part of the result indicated by the dotted box.

Dividing fixed-point numbers

Division is best avoided. On most general-purpose and special-purpose processors, division is considerably slower than multiplication, and dividers are expensive to implement in hardware. This gloomy picture is counterbalanced by the fact that division is also relatively rarely required in signal processing algorithms, and even when it is, an approximate answer is often adequate.

If an algorithm apparently requires a division of two quantities, say a/b, it is often possible to avoid the division by redesigning the algorithm to work with $1/b$ throughout: the division operation then becomes a multiplication of the *dividend* a by $1/b$. Of course, this approach should also be used if b is a constant: never divide by a constant unless you have a good reason for doing so.

If you must divide fixed-point numbers a/b, the first thing you should take care to ensure is that the *divisor* b is not zero. The possibility of b being zero, or even just very small, should ring warning bells: it is a sign of a badly designed algorithm. If b is small it can only be represented in fixed point with low relative accuracy. The *quotient* a/b will then be large but again with low relative accuracy, and almost all the bits in the result will be meaningless.

Now suppose the scaling factors are powers of 2, and that we are dividing x_0 by x_1. Let x_0 be represented in $u_0\mathrm{Q}v_0$ format (in a total of u_0+v_0 bits) and x_1 be represented in $u_1\mathrm{Q}v_1$ format (using a total of u_1+v_1 bits). Then the scaling factors are $\alpha_0 = 2^{-v_0}$ and $\alpha_1 = 2^{-v_1}$. We get the biggest possible result x_0/x_1 by making x_0 as large as possible and x_1 as small as possible, without being zero. This means $x_0 \approx \alpha_0 2^{u_0+v_0}$ (it's actually $\alpha_0(2^{u_0+v_0}-1)$) and $x_1 = \alpha_1$. Then

$$\frac{x_0}{x_1} \approx \frac{\alpha_0}{\alpha_1} 2^{u_0+v_0}.$$

We get the smallest possible non-zero result by setting x_0 as small as possible, without being zero, and x_1 as large as possible. By a similar argument, the result is

$$\frac{x_0}{x_1} \approx \frac{\alpha_0}{\alpha_1} 2^{-(u_1+v_1)}.$$

The full range of possible results therefore requires a total of $u_0+v_0+u_1+v_1$ bits to represent them. This is the same result as we obtained for the multiplication of fixed-point numbers.

This is not the end of the story, however. In the case of multiplication, $u_0+v_0+u_1+v_1$ bits are enough to represent all the possible results *exactly*. For division, this is not the case. Just as some fractions have recurring decimal expansions, such as $1/7 = 0.142857142857\ldots$ (sometimes written $1/7 = 0.\dot{1}4285\dot{7}$ where the dots indicate the part that repeats), some fractions have recurring binary expansions. For example, $1/3$ in binary is $0.010101\ldots_2 = 0.\dot{0}\dot{1}_2$; $1/13 = 0.\dot{0}0010011101\dot{1}_2$. This means that no amount of precision can in general be enough to represent the result of a division exactly. Careful analysis of the likely ranges and precisions of all the numbers involved is required before a suitable result format can be chosen.

If you are implementing a digital signal processing algorithm in hardware, you will probably use a divider that employs an iterative algorithm. You can usually arrange that it provides as many fractional bits of result as you want. Software implementations, however, are constrained by the instructions available: processors that offer division instructions can usually only divide integers, giving an integer result. The division operations on integers available in most programming languages are similarly limited. Fixed-point division can be performed with the help of an integer division instruction and some scaling operations.

Using the same notation as above, the scaling factor of the result of performing a straightforward integer division of the representations of x_0 and x_1 is α_0/α_1. If, as is typically the case, $\alpha_0 = \alpha_1$, the scaling factor of the result is 1: in other words, there are no bits after the binary point. This is usually not the desired outcome, and so it is normally necessary to prescale x_0 so that α_0 is reduced, thus also reducing α_0/α_1. If x_0 is shifted up k bits before the integer division is done, the scaling factor of the result is 2^{-k} and there are k bits after the binary point in the result.

You must of course take care to ensure that overflow cannot occur when x_0 is shifted up.

Look-up tables for division

If you only need to compute a/b to limited accuracy, one approach that can work well in embedded systems or in hardware is to use a look-up table to obtain an approximation to $1/b$, which you then multiply by a.

Exercise 7.11 shows you how to use a couple of extra multiplications to get good approximations to $1/b$ from a small look-up table.

Comparing fixed-point numbers

To compare two fixed-point numbers you must first make their scale factors the same, in the same way as you would before adding or subtracting them. If the scale factors are powers of 2, this simply involves a shift. They can then be compared as if they were ordinary integers. Caution is required when comparing two fixed-point numbers for equality: the process by which the numbers were arrived at may have introduced small errors – see the discussion on rounding below – which means that two numbers that ought to be equal turn out to be slightly different.

Rounding

The results of the arithmetic operations discussed above can all require more bits to represent them than do the operands. Clearly this cannot go on for ever: at some point we will have to discard some bits from our results to make them fit into a new, smaller format.

Discarding bits from the more significant end of a result is safe as long as we know that they will all be zero; otherwise the situation is similar to that of overflow as discussed above.

If we discard bits from the less significant end of a result and the bits discarded are not all zero, there will be two numbers to choose from in the new format, one bigger than the original value and one smaller.

Always choosing one of these options or the other leads to a *systematic error*: in other words, it always modifies the number in the same direction. If the lower alternative is always selected (*rounding down* or *truncation*) the effect is always to reduce the number (or at best leave it alone, in the case where the discarded bits are all zero).

Rounding down is a safe choice in that overflow cannot occur. However, it is possible to turn a non-zero number into zero, which can lead to problems if the result is subsequently used in a division operation.

Figure 7.7(b) shows the effect of using arithmetic operations that systematically round results down on the impulse response of the IIR decaying-average filter discussed in Section 5.7; Figure 7.7(a) shows the impulse response without rounding for comparison. As you can see, rounding causes the response to tail away considerably more quickly. The format used for this example was 0Q8.

An alternative strategy is to *round up* the result: choose the next larger representable value in the new format. This scheme is rarely used.

Yet another strategy is to *round off* the result. Here the choice between the two possible values in the new format is made on the basis of which is closer to the original number: hence the alternative name *round-to-nearest*. In the event of a tie, the original number lying exactly half-way between two adjacent representable values, there are four possible strategies:

(i) Always choose the smaller;
(ii) Always choose the larger;
(iii) Always choose the one whose least significant bit is zero (*round-to-even*);
(iv) Always choose the one whose least significant bit is one (*round-to-odd*).

The second of these is the simplest to implement as it only involves testing whether the most significant of the discarded bits is 1, or, equivalently, adding

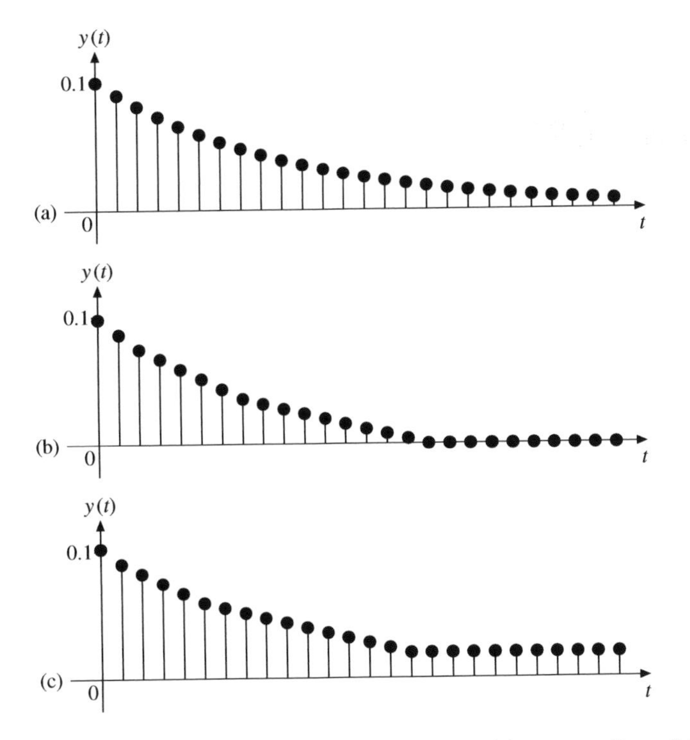

Figure 7.7 Effect of rounding on IIR filter response: (a) no rounding; (b) round down; (c) round to nearest.

one half of the value of the new least significant bit to the number, and then rounding the result down.

Figure 7.7(c) shows the effect of using arithmetic operations that round to nearest on the impulse response of the IIR filter. In this example round-to-even was used in the event of a tie, and the data format used was again 0Q8. The result is perhaps closer to the correct impulse response, but as you can see, will never fully decay to zero. This might be undesirable in some applications.

Table 7.1 gives a number of examples of numbers being rounded from 2Q5 format to 2Q2 format. Note that overflow of the 2Q2 format is possible: the table shows the extra digit in these cases.

Integer division instructions on most processors round the result towards zero.

Effect of rounding FIR filter coefficients

When using FIR filters, the effect of rounding coefficient values is just to introduce a corresponding small error in the output. If one of the calculations in the filter

Table 7.1 *Examples of rounding strategies.*

Original value	Round down	Round to nearest, to even on tie	Round up
00.10000	00.10	00.10	00.10
11.10001	11.10	11.10	11.11
10.00100	10.00	10.00	10.01
11.01001	11.01	11.01	11.10
01.11100	01.11	10.00	10.00
11.11110	11.11	100.00	100.00

implementation is very close to overflow, it is possible that rounding a coefficient will be enough to cause overflow to occur.

It is usually adequate to represent coefficient values to slightly better precision than that used for representing the samples being processed. If very good performance at rejecting certain frequencies is required – in other words, if the magnitude response must be as close to zero as possible over a given frequency range – more precision may be required.

For very demanding applications it is possible to proceed as follows. First design an FIR filter using the techniques of Chapter 5 or using more sophisticated tools, and round the coefficients to the desired precision. Use a high-precision Fourier transform to calculate the response of the filter with rounded coefficients. Then explore alternative filters obtained by perturbing the coefficient values, increasing or decreasing each by one least significant bit, calculating their responses in the same way, and find the one that best meets the design specifications. With shorter filters it is practical to search the space of all possible such modifications to the filter coefficients, and possibly also explore perturbations of more than one least significant bit.

A DC gain of exactly 1 is a common requirement in filter design. The search described above can be modified so that this constraint is satisfied by only examining those sets of filter coefficients that sum to exactly 1.

Effect of rounding IIR filter coefficients

Rounding of IIR coefficient values requires more care. Usually the effect is to alter the frequency response of the filter slightly, but it is possible for rounding to turn a stable IIR filter into an unstable one, especially if it has a sharp response. Figure 7.8 shows the effect of rounding the coefficients of the resonator illustrated

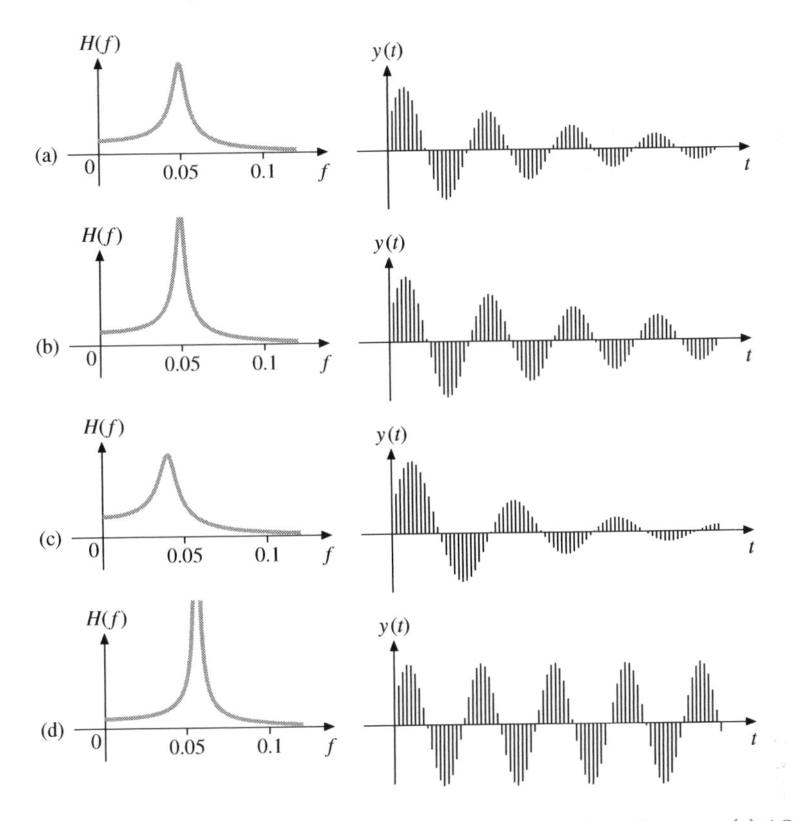

Figure 7.8 Effect of rounding IIR filter coefficients to various formats: (a) 1Q6; (b) 1Q5; (c) 1Q4; (d) 1Q3.

in Figure 5.35(c) to various precisions. As you can see, rounding has a significant effect on the speed of decay of the response of the resonator, as well as on its frequency (the frequency response should be centred on 0.05). In the last example, Figure 7.8(d), the filter has just become unstable: the frequency response goes to infinity and the impulse response will continue at the same amplitude for ever.

7.2 Negative numbers in fixed-point representations

So far we have assumed that all the numbers we are dealing with are greater than or equal to zero. We will now look at how to deal with *signed* values, i.e., those that can be either positive or negative. We will assume that the numbers we are representing are integers; the ideas extend to fixed-point representations by using scaling factors in just the same way as for unsigned integers.

Sign–magnitude representation

The simpler, but less widely used, way to represent signed numbers is to allocate an extra bit, called the *sign bit*, to store the sign of the result. This is called *sign–magnitude* representation.

The main advantage of sign–magnitude representation is its symmetry: there are as many possible positive values as there are negative ones. Multiplication and division are easy to extend from the unsigned case to the signed case: we simply multiply or divide the magnitudes of the numbers and then work out the correct sign for the result, which is the exclusive-or of the signs of the operands.

The first disadvantage of sign–magnitude representation is that there are two ways of expressing 'zero': 'plus zero' and 'minus zero'. This makes comparisons more cumbersome than they ought to be, as it is necessary to treat the two different bit patterns as equal.

The second disadvantage is that addition and subtraction are more complicated than in the case of unsigned numbers, as it is necessary to test for the various combinations of signs of the operands. Exercise 7.13 asks you to work out the details.

Two's complement representation

The second approach to representing signed values is called *two's complement.* This is used for integer arithmetic in almost all processors, and is thus also convenient for fixed-point arithmetic.

Recall that in Section 7.1 we discussed how the odometer on a car might wrap around from 99 999 to 00 000. Now imagine clocking the odometer backwards from 00 004: the last digit will decrement 4, 3, 2, 1, 0 and then all the digits will roll back to 99 999. It is thus reasonable to think of 99 999 as a representation of $-00\,001$, and this leads us to the idea of the two's complement representation.

Imagine that you have a car with a four-digit binary odometer. Clocking it backwards from 0010 (representing $+2$) it will show in succession 0001, 0000, 1111, 1110 and so on. Thus 1111 represents -1 and 1110 represents -2. If we carry on clocking the odometer back we shall eventually get to 0010 again, which we want to use to represent a positive number; at some point we have to draw a line between the negative and positive numbers, and it is conventional to do this at the point where the most significant bit changes. Thus the representations starting with a 1 are negative numbers, and those starting with a 0 are positive numbers (as well as zero itself). Table 7.2 shows the sixteen possible four-bit binary patterns and the numbers they represent when considered as unsigned numbers and as signed numbers using two's complement notation.

Observe that the range of representable numbers is not symmetric: the most negative number in this example is -8, whereas the most positive is $+7$: for two

Table 7.2 *Unsigned and two's complement signed representations.*

Binary pattern	Unsigned value	Signed value
0000	0	0
0001	1	+1
0010	2	+2
0011	3	+3
0100	4	+4
0101	5	+5
0110	6	+6
0111	7	+7
1000	8	−8
1001	9	−7
1010	10	−6
1011	11	−5
1100	12	−4
1101	13	−3
1110	14	−2
1111	15	−1

of the hazards that can arise from this, see Exercises 7.15 and 7.16. On the other hand, we only have one pattern representing zero.

You can see from the table that changing the most significant bit of the four-bit representation from zero to one adds 8 in the unsigned representation: for example, **0**101 represents 5 and **1**101 represents $13 = 5 + 8$. In the two's complement representation, however, changing the most significant bit from zero to one *subtracts* 8: **0**101 still represents 5 but **1**101 represents $-3 = 5 - 8$. One way to think of an n-bit two's complement representation is that the significance of the top bit is -2^{n-1} rather than 2^{n-1} as it is in the case of the unsigned representation.

Notice also that the signed and unsigned integer interpretations of a given n-bit pattern either agree, or differ by exactly 2^n. You can use this fact to verify that adding and subtracting two's complement n-bit integers to give an n-bit result can be done by pretending the values are unsigned and adding or subtracting in the usual way: this is why processors do not have separate signed and unsigned versions of addition and subtraction instructions.

Processors that offer saturation arithmetic generally offer it in both signed and unsigned versions.

As Exercise 7.14 shows, multiplication of two's complement n-bit integers to give a $2n$-bit result cannot be done by simply pretending the values are unsigned: a correction is required. However, as we indicated above, if only the least significant

n bits of the result are required, the result of the multiplication is the same whether the values involved are considered as signed or unsigned.

Division of two's complement quantities is more complicated. The simplest approach is to convert to sign–magnitude notation, divide the magnitudes and convert the result back to two's complement, applying the correct sign.

Fixed-point two's complement

Two's complement fixed-point arithmetic can be expressed in terms of integer operations using scaling factors in exactly the same way as for unsigned fixed-point arithmetic. We need to be aware that because of the presence of the sign bit, the range of magnitudes available is halved. For example, 4Q3 format uses a scaling factor of 1/8. Using two's complement, the most negative number that can be represented is 1000.000_2, whose value is $-64 \cdot \frac{1}{8} = -8.0$; the most positive number we can represent is 0111.111_2, whose value is $63 \cdot \frac{1}{8} = +7.875$. Note once more that the range of values available is not symmetrical about zero.

Converting between two's complement formats

As with unsigned numbers, we frequently need to convert a value from one format to another, often from a wider format to a narrower one.

Again there is the problem of overflow when discarding bits from the most significant end of a number. In the two's complement case the problem has a slight subtlety: it is possible accidentally to change the sign of a number when discarding these bits. We need to ensure (i) that all the bits we are discarding have the same value; and (ii) that the most significant bit remaining in the new value (i.e., its sign bit) has the same value as the discarded bits. For example, we can successfully truncate 11100.01 to 100.01: we have discarded two 1s, and the most significant bit remaining is also a 1. But we cannot truncate 11100.01 to 00.01, since the top bit remaining would not have the same value as the bits discarded: we have changed the sign of the number.

Extending a number at the most significant end requires us to duplicate the sign bit as many times as necessary. In 100.01 the sign bit is a 1: we therefore extend it by padding with 1s to the left, for example to 1110.01. This process is called *sign extension.*

Similar considerations apply to the rounding of two's complement values as to unsigned values. Since all bits in a two's complement number have positive significance apart from the most significant bit, discarding bits from the right-hand end of a number rounds it towards $-\infty$. (When talking about signed numbers it is clearer to say this than the possibly ambiguous 'round down'.) An alternative rounding scheme is 'round towards zero', which rounds positive numbers towards $-\infty$ and negative numbers towards $+\infty$.

The technique of adding a value equal to one half of the new least significant bit and then truncating the result works on two's complement values as well: as you can verify, it rounds to nearest, with ties being rounded towards $+\infty$.

7.3 Floating-point representations

You may have come to the conclusion that keeping track of all the scaling factors, ranges and precisions involved in fixed-point arithmetic is a lot of work; and this is work that you, the signal processing system designer, have to do rather than it being done by the computer. In *floating-point arithmetic* the computer keeps track of the scaling factors for you. Many general-purpose processors include support for floating-point arithmetic and can cope with a very wide range of representable values, offering enough precision for almost any application.

A floating-point number is represented in two parts called a *mantissa* and an *exponent*. The mantissa just contains the binary digits of the number being represented, and the exponent indicates where the binary point is located, usually counting down from the most significant bit of the mantissa. For example, the binary number 101.1101_2 (5.8125 in decimal) would be represented using a mantissa of 1011101_2, and an exponent of $11_2 = 3_{10}$ which indicates that the binary point falls after the third digit from the left of the mantissa.

A floating-point format will usually allocate a fixed number of bits to each of the mantissa and exponent, the mantissa being padded to the right with zeros if necessary to fill up the space. Because the padding is to the right, then (assuming the mantissa is not zero) we can arrange that the most significant bit of the mantissa is a 1. A non-zero floating-point number written with the most significant bit of the mantissa being a 1 is said to be *normalised.*

A normalised number x with an exponent of zero will start $0.1\ldots$ in binary, and therefore $1/2 \leq x < 1$. If the exponent is 1, then $1 \leq x < 2$, and so on; in general if the exponent is u, $2^{u-1} \leq x < 2^u$.

There is no particular reason why the binary point should strictly fall within the mantissa bits: a negative exponent would mean that the binary point is located off the left-hand end of the mantissa and we have to imagine some interpolated zeros. For example, a normalised number with an exponent of -1 would start $0.01\ldots$ and again in general if the exponent is u, $2^{u-1} \leq x < 2^u$.

The number zero does not fall neatly into the above scheme and is usually represented using a special reserved value of the exponent and/or a zero mantissa.

IEEE standard floating point

IEEE 754 is a standard for the representation of floating-point numbers and arithmetic on them that forms the basis of many floating-point implementations. There is much detail in the standard, so we shall only cover the most important aspects of it here.

The two most widely used formats specified by the standard are *single precision* and *double precision*. These differ only in the number of bits allotted to the mantissa and exponent.

A single-precision value occupies a total of 32 bits: one bit for the sign of the number (0 representing a positive number, 1 a negative number); 8 bits for the exponent; and 23 bits for the mantissa. Numbers are usually normalised so that the most significant bit of the mantissa is 1 as discussed above. This bit is not actually stored, and so the mantissa effectively has 24 bits of precision. Single-precision floating-point numbers are thus expressed with a relative error of approximately $2^{-24} \approx 6 \times 10^{-8}$.

A double-precision value occupies a total of 64 bits: one bit for the sign, eleven bits for the exponent and 52 bits for the mantissa. Again, the most significant bit of the mantissa is 1 and is not stored, giving effectively 53 bits of mantissa precision. The relative error in double-precision floating-point numbers is thus approximately $2^{-53} \approx 10^{-16}$.

In each case the exponent is stored with a *bias*: the stored value is the actual exponent plus an offset. In the single precision case, the bias is $+127$; for double precision, it is $+1023$. Table 7.3 shows some examples of numbers converted into the single-precision IEEE format, including the representable numbers with largest and smallest magnitude with a normalised mantissa. The leading 1 of the mantissa (the bit that is not actually stored) is shown as '(1.)' in the table.

The bias in the exponent makes it possible to reserve the cases where the exponent field is all zeros or all ones for special values: a couple of these are shown in Table 7.4.

If the exponent and the mantissa fields both consist entirely of zero bits, the number zero is represented. As in the case of the sign–magnitude representation for fixed-point values, there are thus two bit patterns corresponding to the value zero, one with the sign bit clear and one with it set. These are referred to as $+0$ and -0 respectively. The IEEE standard states that the two values should compare as 'equal' even though they have different bit patterns: beware!

If the exponent field consists entirely of ones and the mantissa entirely of zeros, an 'infinity' is represented. As with zero, there are two infinities: $+\infty$ and $-\infty$. Infinities are produced when overflow occurs or when dividing by zero.

If the exponent field consists entirely of ones and the mantissa is not composed entirely of zeros, the thing represented is called a 'not-a-number', or 'NaN'. These

Table 7.3 *Example IEEE 754 floating-point numbers.*

Value or name	Base 2 form	Sign bit	Mantissa bits	Biased exponent
17	$1.0001_2 \times 2^{100_2}$	0	(1.)000100...	10000011
-17	$-1.0001_2 \times 2^{100_2}$	1	(1.)000100...	10000011
6.022×10^{23}	$1.11111_2 \times 2^{1001110_2}$	0	(1.)111111...	11001101
3.403×10^{38}	$1.11111_2 \times 2^{1111111_2}$	0	(1.)111111...	11111110
1.175×10^{-38}	$1.00000_2 \times 2^{-1111110_2}$	0	(1.)000000...	00000001

Table 7.4 *IEEE 754 special values.*

Value or name	Base 2 form	Sign bit	Mantissa bits	Exponent bits
$+0$	0_2	0	000000...	00000000
-0	0_2	1	000000...	00000000
$+\infty$	$+\infty$	0	000000...	11111111
$-\infty$	$-\infty$	1	000000...	11111111
NaN		0 or 1	non-zero	11111111

are used to flag errors such as the division of zero by zero or taking the square root of a negative number. Any operation on a NaN returns a NaN which causes any such error to propagate through a calculation. The value in the mantissa field is sometimes used to indicate the type of error that occurred, to help with debugging.

The IEEE standard also describes 'subnormal' numbers which fall between the smallest normalised non-zero floating-point number and zero. These are represented using a zero exponent field and a non-zero mantissa field. Many implementations which claim to be compliant with the standard do not in fact support subnormal numbers.

Floating-point formats with exponents that represent powers of numbers other than 2 do exist, but are rare in practice.

Floating-point arithmetic

Algorithms for performing arithmetic on floating-point numbers essentially follow the methods discussed above for fixed-point arithmetic, with the scaling factors involved being automatically chosen by normalisation of the results. However, the details, especially of rounding algorithms, are complicated and not particularly instructive, and so we shall not discuss them here.

The IEEE standard specifies precisely what result should be obtained for any given arithmetic operation on floating-point numbers. The basic rule is that we pretend that the arguments to the operation are exactly represented, calculate the exact result of the operation on those arguments, and then round this result to a representable value. For example, 2 and 3 can both be represented exactly as IEEE single-precision numbers. The standard says that the result of dividing these numbers should be a rounded version of the value 2/3 (which is not itself exactly representable). The standard specifies several rounding modes: many implementations support only round-to-nearest with rounding to even on ties.

One might hope that this would imply that all computers whose floating-point arithmetic complies with the IEEE standard would, given identical input, give identical results from a given sequence of operations. Sadly this rarely happens in practice, as many processors work internally to higher precision than strictly necessary, only rounding results to the desired final precision when they are stored in memory. The internal precision used varies, and in any case postponing rounding in this way – rather than doing it immediately after each operation – can lead to a different result from what might be expected, although the difference is often small.

Furthermore, many high-level languages do not specify the order in which operations should occur. For example, if you write the expression $a+b+c$, a compiler may produce code which evaluates $a+(b+c)$ or it may produce code which evaluates $(a+b)+c$. These do not necessarily produce the same result. The compiler may even generate the two alternatives at different points in the same program.

Block floating point

Block floating point is a compromise between fixed-point and floating-point representations, where a single exponent is used as the scaling factor for a vector of mantissas so that all are scaled by the same amount. NICAM stereo television sound uses a scheme like this to cover a wide amplitude range using a relatively small total number of bits, transmitting only one exponent for a block of 32 samples.

Block floating point is also sometimes used in fast Fourier transform algorithm implementations, with a single exponent being used for the entire vector of samples. As you can verify by inspection of Figure 4.14 the maximum magnitude of the numbers involved at any stage in the calculation can only be twice as big as the maximum magnitude in the previous stage, and so the exponent can only increase by at most 1 at each stage. Using block floating point means that good relative accuracy can be achieved whether the signal being transformed has a single dominant high-amplitude frequency component or if it has a large number of low-amplitude frequency components.

7.4 Choosing between fixed point and floating point

Most computer programming languages can represent numbers either as floating-point values or as integer values, often with several alternatives for each. For example, the C programming language includes `float` and `double` types for floating point, and `short`, `int` and `long` types for integers. The integer types can also be used for fixed-point arithmetic as we have seen.

The decision as to whether to implement a signal processing system using fixed-point or floating-point arithmetic depends on a number of factors.

Working with fixed point demands considerable effort on the part of the designer in keeping track of scale factors through calculations, and checking what ranges and precisions of values are required. A single mistake can lead to a fault that can be very hard to diagnose. In most embedded systems, however, fixed point will be much faster than floating point and the resulting code size will be much smaller. This all makes for a better-performing, cheaper system as a reward for the design effort.

Conversely, it is much quicker to design a system based around floating-point arithmetic. For experimenting with algorithms, or doing most of the exercises in this book, floating point is usually the right choice. So much software is written using floating-point arithmetic that general-purpose processors are designed with a large part of their hardware dedicated to performing these operations quickly, even to the point where floating-point arithmetic can be faster than integer arithmetic.

Hardware signal processing systems almost invariably employ fixed-point arithmetic because floating-point hardware, especially hardware conforming to the IEEE standard, is complicated, power-hungry and expensive and can be difficult to use in some applications since operations may not take a predictable amount of time.

In short, floating point is ideal if you are experimenting with algorithms or if you are designing a system with relaxed constraints of cost, size and power. Otherwise, fixed point is likely to be the best solution.

Exercises

7.1 Convert $x = 3.5625$ to 4Q5 fixed-point notation.

7.2 Generate a vector of 256 points containing one complete cycle of a sine wave of amplitude 1 (i.e., the sample values should be in the range from -1 to $+1$). Simulate the effects of (a) clipping and (b) wrap-around on this waveform in each case by modifying the values of samples whose absolute value exceeds 0.9. Plot your modified waveforms to check that they resemble Figure 7.3 and then plot the magnitudes of the Fourier components of the original waveform and each of the two modified waveforms. Listen to the two waveforms. Which is the more distorted?

7.3 Write and test a set of four routines to add, subtract, multiply and divide fixed-point numbers in 8Q8 fixed-point notation using only integer arithmetic.

7.4 Write and test an 8-point FIR filter routine with input samples and coefficients expressed in 2Q6 fixed-point notation, using only integer arithmetic.

7.5 Suppose you are using a 32-bit processor whose only multiply instruction produces the least significant half of the 64-bit integer product of two 32-bit integer operands. Can you use a single multiplication of this form, followed by a shift, to multiply two unsigned numbers in 14Q14 format to produce a 14Q14 result? (Assume that you know the answer will be in the range representable by that format.) What is the answer to this question for 12Q12 format? For 10Q10? For 10Q11? For 10Q12? What is the condition that u and v must satisfy for the answer to the question to be 'yes' for format uQv?

7.6 You are programming a 32-bit processor which features a 'fractional multiply' instruction that performs a fixed-point multiplication of two unsigned numbers in 1Q31 format, returning the result in the same format. Show how to use this instruction to carry out a multiplication with operands and result in 8Q10 format.

7.7 You wish to implement an 'automatic gain control', a system which scales its input by a factor which it continuously adjusts so that the long-term average of the output amplitude is 1, whatever the average amplitude of the input. Such devices are used in radio receivers to make them insensitive to the strength of the incoming signal.

One approach is to feed the absolute value of the input signal to an IIR decaying-average filter with a long time constant to calculate its average amplitude. The output of the system is then its input divided by the output of the filter, as shown in Figure 7.9. What happens to this system if its input is zero? Discuss modifications to avoid this problem.

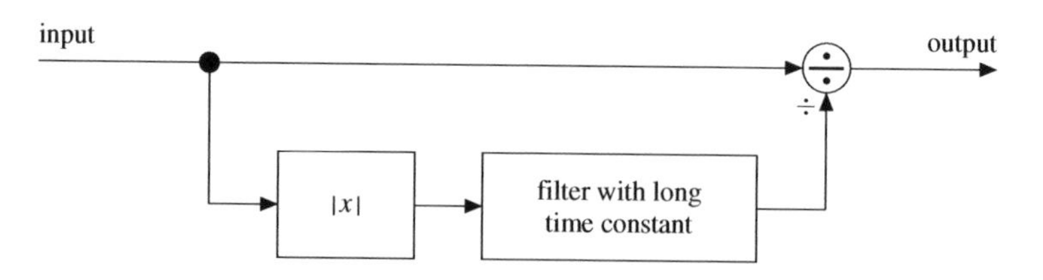

Figure 7.9 Automatic gain control.

Discuss the design an alternative automatic gain control using a multiplier as a scaling device rather than a divider. (**Hint** for one possible approach: suppose you maintain a current multiplicative scaling factor k; how could you detect that k is too high or too low?)

7.8 Express $1/3$ as accurately as possible in fixed-point 1Q7 format. Multiply the resulting integer by 3. Is the answer equal to 1 in fixed-point 1Q7 format? For what values of n can you express $1/n$ exactly in 1Q7 format?

7.9 Some processors (and programming languages) have a 'remainder' or 'modulo' feature which calculates the integer remainder of the integer division a/b. In other words, it calculates a number r such that $a = qb + r$ and $0 \leq r < b$. (Implementations differ confusingly in the interpretation of 'remainder' for negative a and/or b; for now we shall just assume that all the numbers involved are positive or zero.) Show how to use the remainder from an integer division to calculate some fraction bits of a (fixed-point) result. Demonstrate your method using the example $355 \div 113 = 3$ remainder 16.

7.10 Determine by experiment how your favourite programming language interprets the ideas of the 'quotient' and 'remainder' of the integer division x/y when one or both of x and y is negative.

7.11 Show that if $ab = 1 + \epsilon$, where ϵ is small compared with 1, then $ab(2 - ab)$ is closer to 1 than ab is, and thus that $b(2 - ab)$ is a better approximation to $1/a$ than b is. Show that if b is accurate to k bits as an approximation to $1/a$ (i.e., that the relative error in the approximation is 2^{-k}), then $b(2 - ab)$ is accurate to approximately $2k$ bits. Note that the improved approximation is obtained using only subtraction and multiplication operations. (This method is called a *Newton–Raphson iteration*, and the doubling of the number of bits of accuracy in the result when the iteration is applied is called *quadratic convergence*. Similar methods exist for evaluating functions such as square root.)

Write a program that constructs a look-up table with 16 entries to calculate approximately the reciprocal of a fixed-point value a represented in 4Q12 format, returning the result in the same format. Assume that $a \geq 1$. Plot the relative error in the result against a from 1 to 15 in steps of 0.1.

Add a step that improves this approximation using a Newton–Raphson iteration and plot the relative error in the refined result against a over values of a from 1 to 15 in steps of 0.1. What is the worst-case relative error?

Add another Newton–Raphson iteration and plot the relative error again. Roughly how big a look-up table would you need in order to obtain this level of accuracy without the Newton–Raphson method?

7.12 For each of the following rounding schemes:

(i) Round-to-nearest with round-to-even on ties;
(ii) Round-to-nearest, rounding up on ties;
(iii) Round-to-nearest, rounding down on ties;

construct an example of a number represented in 1Q7 format such that when it is rounded to 1Q5 format, and then this rounded result is rounded to 1Q3 format, the result is different from the result of rounding the original number to 1Q3 format in a single step.

(The moral of this exercise is: never round a number twice if you can possibly avoid it.)

7.13 Describe how to add two integers in sign–magnitude representation. More formally, an integer n is represented in sign–magnitude notation as $n = (s, m)$ where $m \geq 0$, and $n = +m$ or $n = -m$ according as $s = 0$ or $s = 1$ respectively; given two signed integers n_0 and n_1, represented in sign–magnitude notation as $n_0 = (s_0, m_0)$ and $n_1 = (s_1, m_1)$, describe an algorithm to calculate a sign–magnitude representation of n, where $n = n_0 + n_1$.

7.14 Suppose x and y are 16-bit two's complement integers and you multiply them together using a multiply instruction designed for unsigned numbers, giving a 32-bit result. What is the error in the result (viewed as an unsigned number), in terms of x and y, if both x and y are positive? If x is positive and y is negative? If y is positive and x is negative? If both are negative? (**Hint:** Use the fact that the interpretations of signed and unsigned values agree for positive numbers and differ by exactly 2^{16} for negative numbers.)

Show that correcting the result using the error you have calculated above never affects the least significant 16 bits of the result.

7.15 You write a simple function to calculate the absolute value of a two's complement integer. It tests whether the number is negative, and, if it is, subtracts it from zero. When you test the function you find that occasionally it seems to return a negative answer. What is happening?

7.16 What are the smallest (i.e., most negative) and largest (i.e., most positive) results you can get by multiplying two 16-bit two's complement integers? How many bits are required to represent the product of two n-bit two's complement integers as a two's complement integer?

7.17 Construct an example of three numbers a, b and c where $(a+b)+c \neq a+(b+c)$ when evaluated using IEEE single-precision floating-point arithmetic.

7.18 How would you write a function to add up all the elements in a vector of floating-point numbers passed to it in order to achieve maximum accuracy in the final result? Assume that the application is not speed-critical.

Part II

Applications

8
Audio

In this chapter we will look at some practical examples of the digital processing of audio signals.

8.1 The ear

To a reasonable approximation, the average human ear can hear signals with frequency components from about 20 Hz to about 20 kHz. Apart from at the extremes of this range, its sensitivity to frequency is approximately logarithmic: a step in frequency from say 100 Hz to 110 Hz is perceived as similar to the step from 200 Hz to 220 Hz or 1000 Hz to 1100 Hz. Musicians use a logarithmic scale to express pitch: an *octave* is a factor of 2 in frequency.

The ear is not very sensitive to the phase of the frequency components of audio signals: it is only their magnitudes that are important.

The ear's sensitivity to amplitude is also approximately logarithmic: doubling the amplitude of a signal at a given frequency will sound like the same step in volume whether the original signal was loud or quiet. The ear's sensitivity to amplitude varies significantly with frequency. It is most sensitive to quieter sounds in the region from around 500 Hz to 5 kHz.

8.2 Sample rates and conversion

Ordinary audio (so-called 'red book', from the colour of the cover of the book that contained the original specification) compact discs store data sampled at a rate of 44.1 kHz. Each sample is stored directly as a 16-bit integer. (There is in fact one sample for each of the left and right stereo channels, but we shall ignore that for now.) The Nyquist limit at this sample rate (see Chapter 2) is $1/2 \times 44.1 = 22.05\,\text{kHz}$, which is above the highest frequency audible by the average human ear. As far as the range of frequencies is concerned, then, the format used in ordinary compact discs is more than adequate.

Suppose we design a compact disc player that outputs the sequence of samples directly to a digital-to-analogue converter. The output of the converter will include aliases which, although themselves not audible, we should try to remove since they may cause a subsequent amplification stage to introduce audible distortion.

A sine wave at 20 kHz sampled at 44.1 kHz will have an alias at $44.1 - 20 = 24.1$ kHz, and so the anti-aliasing filter needs to pass frequencies up to 20 kHz and block frequencies above 24.1 kHz: see Figure 8.1(a). It is difficult to build a filter using ordinary analogue electronic components to meet this rather tight specification, especially if we want to avoid using any electronic components that must be adjusted during manufacture to compensate for the tolerances of other components. Such designs are expensive to build and prone to drift, meaning that their performance will degrade over time. The specification can, however, be met fairly easily using a digital FIR filter: depending on the exact specification, a few tens of coefficients will be needed.

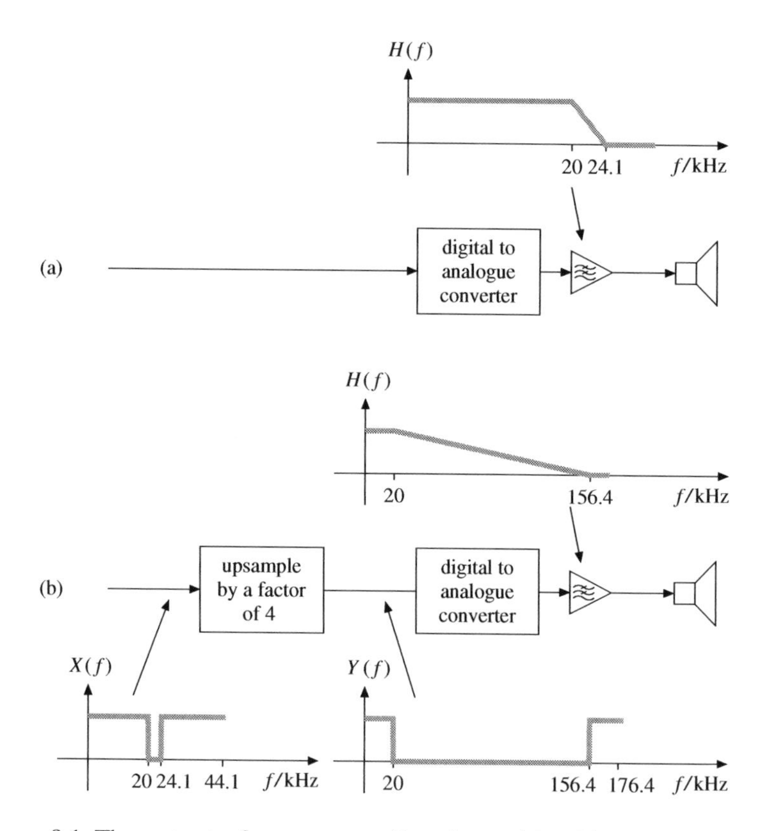

Figure 8.1 The output of a compact disc player (a) without oversampling; (b) with oversampling.

A strategy adopted in many compact disc players is *oversampling*: digitally convert the signal sampled at 44.1 kHz to a signal sampled at say $4 \times 44.1 = 176.4$ kHz, and pass these samples to the digital-to-analogue converter. Our 20 kHz sine wave will now have an alias at $176.4 - 20 = 156.4$ kHz. The specification for the anti-aliasing filter at the converter output is now much more relaxed as it can fall off smoothly between 20 kHz and 156.4 kHz: see Figure 8.1(b).

A pleasant and rather surprising consequence of oversampling is that, with a little extra work, we can use fewer bits per sample to represent the oversampled signal and achieve the same signal quality. This allows us to use a lower-precision (and hence cheaper) digital-to-analogue converter. The idea can be taken to the extreme of using a sample rate of several megahertz and a 'one-bit digital-to-analogue converter', as we shall see at the end of this chapter.

Some people claim to be sensitive to the phase of the frequency components towards the upper end of the audible range. They argue that the 44.1 kHz sample rate used by compact discs is too demanding on the anti-aliasing filters in a system, and that these filters inevitably modify the phases of the higher frequency components. They use this as an argument to justify the use of higher sample rates such as 48 kHz, 96 kHz, or even 192 kHz. Leaving aside the question of whether these people really can hear the effect of phase on these frequencies which themselves are almost inaudible, the use of digital upsampling methods means that phase distortion in modern systems is negligible and that a sample rate of 44.1 kHz is entirely adequate to record the amplitude and phase information of all frequency components of an audio signal up to 20 kHz.

A marginally better argument in favour of using a sample rate of, for example, 48 kHz is that it has a simple ratio to sample rates used in other applications: telephony, for example, commonly uses a sample rate of 8 kHz, usually regarded as the minimum rate that can be used for intelligible speech. As we saw in Chapter 5 the filters required to convert between sample rates are more straightforward if the ratios involved are simple.

A more reasonable criticism of the compact disc format is that the sample values are stored linearly as 16-bit integers. As we discussed in Chapter 3, it is possible to get better perceived quality from a given number of bits – or, equivalently, the same quality with fewer bits – if a non-linear quantisation scheme along the lines of mu-law or A-law is used. Sixteen bits are enough for audio in almost any application: if you turn up the volume on your amplifier so that the loudest parts of the music from a compact disc are at the threshold of causing pain to your ear, the noise introduced by the quantisation to 16 bits is likely to be lower in amplitude than the background noise in any typical environment. Commercially produced compact discs are nevertheless often processed using a noise-shaping technique like the ones we will describe at the end of this chapter, giving the effect of finer quantisation in the frequency bands where the ear is most sensitive.

8.3 Audio in the frequency domain

It is often convenient to process audio in the frequency domain. Indeed, many items of audio equipment have a display which shows an (albeit crude) representation of the signal in the frequency domain as a row of bars that flicker up and down, usually called a 'graphic equaliser display'. The display shows the magnitudes of various frequency components of the signal, usually at logarithmically spaced frequencies.

When audio is processed digitally in the frequency domain, the signal is usually divided into overlapping blocks each a power of 2 samples long. The blocks overlap by half their length so that each sample is included in two adjacent blocks, as shown in Figure 8.2. The waveform here is from the beginning (the 'attack' part) of a note played on a piano.

The block length is chosen to be a power of 2 primarily so that it is easy to apply the fast Fourier transform algorithm to the block. As we said in Chapter 4, it is essential to window the block before taking the transform. Beyond the power-of-2 constraint, choosing the block size is a compromise. If we choose long blocks the Fourier transform will have many points and therefore will analyse the signal into a large number of closely spaced frequency components; however, the resolution of the transform in time will be poor, and it will effectively average out any variations in the amplitude of each frequency component over the period of time occupied by the block. On the other hand, a short block size will give poor resolution in frequency but good resolution in time. In fact, the resolution in frequency is exactly inversely proportional to the resolution in time. In some applications which must cope with a wide variety of types of audio signal (including, as we shall see below, MP3 compression), the system can dynamically change its block size to suit the nature of the signal.

Notice that the choice of block size does not have a significant effect on the computational load involved in taking the Fourier transform, as long as a fast algorithm is used. As we saw in Exercise 4.11, a transform of size $2N$ takes only slightly more than twice as long as a transform of size N. This is almost exactly cancelled out by the fact that, if our block size is $2N$ rather than N, we will only need half as many transforms over a given number of samples.

Popular choices of block size for a sample rate of 44.1 kHz are $2^{10} = 1024$ samples and $2^{11} = 2048$ samples. These correspond to block lengths of 23.2 ms and 46.4 ms respectively. Most audio signals do not change significantly in nature over time periods of that size.

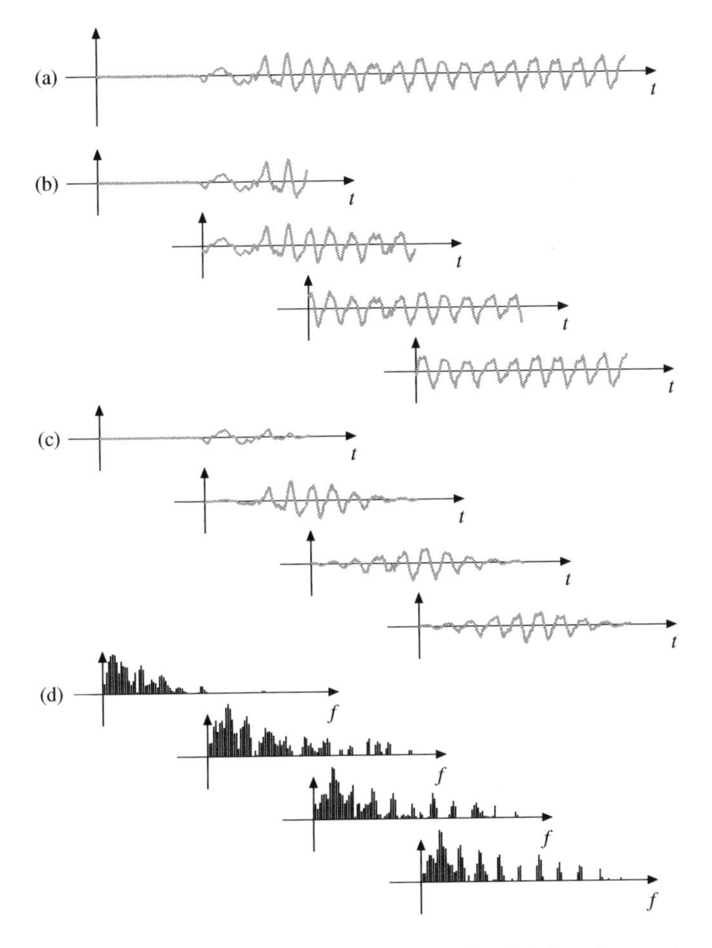

Figure 8.2 Taking Fourier transforms of overlapping blocks of an audio signal: (a) original signal; (b) overlapping blocks; (c) windowed blocks; (d) magnitudes of resulting Fourier transforms.

One way to present the results of transforming the overlapped blocks is as a *spectrogram*. Here the amplitude is represented by a range of colours or shades of grey, again usually on a logarithmic scale. Figure 8.3 shows a spectrogram of a longer version of the piano note attack signal used in Figure 8.2. Here we have used black to represent maximum amplitude and white to represent an amplitude of zero. The vertical axis is frequency, with the higher frequency components towards the top of the page, and the horizontal axis is time. Each column of pixels in the spectrogram corresponds to the result of one Fourier transform. As you can see, there is a strong frequency component towards the bottom of the spectrogram (just below 500 Hz in this example) and a number of other strong frequency components equally spaced in frequency above it. The waveforms produced by most musical

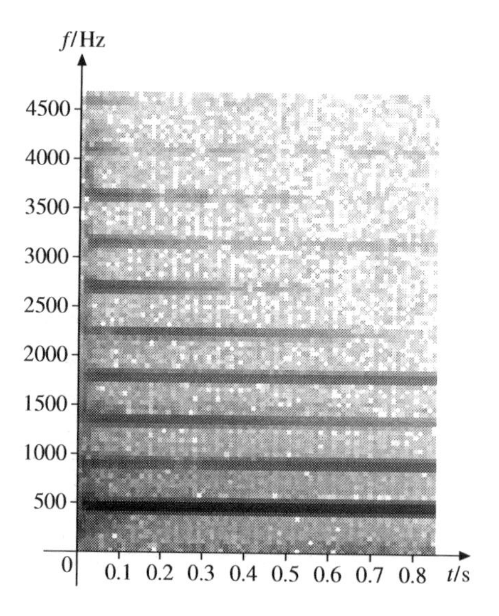

Figure 8.3 Fourier transforms of successive overlapping blocks of an audio signal stacked along a time axis.

instruments that have a definite 'pitch' (but not, for example, most types of drum) exhibit this pattern. The lowest frequency is called the *fundamental frequency*, and the frequency components at integer multiples of the fundamental are called *harmonics*. Notice how the note starts with many higher harmonics present but that these harmonics die away shortly after the note starts, leaving a purer tone.

8.4 Compression of audio signals

The data format used on compact discs is *uncompressed*: the samples are simply stored directly. *Compression* is the process of converting the sample data into a more compact representation; the reverse process is called *decompression*. In the case of audio signals, the idea is that the result of compressing the samples and then decompressing them sounds as similar as possible to the original.

Compression schemes are divided into the *lossless*, where the original samples are guaranteed to be recovered exactly, and the *lossy*, where they might not be recovered exactly. Strictly speaking, 'lossless compression' is a contradiction in terms. If a scheme could guarantee to compress any sequence of data and recover it exactly from the compressed form, we would simply iterate the compression until there was nothing left of our data; under the terms of the guarantee we could then iteratively decompress this 'nothing' to recover our original sequence, which is plainly impossible. There will always be some cases where the output of a lossless compressor will be bigger than its input. 'Lossless compression' is thus usually used to refer to a scheme which almost always manages to compress the data it encounters, and which perhaps gives some guarantee about its worst-case performance.

We shall only consider lossy audio compression algorithms here. We shall look at three algorithms in increasing order of sophistication. As the algorithms get more sophisticated, the *compression ratio* (the size of the input data divided by the size of the output data) improves for a given quality of reconstructed result. To achieve these improvements the algorithms take advantage of increasingly subtle aspects of the way we hear sounds.

Compression schemes are often most easily understood by looking at the action of the decoder first. Some standardised schemes are specified solely in terms of the decoder action, leaving the design of the encoder to the implementers. There are then often tradeoffs between the complexity of the encoder and the quality of the final result. Many schemes are also designed so that the decoder is much simpler than the encoder, usually because they are used in situations either (a) where a signal will only be compressed once, but decoded many times; or (b) where decoding must happen in real time, but encoding need not. In the case of a recording of a piece of music for consumer distribution, both these conditions normally apply. The original recording can be made without compression, and then compression subsequently applied 'off-line', i.e., without having to occur in real time. It is desirable, however, both that the decompression can be done in real time so that a playback device does not need to have space to store the uncompressed signal, and that the playback device can be simple and hence cheap. Low power consumption is also desirable in applications such as portable audio players, which again is an argument for simplicity in the decoder.

Delta pulse-code modulation

Delta pulse-code modulation, or DPCM, is a simple way to compress audio signals. Its compression performance is rather limited, but it is very simple to implement and is suitable for extremely resource-constrained systems.

Roughly speaking, instead of encoding each sample separately, we encode the difference between each sample and the last, but using only a limited selection of possible difference values. The difference values are called *deltas*.

As noted above, it is often clearest to start with a description of the decoder. The block diagram of a DPCM decoder is shown in Figure 8.4. Its input is a series of numbers, which it uses to index into a look-up table of so-called *delta* values called an *inverse quantiser*. The *integrator* keeps a running total of the delta values and it is this accumulated running total that forms the decoded output. Of course, some of the delta values will be positive and some negative, and the values chosen are usually symmetric about zero. (We have omitted from the figure the machinery required to initialise the accumulator.)

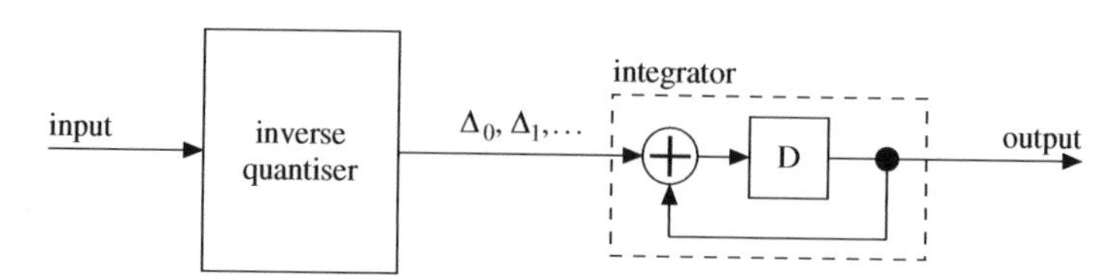

Figure 8.4 DPCM decoder.

How would we encode a signal to drive this decoder? A simple approach would be to take the difference between consecutive input samples using a *differentiator* and quantise each difference to the nearest available delta value, as shown in Figure 8.5. The quantiser can either be a look-up table, or, if the set of delta values is chosen conveniently, it may be possible to implement it as a simple function. However, we can do slightly better than this by trying at each step to make up for any error introduced into the decoder's output by previous steps. To evaluate this error, the encoder needs to include a copy of the decoder: this is a common theme in compression algorithms.

Figure 8.6 shows an encoder for DPCM using this more sophisticated strategy. The embedded decoder is indicated by the dotted box. The subtractor at the left works out the difference between the decoder's output for the previous sample

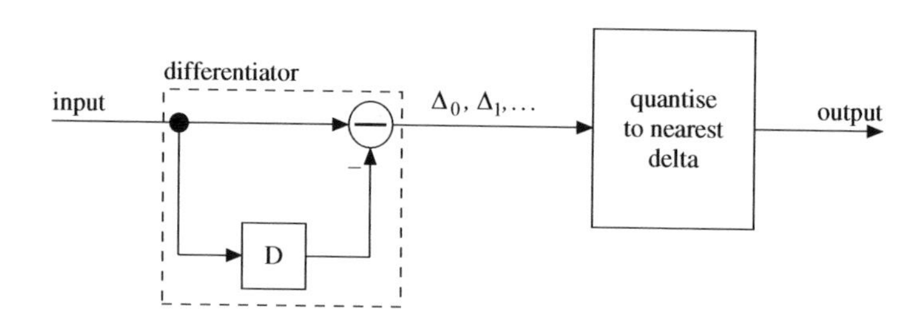

Figure 8.5 A simple DPCM encoder.

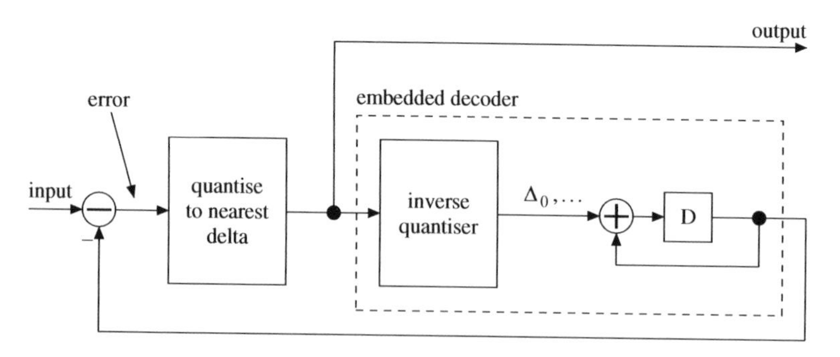

Figure 8.6 A more sophisticated DPCM encoder.

and the actual value of the current sample. Its output is quantised to the nearest available delta value, which will make the next sample output by the decoder as close as possible to the current input sample. Note that there is a delay of one sample through the system. Most compression schemes introduce some delay; schemes that achieve higher compression ratios generally introduce larger delays.

Figure 8.7 shows a slightly contrived example of DPCM in action where only four different deltas are available: +4, +1, −1 and −4. The plot shows the original signal (grey) and the result of passing it through an encoder like the one in Figure 8.6 followed by a decoder (black samples). The short horizontal lines show the values from which the encoder could select at each sample.

The weaknesses of the scheme are apparent: when the signal changes quickly, the decoder output cannot keep up because the biggest available delta value is not large enough; and when the signal is flat, as at the beginning of the signal in the example, the system is forced to used deltas of +1 and −1 alternately, introducing oscillation into the output.

The first weakness can to some extent be overcome using a careful choice of delta values. The second can be overcome by including zero as an available delta value, although this brings with it the slight inconvenience of making the total number of different delta values odd.

Adaptive delta pulse-code modulation

Adaptive delta pulse-code modulation, or ADPCM, overcomes some of the disadvantages of DPCM by in effect adjusting the set of available delta values continuously in response to the signal. The idea is outlined in Figure 8.8. Again the encoder includes an exact copy of the decoder (and so we have not shown the decoder separately).

Looking first at the decoder part, the difference between the arrangement here and that of the DPCM decoder in Figure 8.4 is the addition of the filter and multiplier. The filter simply analyses which delta values are being used. If the values which are large in absolute value are being used more than the smaller ones,

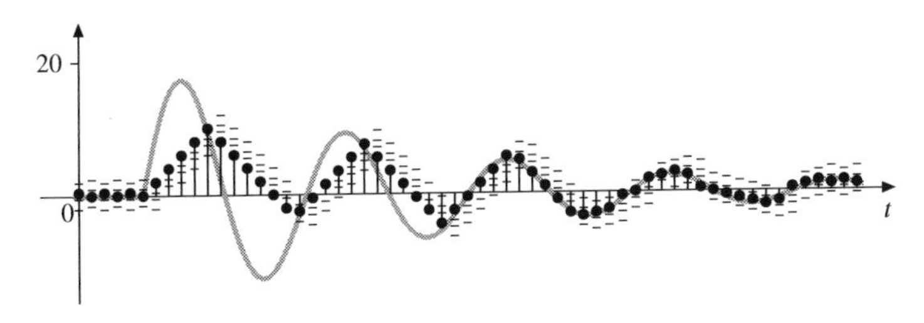

Figure 8.7 Action of a DPCM encoder and decoder on an example signal.

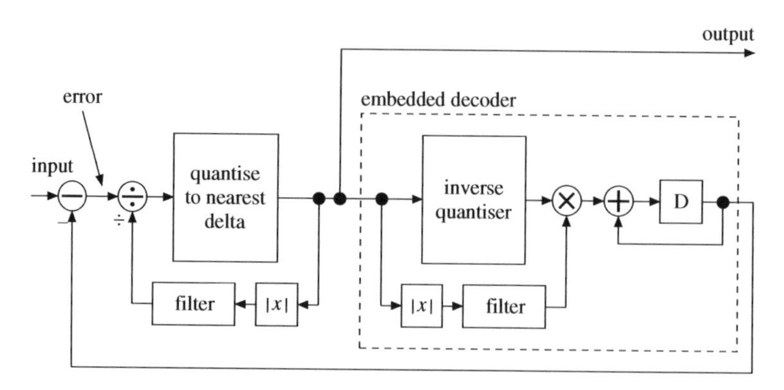

Figure 8.8 An ADPCM encoder.

it increases the multiplication factor at the output of the table: in effect, it scales up the set of available delta values. Conversely, if the smaller-magnitude delta values are mostly being used it decreases the multiplication factor, effectively scaling the set of available delta values down. The 'filter' will typically be based around some kind of decaying-average IIR filter acting on the absolute values of its input, possibly with some non-linear scaling on its input and output. The characteristics of the filter will be designed to match the characteristics of the expected type of signal, and as usual care is needed to ensure that the filter will always be stable.

The remainder of the encoder is complementary to the decoder. The subtractor at the input, which calculates the error in the decoded output, is complementary to the integrator at the output of the decoder and the divider is complementary to the multiplier. The two filters are identical and are fed with the same inputs, and so their outputs will be identical. (In practice, of course, the encoder only calculates the filtered absolute value of the output signal once, using the result both to scale the output of the inverse quantiser and to inversely scale the input to the quantiser.) The scaling of the delta values will therefore be the same in both the quantiser and the inverse quantiser: there is enough information in the encoded data to recover the changing scale factor.

Figure 8.9 shows the performance of this ADPCM encoder on the same waveform as Figure 8.7 and using the delta values of $\pm\alpha$ and $\pm 4\alpha$, where α is the scaling factor at the output of the filter. The delta values are scaled by a variable factor as a result of the action of the filter, as shown by the fine horizontal lines which indicate the available values from which the encoder chose at each sample time. As you can see, the lines become closer together when α decreases, which is when the signal amplitude is small; the lines are further apart when α increases, which is when the signal amplitude is large. It takes a few samples for α to

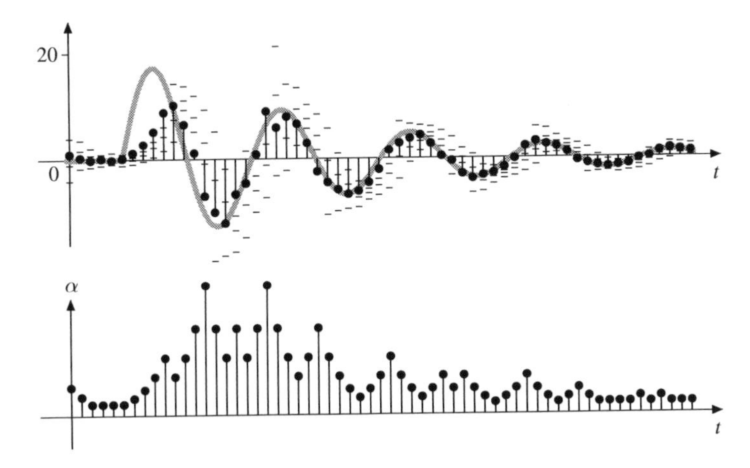

Figure 8.9 Action of an ADPCM encoder and decoder on an example signal.

adapt when the signal suddenly changes in amplitude, but after that it tracks the signal well.

ADPCM is primarily aimed at encoding speech signals. There are various standards for ADPCM implementations in telephony applications, which are slightly more complicated than the version we have presented here. A common refinement is the replacement of the simple integrator at the output of the decoder with an FIR filter, an IIR filter or a combination of the two. This filter can itself also be designed so that it changes its characteristics adaptively.

Telephone-quality speech is usually sampled at 8 kHz with 8 bits per sample using A-law or mu-law quantisation (see Chapter 3), giving a total data rate of 64 kbit/s. ADPCM compression can reduce this by a factor of 2 to 32 kbit/s, normally with no loss in perceived quality. This corresponds to four bits of compressed data per input sample.

MP3 compression

MP3 is the popular name for one of the audio compression algorithms used in the MPEG-1 and MPEG-2 video compression schemes (see Chapter 10). These standards describe three audio compression schemes, called, in increasing order of sophistication, 'Layer 1', 'Layer 2' and 'Layer 3'. Layer 3 offers the best compression ratio, and it is this to which the '3' in 'MP3' refers.

MP3 is specified solely in terms of the decoder, and there is considerable variation between encoders in both quality and speed: the two are to some extent inversely correlated. The standard does suggest one way an encoder might work, and most encoders are based on the recommended method.

A good MP3 encoder can produce a compressed data stream at a rate of 128 kbit/s which, when decompressed, is indistinguishable in quality to the average listener from a recording on a compact disc. (The exact figure is the subject of heated debate among audio enthusiasts.) We will outline the processing involved in a typical encoder with an input sample rate of 32 kHz; because of the complexity of the standard we will omit some of the finer details.

Most MP3 encoders employ a so-called *perceptual model*: in other words, they exploit particular characteristics of human hearing. We mentioned at the beginning of this chapter that the ear is most sensitive to frequencies in the region from around 500 Hz to 5 kHz: an MP3 encoder will in practice incorporate a complete sensitivity curve for the ear, giving the relative sensitivities to each frequency. A typical such curve might look like the one in Figure 8.10.

The other feature of the perceptual model that we shall consider here is called *frequency masking*. Suppose a sound is composed of just two frequency components, one rather louder than the other. The quieter one will only be heard if it is sufficiently far away in frequency from the first: in effect, the ear's sensitivity is considerably reduced in a range of frequencies around the loud tone. Figure 8.11 shows an example of what the sensitivity curve of the human ear might look like in the presence of a 1 kHz tone: you can see how the sensitivity is reduced (i.e., the quietest perceptible tone is louder) in the frequency area around 1 kHz.

The aspects of the perceptual model above were described in terms of their effects in the frequency domain. A typical MP3 encoder, as shown in Figure 8.12, will in fact transform its input into the frequency domain twice, in slightly different ways. The first uses an ordinary Fourier transform whose output is used for the perceptual model. The second is more complicated and is used for the actual encoding process. The results of the two processes are rather similar; one of the

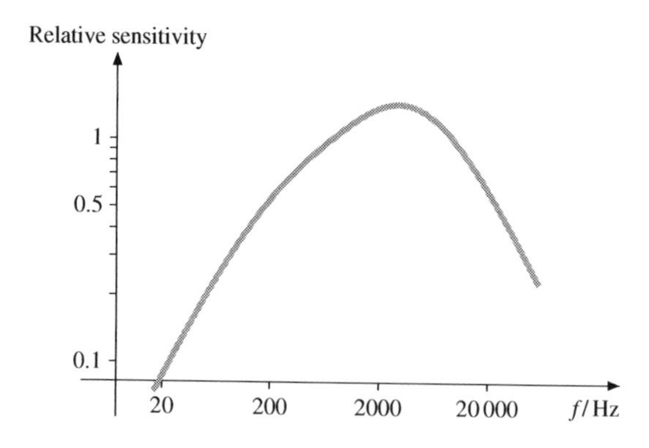

Figure 8.10 Typical sensitivity of the human ear to different frequencies (logarithmic scales).

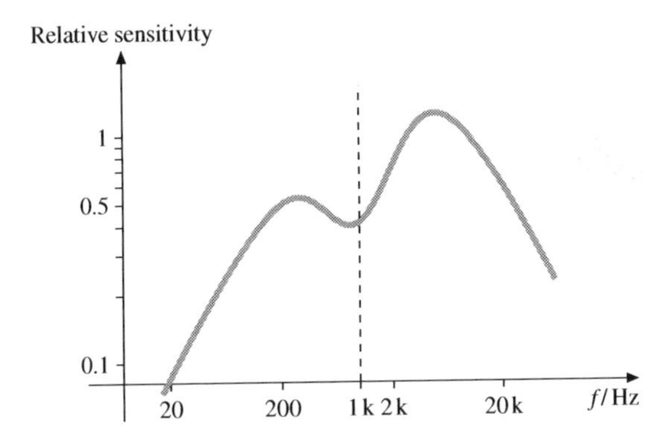

Figure 8.11 Typical sensitivity of the human ear to different frequencies in the presence of a tone at 1 kHz.

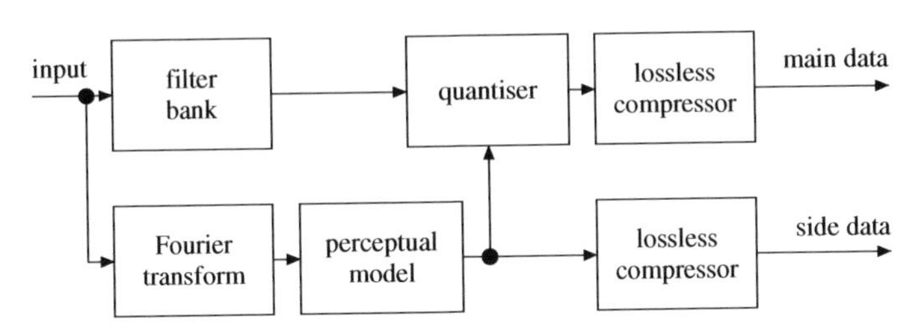

Figure 8.12 Block diagram of a typical MP3 encoder.

reasons for using this arrangement is to make MP3 more similar to earlier audio compression standards.

The perceptual model analyses the Fourier transform of a segment of the input signal. It identifies the strongest frequency components by finding local maxima and calculates their masking effect. The extent of the masking effect depends on the frequency and the sharpness of the peak. The result of this process is a curve which tells us the relative importance of the various frequency components in the input signal.

We now turn to the second method by which the input signal is transformed into the frequency domain. Here the signal is first passed through a set of 32 band-pass FIR filters, collectively called a *filter bank*: see Figure 8.13. Each filter is designed to pass a frequency band of the same width, and together the filters cover the band from 0 Hz to the Nyquist limit of 16 kHz. Numbering the filters from 0 to 31, filter i passes frequencies in the range $500i$ Hz to $500(i+1)$ Hz and blocks all other frequencies. The filters are not perfect, and so there is a small amount of overlap in their responses, as sketched in Figure 8.14.

Since the output of each filter only occupies a narrow frequency band, it can be sub-Nyquist sampled (see Chapter 2). As you can verify, it is possible to

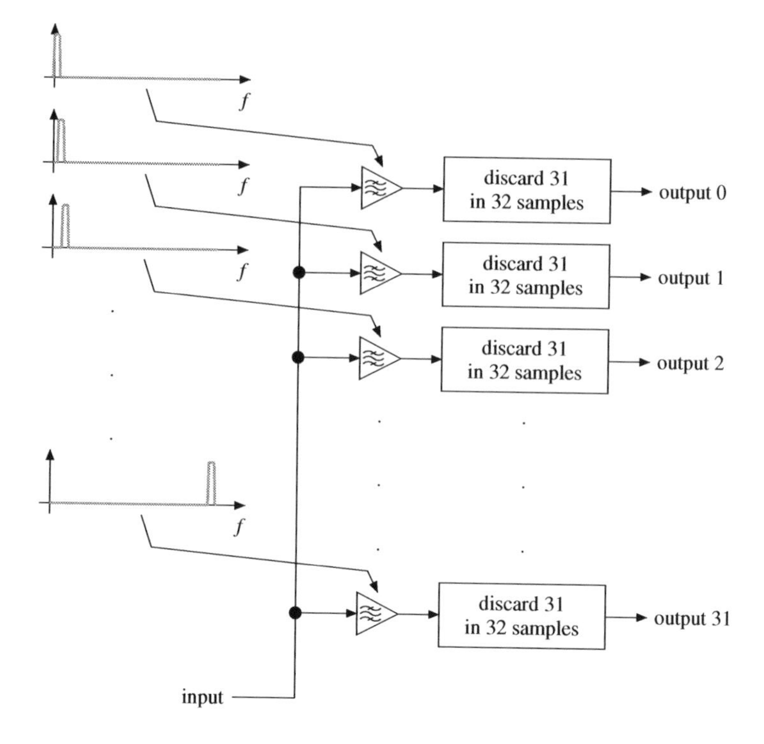

Figure 8.13 MP3 encoder filter bank.

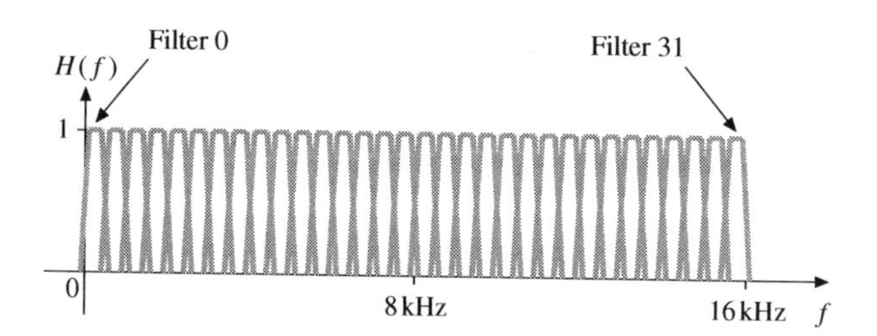

Figure 8.14 Superimposed responses of band-pass filters in the filter bank of an MP3 encoder.

sample each output at 1 kHz without any aliasing being introduced other than as a result of imperfections at the edges of the filter responses. Since 1 kHz exactly divides the input sample rate of 32 kHz, sub-Nyquist sampling can in this case be achieved by simply discarding 31 out of every 32 consecutive samples. As in the case of interpolation filters, discussed in Chapter 5, when reducing the sample rate in this way there is no need for the filter to compute samples which would then immediately be discarded.

The output of the filter bank thus consists of 32 signals each sampled at 1 kHz for a total sample rate of 32 kHz, the same as the input: as long as the filters are carefully designed, the filter bank has not destroyed any information.

Overlapping blocks from the output of each filter are now windowed (see Section 4.7) and converted to the frequency domain using a version of the discrete cosine transform. This is similar to the Fourier transform but works with real input and output data. It is most commonly used in image processing, and so we shall defer a description of it until Chapter 9.

The encoder can choose to make the blocks either 36 samples long, overlapping by 18 samples, or 12 samples long, overlapping by 6 samples. It will usually choose the former when the spectrum of the sound is changing slowly and the latter where it is changing quickly.

The input signal is now represented in the frequency domain. We compare this representation with the threshold curve produced by the perceptual model and discard any frequency components falling below the curve. The amplitudes of the remaining components are now quantised (logarithmically, because amplitude is perceived approximately logarithmically) using more bits for those components where the threshold is low and fewer bits where the threshold is high. The resulting bits are then further compressed using a lossless compression scheme. This forms one part of the output of the decoder, called the *main data*. The output also contains a compressed version of the other information the decoder will need to reconstruct the signal, such as the quantisation levels that have been chosen by the perceptual model. This second part of the output data is called the *side information*.

The decoder reverses the steps in the encoder, although it does not need the perceptual model: it simply needs to be told the results of decisions the encoder made on the basis of the model.

For each block it first decompresses the side and main information. It uses these to reconstruct an approximation to the frequency-domain representation of the input signal. It then inverts the discrete cosine transform to reconstruct the windowed time-domain samples from the output of each filter in the filter bank.

Successive blocks of these samples can now be overlapped and added to build up a continuous stream of time-domain samples corresponding to the output of each filter in the encoder's filter bank. The effect of the sub-Nyquist sampling is undone by shifting each of these streams back to its original frequency band, using a technique we shall describe in Chapter 11, and the results added together to reconstruct the original signal.

8.5 Pitch extraction

We will now look at a particular audio signal processing problem and see how it can be solved in different ways.

The problem is to extract pitch information from an audio signal. For example, the signal might be a recording of a person singing or of a musical instrument being played, and we wish to work out the notes of the tune. Alternatively, it might be a recording of speech, and we would like to extract the *intonation*: how the pitch of the speech rises and falls to convey emphasis and other meaning. In British English, for example, the pitch of the voice usually rises at the end of a question and falls at the end of a statement.

Figure 8.15 shows four waveforms. The top waveform is a recording of the note A above middle C on a piano (this is the note that orchestras tune to, with a nominal frequency of 440 Hz). The second waveform is the same note played on a flute; the third is the same note played on an oboe; and the last is the note sung by a human voice using the 'ah' vowel. In each case the segment of waveform shown is taken from the middle of a note because for most instruments the waveform is steadiest there. On each waveform the two dashed lines indicate a subsegment of

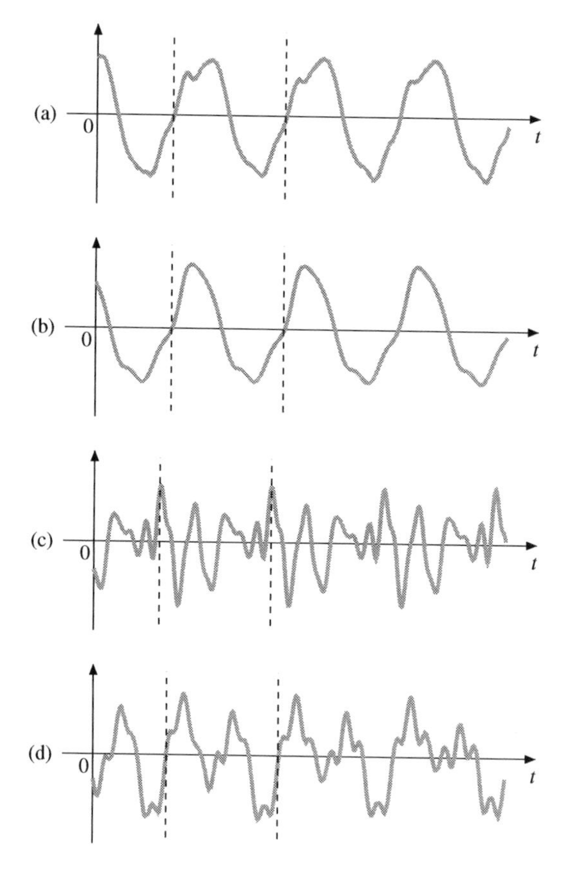

Figure 8.15 Audio waveforms: (a) piano; (b) flute; (c) oboe; (d) human voice singing 'ah'.

length 1/440 seconds. If the waveform were perfectly periodic it would consist of this segment repeated indefinitely.

Many different algorithms have been proposed to solve the problem of pitch extraction; we will outline a few of them and see how they perform on these waveforms.

Time-domain methods

The simplest approaches to pitch extraction are time-domain methods, i.e., ones that analyse the waveform samples directly.

The first idea we might consider is to look at the *zero-crossings* of our signal: the points in time where one sample is negative and the next is positive (a positive-going zero-crossing) or where one sample is positive and the next is negative (a negative-going zero-crossing). We can then measure the time between two successive crossings in the same direction and report that as the period T and its reciprocal as the frequency f. Figure 8.16 shows this process on our four example waveforms. In each case a pair of consecutive zero-crossings has been chosen at random and the implied frequency is shown.

The disadvantage of this method is that it does not cope with the case where there is a spurious extra zero-crossing in the middle of a period.

A slightly more sophisticated approach avoids, and even exploits, this problem. Let $u_0, u_1, u_2, \ldots$ be the set of times where there is a positive-going zero-crossing, in ascending order. Consider the set of all possible differences between these times $u_j - u_i$, where $j > i$, and make a histogram of the differences, as illustrated in Figure 8.17. The histogram has a peak at around $t = 1/440\,\mathrm{s}$, as we would expect. It also has a peak at twice that value, which is not unreasonable: after all, if a waveform has a certain period T, then it also can be thought of as repeating every $2T$. We can ensure our pitch extractor chooses T over $2T$ by a crude method such as multiplying all the histogram heights by a decreasing function of period t such as $1/t$: the peak at $2T$ will now have to be twice as big as the one at T to be chosen.

The above two methods are both simple and do not require a great deal of processing power or storage: in a real-time system the zero-crossings could be found 'on the fly' and so there would not even be any need to store the input waveform. Both methods would be suitable for use in a low-cost embedded system such as a tuner for guitars or other musical instruments.

Frequency-domain methods

Neither of the above time-domain methods gives particularly good results on more complex waveforms, especially those, such as speech waveforms, that change over time. For such waveforms we need a greater level of sophistication.

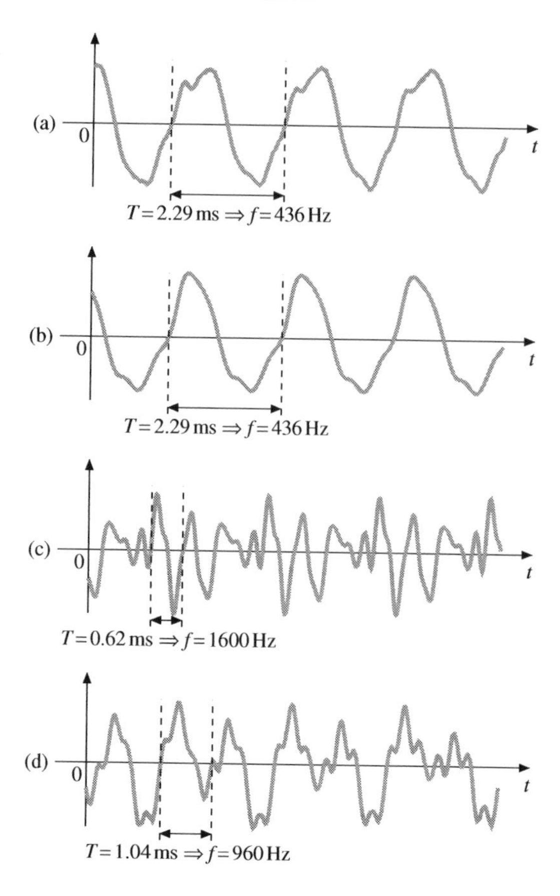

Figure 8.16 A simple time-domain method for pitch extraction illustrated on four audio waveforms: (a) piano; (b) flute; (c) oboe; (d) human voice singing 'ah'.

In Chapter 4 we saw how to use a windowing function to allow us to apply a Fourier transform to a signal that is not necessarily periodic. In this problem, we do not know the period of the signal before we start, and so we must treat it initially as non-periodic. Figure 8.18 shows the result of applying a Hamming window to each of our four example signals and taking a Fourier transform. Only the magnitude of each frequency component is shown.

In each case, the Fourier transform shows a number of peaks. The first peak is at the fundamental frequency of the signal, in this case 440 Hz, with successive peaks at multiples of that frequency. Usually the amplitudes of these higher frequency components fall off gradually, and we can estimate the pitch of the original waveform by looking for the highest magnitude peak. This is not always good enough, however, as sometimes the highest peak is actually one of the harmonics. In the case of the oboe the fundamental is much weaker than the third and fourth harmonics, at about 1320 Hz and 1760 Hz respectively.

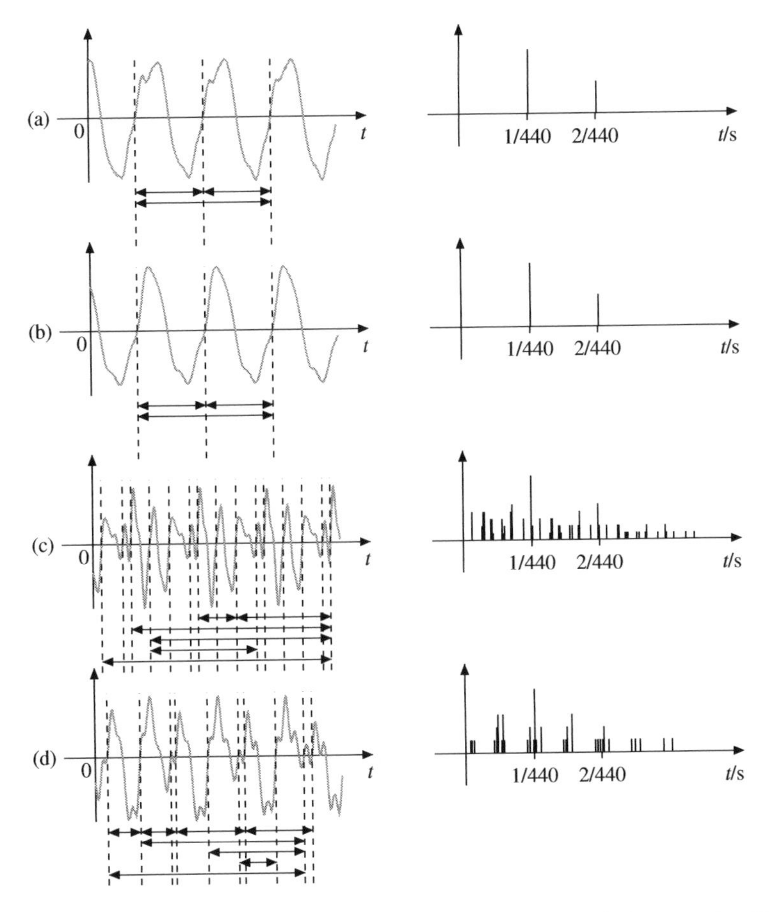

Figure 8.17 A more sophisticated time-domain method for pitch extraction illustrated on four audio waveforms: (a) piano; (b) flute; (c) oboe; (d) human voice singing ‘ah’.

The next level in sophistication is to analyse the Fourier transform further. Looking at the transforms in Figure 8.18 it would seem reasonable to say that the transform consists of a number of spikes at regularly spaced frequencies (the harmonics of the frequency we are interested in) which fit within an overall slowly varying envelope. Seen in this way, the amplitudes of the frequency components arise as the product of the spikes and the envelope function.

As we saw in Chapter 5, multiplication in the frequency domain is equivalent to convolution, or filtering, in the time domain. When you sing, passing breath through the vocal cords produces a signal which has a spectrum of spikes, much like $|E(f)|$ in Figure 8.19(a): the signal is called the *excitation waveform*. The vocal tract acts as a filter with response like that shown as $|H(f)|$ in Figure 8.19(a). The product of $|E(f)|$ and $|H(f)|$ gives $|A(f)|$, which is the spectrum of the sung

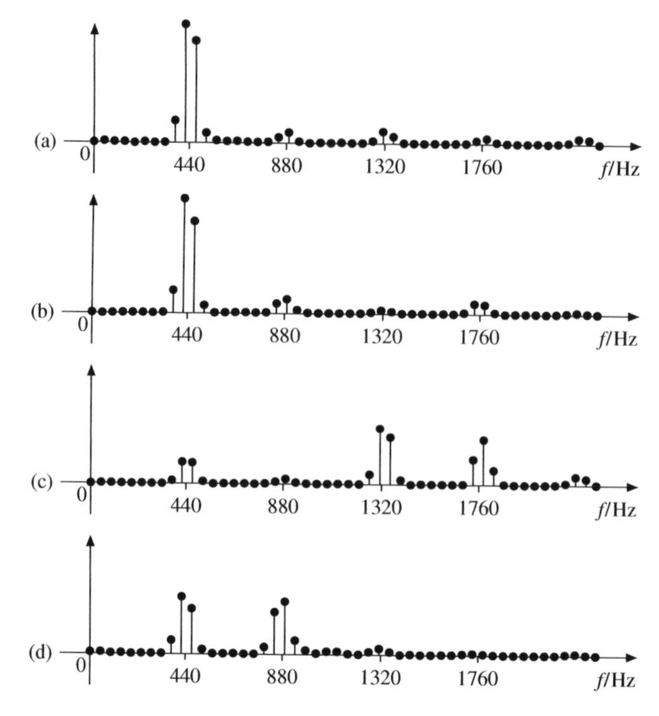

Figure 8.18 Spectra of four audio waveforms: (a) piano; (b) flute; (c) oboe; (d) human voice singing 'ah'.

note. You can hear the approximate impulse response of the filter formed by your vocal tract by shaping your mouth as if you were about to sing a note and (gently!) flicking your finger either against the side of your cheek or, slightly more painfully, against your neck near to where your vocal cords are: you may need to experiment to get a good result. Try changing the shape of your mouth as if you were about to sing different vowel sounds, such as 'ah', 'ee' or 'oo', and hear the difference between the impulse responses.

Low-cost speech synthesisers use an IIR filter to model the vocal tract. A spiky excitation signal is fed into the filter, and the parameters of the filter adjusted continuously to produce the speech waveform.

If we take the logarithm of $|A(f)|$ we obtain the plot in Figure 8.19(b). Since $|A(f)| = |E(f)||H(f)|$, $\log|A(f)| = \log|E(f)| + \log|H(f)|$: we have the sum of a periodic signal and a non-periodic signal, and our ultimate goal, the pitch of the original waveform, is the interval between the spikes in the periodic waveform. To separate out this periodic component from the non-periodic component we can take another Fourier transform. Were it not for the logarithm operation, this would get us back very close to where we started because the forwards and

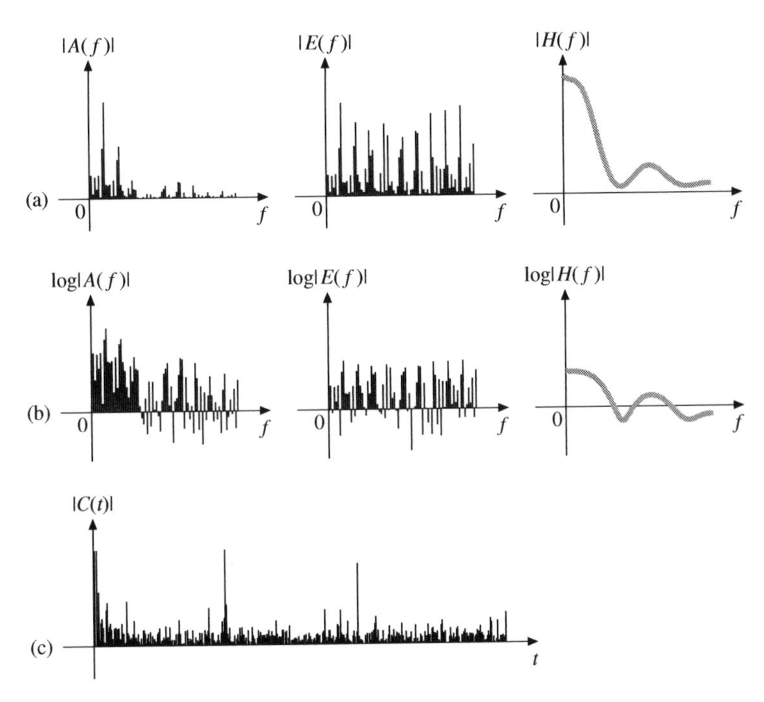

Figure 8.19 Calculating the cepstrum: (a) spectrum of original signal; (b) logarithm of spectrum; (c) the cepstrum is the Fourier transform of the logarithm of the spectrum.

inverse Fourier transforms are very similar in form. However, with the intervening logarithm operation the result is rather different: see Figure 8.19(c).

Because the procedure is similar to calculating a spectrum and then reversing the process, the result is called a *cepstrum.* (Strictly speaking, what we have calculated above is called the *real cepstrum.*) In a rare outbreak of humour in the field of signal processing, the various quantities involved in the cepstrum also have similarly jokey names: for example, the horizontal axis, which has units of time, is called the *quefrency* (as opposed to 'frequency') axis. 'Magnitude' becomes 'gamnitude', 'harmonic' becomes 'rahmonic', and so on.

The slowly varying information in $|H(f)|$ is compressed into the low-quefrency end of the cepstrum. The spikes in $|E(f)|$ have become the regular peaks in the cepstrum. The cepstra of our four example waveforms are shown in Figure 8.20. Observe how the cepstrum has strong peaks at the fundamental frequency even if that frequency component was relatively weak in the original signal.

Cepstral analysis works well in cases where the fundamental of the original signal is weak or missing, and is widely used in speech processing as it is one of the most reliable ways to extract intonation information.

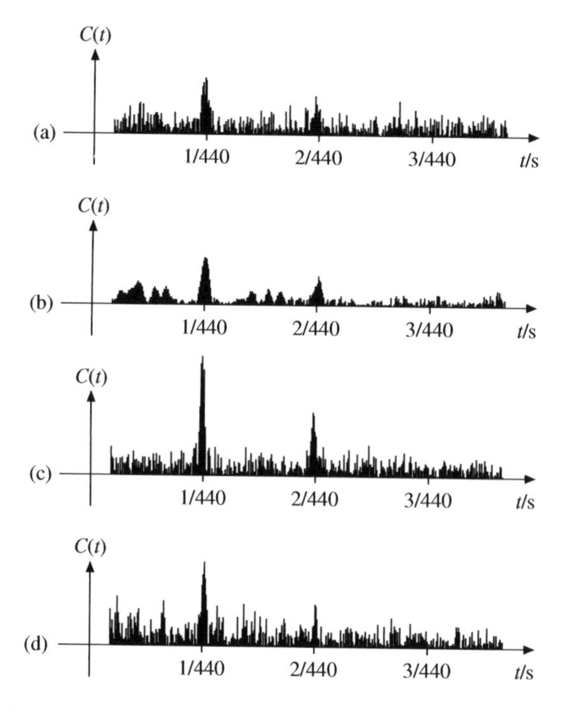

Figure 8.20 Cepstra of example waveforms: (a) piano; (b) flute; (c) oboe; (d) human voice singing 'ah'.

A likelihood method

We will now look at an example of how we can apply the probability-based techniques we discussed in Chapter 6 to the problem of pitch extraction. If we write T_0 for the fundamental period of the segment W of waveform we are analysing, we can use a likelihood method to make a plot of $P(T = T_0 \mid W)$ against T. In other words it will tell us how confident we should be that the fundamental period is T for any given T. We can then either simply take the maximum value of this function, or we can retain the full information in the distribution and continue to process it, for example by combining results from different waveform segments.

We start as we did in the time-domain pitch extraction method discussed above where we calculated the set of time differences between the positions where the signal has zero-crossings, giving rise to a histogram like the ones in Figure 8.17.

The first step in a Bayesian analysis is to design a model. Imagine we already know the actual period of the waveform, T_0. We pick two zero-crossings with the same polarity at random: let their times be t_0 and t_1. We can write the time difference between t_0 and t_1 as a whole number of periods k plus or minus some offset t as $t_1 - t_0 = kT_0 + t$, where we can ensure that $-\frac{1}{2}T_0 < t \leq +\frac{1}{2}T_0$. Note that (assuming all the times are represented as integer sample indices) there are T_0 possible different values of t.

Now there is a chance that the two crossings we have chosen will be at corresponding points on the waveform a whole number of periods apart and so t will be approximately zero: call this 'case 0' or C_0 for short. On the other hand, there is also a chance that the crossings are not at corresponding points on the waveform and therefore not a whole number of periods apart, in which case we know essentially nothing about t (other than that $-\frac{1}{2}T_0 < t \leq +\frac{1}{2}T_0$). Call this 'case 1', or C_1.

To simplify matters, let us suppose these two alternatives are equally likely, so that $P(C_0) = P(C_1) = 1/2$. In the first case we will assume that t is distributed as a Gaussian with mean zero and standard deviation $\sigma = \frac{1}{50}T_0$; in the second case we will assume that all values of t are equally likely, and so $P(t \mid T_0, C_1) = 1/T_0$. Now we apply the rules of conditional probability and obtain:

$$
\begin{aligned}
P(t \mid T_0) &= P(t \mid T_0, C_0)P(C_0) + P(t \mid T_0, C_1)P(C_1) \\
&= \frac{1}{2} \cdot \frac{1}{T_0} + \frac{1}{2} \cdot \frac{1}{\sigma\sqrt{2\pi}} \mathrm{e}^{-\frac{1}{2}(t/\sigma)^2} \\
&= \frac{1}{2}\left(\frac{1}{T_0} + \frac{1}{\sigma\sqrt{2\pi}} \mathrm{e}^{-\frac{1}{2}(t/\sigma)^2}\right).
\end{aligned}
$$

Figure 8.21 shows a sketch of this function. (The function is not quite normalised, the area under the curve being very slightly less than 1: why?)

Now we can consider the case where we have a collection $\{t_i\}$ of N values of t. We shall assume that the values are independent of one another (although, if they are generated by exhaustively enumerating all the time differences between zero-crossings of the same polarity in a given waveform segment, they will not be). We then have

$$P(\{t_i\} \mid T_0) = \prod_i P(t_i \mid T_0).$$

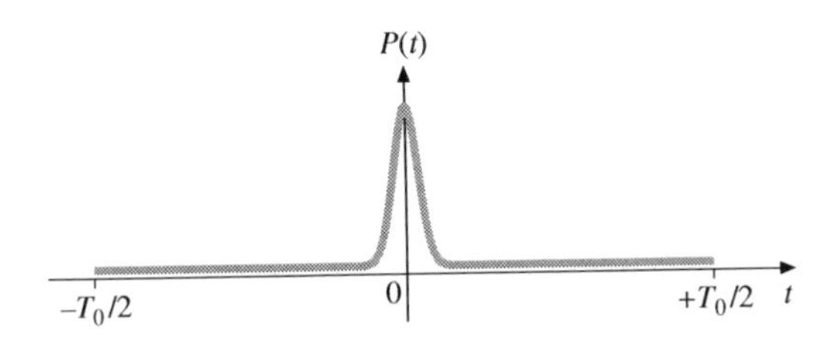

Figure 8.21 Modelled distribution of t.

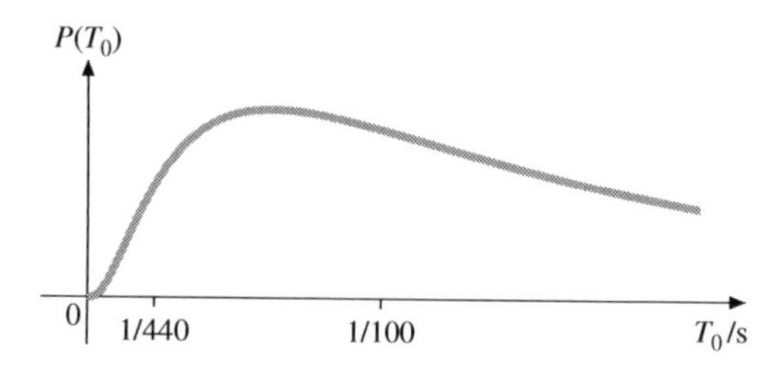

Figure 8.22 Prior distribution on T_0.

What we want to evaluate is $P(T_0 \mid \{t_i\})$ for a range of values of T_0. Bayes' theorem (Equation 6.2) tells us that

$$P(T_0 \mid \{t_i\}) = \frac{P(\{t_i\} \mid T_0)P(T_0)}{P(\{t_i\})}. \tag{8.1}$$

Since each t_i can take on T_0 different values, there are T_0^N possible different sequences of $\{t_i\}$. If we assume all these sequences are equally likely, then we have $P(\{t_i\}) = 1/T_0^N$.

Recall that $P(T_0)$ is called the *prior distribution* on T_0. It describes the information we have about T_0 before starting our analysis. For example, we might know that we are dealing with a speech waveform, and therefore that the fundamental frequency is likely to be in the range from 100 Hz to 250 Hz and thus that T_0 is likely to be between 4 ms and 10 ms. Or we may know that the signal is a soprano singing, in which case we would expect the frequency to be rather higher, and T_0 correspondingly lower. An advantage of the Bayesian approach is that it lets us express this kind of information precisely, in the form of a prior distribution on the values of T_0.

Figure 8.22 shows a suitable prior for T_0 in this case: the distribution shown covers the middle range of most musical instruments and human voices.

We are now in a position to use this prior distribution to evaluate the posterior probability of T_0 for each of the four waveforms shown in Figure 8.18 using the prior distribution in Figure 8.22 and Equation (8.1). Figure 8.23 shows the results, plotted using a logarithmic vertical probability axis because the ratios involved are so great. As you can see, the maximum in the posterior probability distribution gives the correct answer in each case.

8.6 Delta–sigma conversion

Delta–sigma conversion (also known as *sigma–delta conversion*) is a technique for converting signals between analogue and digital forms, in either direction. The technique is widely used in audio applications, for reasons we shall see later. Variations on the technique go by names such as 'one-bit conversion', 'bitstream conversion' and, more loosely, 'oversampling'.

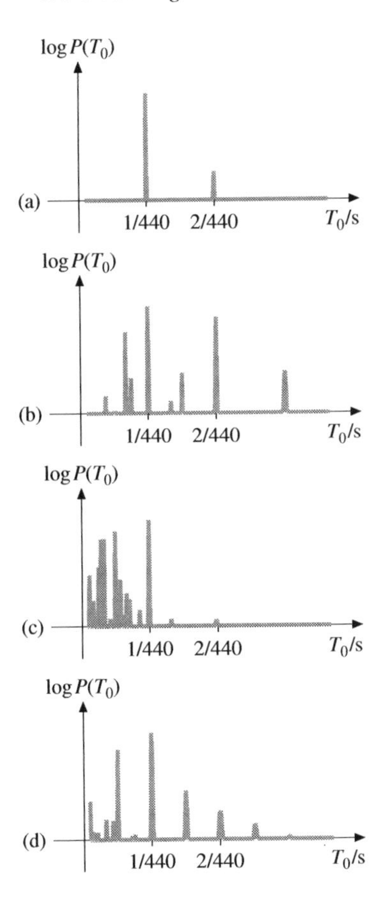

Figure 8.23 The Bayesian pitch extraction method illustrated on four audio waveforms: (a) piano; (b) flute; (c) oboe; (d) human voice singing 'ah'.

In this section we will look at how to make a delta–sigma digital-to-analogue converter. Similar ideas can be applied to analogue-to-digital conversion.

The underlying idea is to convert a stream of samples at one sample rate, say f_s, to a stream of samples at a higher rate kf_s where the new samples are quantised more coarsely than the original ones. Taken to the extreme, we quantise to a single bit, giving rise to the names mentioned above. The factor k, usually an integer, is called the *oversampling factor*.

Suppose we start with 16-bit samples at a rate of 44.1 kHz from a compact disc. To play these samples back we could convert them to analogue form using a high-precision, and therefore expensive, 16-bit digital-to-analogue converter, followed by an (also expensive) anti-aliasing filter with a sharp cutoff. Delta–sigma conversion lets us replace this arrangement with some digital processing (which is cheap and does not need adjustments during manufacture) to produce a 1-bit output sampled at perhaps $k = 256$ times higher a sample rate: $256 \times 44.1\,\text{kHz} = 11.3\,\text{MHz}$.

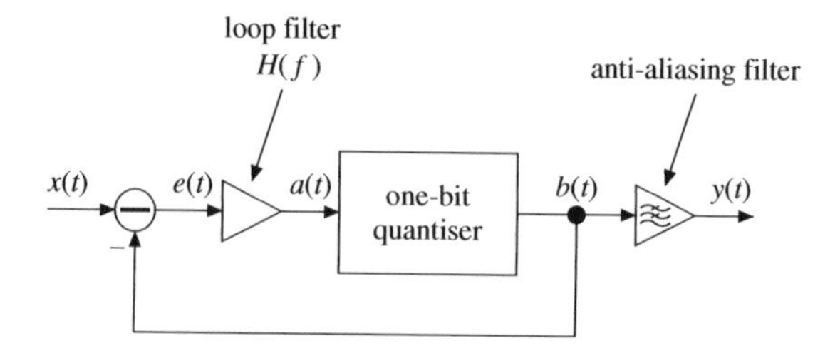

Figure 8.24 Block diagram of a delta–sigma converter.

This output, which we can equally well regard as digital or analogue, will still require an anti-aliasing filter, but because of the much higher sample rate, this filter can be very simple, and thus cheap.

Figure 8.24 shows an example of a delta–sigma converter followed by an anti-aliasing filter. The (digital) input signal is $x(t)$, which is converted to a stream of bits $b(t)$. The final (analogue) output of the converter, after the anti-aliasing filter, is $y(t)$. We will assume that $x(t)$ contains no frequency components outside the band passed by the anti-aliasing filter. Note that all these signals, including $x(t)$, are sampled at the higher sample rate kf_s.

If the signals $y(t)$ and $x(t)$ match closely, so must their Fourier transforms. Since the Fourier transforms of $b(t)$ and $y(t)$ agree over all the frequency components passed by the anti-aliasing filter, any differences between them will have to be in the frequency components blocked by that filter. We therefore want the error signal $e(t) = x(t) - b(t)$, calculated by the subtractor on the left, to have no frequency components among those frequencies passed by the anti-aliasing filter; the remaining frequency components, however, can take on any values at all. To reflect this, we pass the error signal through a filter, called the loop filter, whose frequency response $H(f)$ is designed to remove the frequency components in which we are not interested. Its output is the filtered error signal $a(t)$, which is then quantised to one bit to produce the bit stream $b(t)$.

The signals inside the delta–sigma converter when fed with a sine wave are shown in Figure 8.25. As you can see, the higher-amplitude unwanted components in $B(f)$ tend to be at higher frequencies, while the unwanted components at lower frequencies are of a lower amplitude. The requirements on the anti-aliasing filter at the output are therefore not very demanding.

In a simple delta–sigma converter the loop filter is simply an integrator, which can be thought of as a decaying-average filter with infinite time constant. The name 'delta–sigma' comes from the subtractor (suggesting 'delta') followed by the integrator (suggesting 'sigma').

The error signal is also referred to as the 'noise'. The error signal filter is then called a *noise shaping* filter.

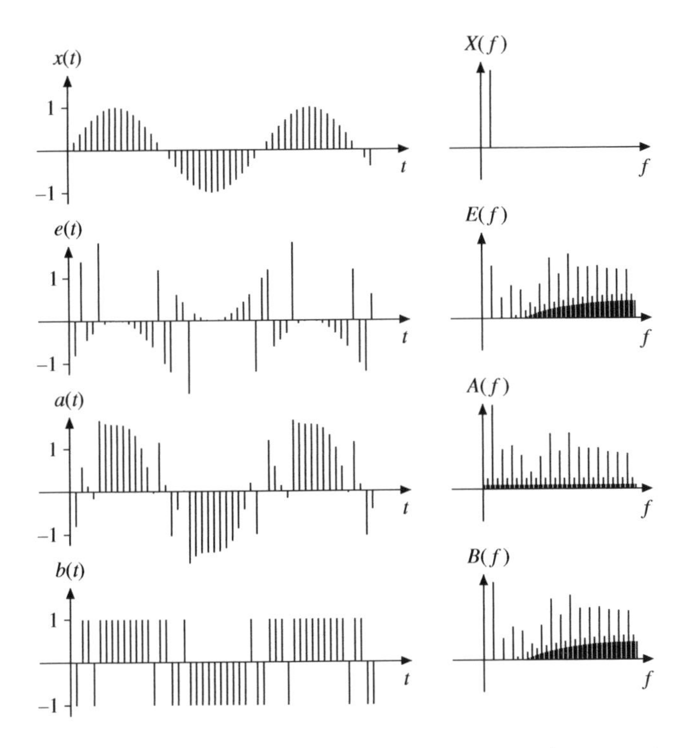

Figure 8.25 Signals inside a delta–sigma converter operating on a sine wave.

The origin of the error signal is the quantiser. The 'noise' produced by quantisation is usually strongly correlated with the signal being quantised, and is usually very far from being white; it is perhaps better characterised as 'distortion'. For audio applications distortion is definitely not desirable, and the output of the delta–sigma converter would sound better if the noise were more nearly white. This can be achieved by adding extra white noise to $a(t)$ before it is quantised; perhaps counter-intuitively, this reduces the perceived distortion in the final output by making the spectrum of the error signal more uniform.

In more sophisticated versions the filter is carefully designed for a given application. In particular, it can be designed to have a response complementary to that of the anti-aliasing filter. It then 'shapes' the noise in $b(t)$ so that it is removed as effectively as possible by the anti-aliasing filter.

If a delta–sigma converter is to be used in an audio application it is possible to adjust the characteristics of the noise-shaping filter so that the noise is moved away from the frequency bands where the ear is most sensitive.

Exercises

8.1 How many bits are needed to represent the samples of one second of stereo audio in the format used by a compact disc? (The disc carries other data, such as track information, which you can ignore for the purposes of this part of the question.)

It is said that the maximum recording time of approximately 74 minutes for compact discs was chosen because that was the time taken by the conductor Herbert von Karajan to get through a somewhat leisurely performance of Beethoven's Ninth Symphony (his longest). How many bytes of sample data is this? How does this compare with the 650 MB (megabyte) capacity claimed by CD-ROMs? What do you think is the reason for the difference?

8.2 If a 128 kbit/s MP3-compressed recording is indistinguishable from a recording on a compact disc, what compression ratio is being achieved?

8.3 A friend offers you his lossless compressor for audio signals. You run it on your collection of 1000 files, each 10 MB long. To your dismay you discover that although it has compressed 99% of the files each by a factor of exactly 10, the remaining 1% of your files have each increased in size by a factor of 100, making the total size of the files in your collection slightly larger. Suggest a simple improvement your friend could make to his compression algorithm and calculate the overall compression ratio you will achieve on your collection of files after your friend has implemented your suggestion.

8.4 Obtain an uncompressed audio file of some music. Choose a range of delta values of the form $\pm 2^n$ for a suitable range of n for use in a DPCM compression scheme to suit the range of your audio samples, and explain your choice. Show how to implement the quantiser in Figure 8.5 without using a large look-up table.

Using these delta values, implement a DPCM decoder for audio signals along the lines of Figure 8.4. Implement an encoder like the one in Figure 8.6. Encode and decode your audio file and compare the result with the original signal. Ensure that there is no overall offset in amplitude between the original and the output of the decoder: you may find it easiest to set the first few samples of the original file to zero as well as being careful about how you initialise the accumulator in the decoder.

What compression factor have you achieved?

8.5 Devise a reasonable set of 32 delta values suitable for use in a DPCM encoder with signed 16-bit input samples such that the quantiser in Figure 8.5 is as simple to implement as you can manage.

8.6 Consider a DPCM scheme with four different delta values available: +4, +1, −1 and −4. Construct a periodic signal $s(t)$ such that if $s(t)$ is fed to the input of the encoder of Figure 8.5, and the output of the encoder is then decoded using the decoder of Figure 8.4, the resulting output increases without limit.

8.7 In the simple time-domain pitch extractor we first found the positive-going zero-crossing points of the signal by looking for a negative sample followed by a positive sample. Suggest how you could use the values of these samples to estimate the position of the zero crossing to a resolution of better than one sample time.

8.8 Write a program to simulate a simple delta–sigma converter with an integrator as the filter element. Make the signal $b(t)$ take on the values $+1$ or -1. Use your program to reproduce the plots of $E(f)$, $A(f)$ and $B(f)$ in the first row of Figure 8.25 using an input sine wave with an amplitude of 0.5.

9
Still images

Images fall broadly into two classes: *bilevel* and *continuous tone*. In the former case, each pixel in the image can be one of only two colours, typically black and white: most line drawings and printed pages of text fall into this category. Photographs are examples of continuous-tone images: they typically contain a wide range of intensities and colours, and often feature areas over which the intensity or colour changes smoothly. A bilevel image can be thought of as a special case of a continuous-tone image.

The majority of the algorithms we shall discuss are designed to operate on continuous-tone images, although they can of course also be applied to bilevel images.

9.1 Luminance and chrominance

Continuous-tone images can be *greyscale* or *colour*. In a greyscale image each point has associated with it a single real value, called the *luminance* at that point. A colour image can be represented by associating three values with each point, giving the amount of each of the three primary colours (red, green and blue) at that point. We say that a picture is represented using the red, green and blue *channels*.

An alternative representation for colour images is as luminance and *colour difference* or *chrominance* values. The luminance value is simply the intensity of a greyscale version of the same image, and the chrominance values, two numbers for each point on the image, give the extra information needed to encode the colour. The conversion from red, green and blue values (R, G and B) to luminance (Y) and chrominance values (C_B and C_R) can be done by taking linear combinations as follows:

$$Y = +0.299R + 0.587G + 0.114B$$

$$C_B = -0.169R - 0.331G + 0.500B$$

$$C_R = +0.500R - 0.419G - 0.081B.$$

Notice that if the values of R, G and B are in the range from 0 to 1, Y will also be in this range. Notice also that C_B and C_R can be negative. The values of the coefficients we have given here are chosen to reflect the human eye's sensitivity to different colours. For example, in the expression for Y, the coefficient for the green component is larger than that for blue or red because the eye is more sensitive to green light. The values shown are those used in several image compression and digital television standards. Different sets of values are used in other applications, but the underlying idea is the same.

We can write these expressions in matrix form

$$\begin{pmatrix} Y \\ C_B \\ C_R \end{pmatrix} = \begin{pmatrix} +0.299 & +0.587 & +0.114 \\ -0.169 & -0.331 & +0.500 \\ +0.500 & -0.419 & -0.081 \end{pmatrix} \begin{pmatrix} R \\ G \\ B \end{pmatrix}$$

or, inverting the matrix,

$$\begin{pmatrix} R \\ G \\ B \end{pmatrix} = \begin{pmatrix} +1.000 & 0.000 & +1.402 \\ +1.000 & -0.344 & -0.714 \\ +1.000 & +1.772 & 0.000 \end{pmatrix} \begin{pmatrix} Y \\ C_B \\ C_R \end{pmatrix}.$$

The advantage of separating the luminance and chrominance information in this way is that the human eye can resolve changes in luminance much better than it can changes in chrominance: this is why bright yellow text on a dark blue background is much easier to read than bright yellow text on a light blue background. We can therefore store and process the chrominance parts of the signal at a lower resolution than the luminance part without visibly degrading the image.

Image processing algorithms are often described as if they were acting on a greyscale image. They can in many cases be extended to full-colour images by processing the components independently, whether represented in terms of red, green and blue channels or in terms of one luminance and two chrominance channels.

9.2 Gamma

Greyscale images are often encoded using a non-linear relationship between Y, the number representing the luminance value, and I, the actual brightness or light intensity in the image. The relation between these quantities is usually a power law of the form

$$I = Y^{\gamma}$$

(or a slight modification thereof), where I and Y range from 0 to 1 and where γ is typically between 2 and 2.5. The form of the relationship originates in the non-linear response of a cathode ray tube (CRT) display to the voltage applied to it.

Practically all image processing operations work on the assumption that light intensity is represented linearly: for example, it only makes sense to add two luminance values together if they are linearly related to actual image brightness. It is therefore necessary to find out what value of γ has been used during encoding so that the effect can be undone using the above expression before the image is processed. Before re-encoding the result of an image processing operation the inverse relation

$$Y = I^{1/\gamma}$$

can be applied: this is called *gamma correction*. In colour images the red, green and blue channels should have gamma correction applied to them separately.

Gamma correction has the benefit that the encoded Y values are approximately on a perceptually linear scale, which means the encoding is efficient for images intended for human viewing.

9.3 An image as a signal

Images differ from most of the other signals that are discussed in this book in that their domain is two-dimensional: the intensity of a pixel is a function $u(x, y)$ of its two coordinates, x and y, within the image. Images are usually sampled on a regular rectangular grid of points, with a certain sample rate along each axis. The sample rate is expressed in pixels per unit length. A printer or scanner, for example, will have its imaging resolution expressed in dots per inch or a similar unit.

We shall generally assume that the pixels are square: that is, that the sample rates in the x and y directions are the same. Most scanners and cameras are capable of producing images with square pixels.

Most of the signal processing operations we have seen on signals with one-dimensional domain can be extended to the two-dimensional case, as we shall now see. For simplicity we shall show examples of greyscale images using integer sample values with zero denoting black, rather than real numbers.

The Fourier transform of an image

A two-dimensional Fourier transform can be defined in terms of one-dimensional Fourier transforms. We simply take the Fourier transform of each row of the image, and then take the Fourier transform of each column of the result: see Figure 9.1. (Alternatively, these two batches of operations can be carried out in the reverse order: the result is the same.) The resulting Fourier transform has the same dimensions as the original image. The original image is said to be in the *spatial domain* and its Fourier transform in the *frequency domain*.

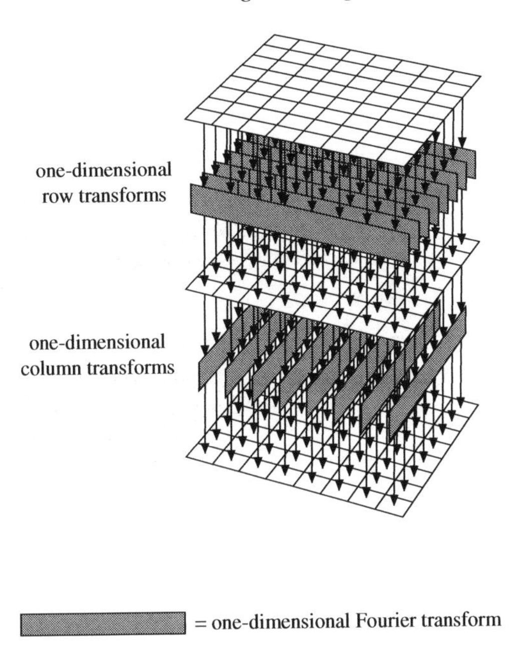

Figure 9.1 The two-dimensional Fourier transform: transform the rows of the image, and then transform the columns.

The frequency resolution along each axis is, as you might expect by analogy with the one-dimensional case, equal to the sample rate of the original image along that axis divided by the number of pixels along that axis. We shall look at the interpretation of the individual frequency components below.

Note that although the image is real, the result of the first batch of transforms, and hence the input to the second batch of transforms and the final result, will be complex.

The two-dimensional inverse transform can be calculated in the analogous way: take the inverse transform of each row, and then the inverse transform of each column of the result.

It is possible to use two-dimensional Fourier transforms to perform analysis of images in the frequency domain, in much the same way as one-dimensional transforms are used to analyse audio signals in the frequency domain. In the one-dimensional case we made the assumption that the section of signal being transformed was one period of a periodic waveform. In two dimensions the assumption is that the square or rectangular region being transformed is one tile of an infinite pattern that repeats along both the x and the y axes. Images (except, perhaps, specially constructed segments of textures designed to be tiled regularly) are unlikely to exhibit periodic patterns like this, and so Fourier transforms are less often appropriate in image processing applications. We will look below at a variation on the Fourier transform, called the discrete cosine transform, that is more often useful; meanwhile, we will describe how to filter an image, and how the job can be accelerated using Fourier transforms.

Figure 9.2 shows some example synthetic images and their Fourier transforms. Only the magnitude of each Fourier component is shown. In each case the origin is at the centre of the image on the left, which you should imagine as one tile in a repeating pattern. Thus, for example, Figure 9.2(j) is actually a disc just like Figure 9.2(b) but with a different centre. The axes of the images are x and y (with the y axis running upwards); the axes of the transform are f_x and f_y.

Some useful properties of two-dimensional transforms are apparent from these examples, most of which are parallels of the situation in the one-dimensional case:

(i) All the transforms have 180° rotational symmetry (this is a consequence of the original image being real rather than complex);

(ii) A circularly-symmetric image has a circularly-symmetric transform: see Figure 9.2(b) and (c);

(iii) An image with discontinuities (sharp edges) will have frequency components that tail off more slowly than one with smooth edges: compare Figure 9.2(b) and (c);

(iv) An image where all columns are the same (i.e., which just consists of horizontal stripes) has all its non-zero frequency components in the line $f_x = 0$: see Figure 9.2(d);

(v) An image where all rows are the same (i.e., which just consists of vertical stripes) has all its non-zero frequency components in the line $f_y = 0$: see Figure 9.2(e);

(vi) More generally, an image which just consists of stripes at right angles to a particular direction has all its frequency components along a line in that direction from the origin: see Figure 9.2(f), where we have chosen (1,2) as the direction vector;

(vii) An image that does not tile correctly, even if only by a small amount, can have many non-zero frequency components: compare Figure 9.2(f) and (g);

(viii) An image can be translated horizontally, vertically, or some combination of the two, without affecting the magnitudes of the Fourier coefficients (although the phases do change): compare Figure 9.2(b), (h), (i) and (j);

(ix) The Fourier transform of random noise looks like (different) random noise (see Figure 9.2(k));

(x) Rotating an image rotates its transform: see Figure 9.2(l), (m) and (n).

As with the one-dimensional transforms shown in Figure 4.7, we can formulate more precise versions of these statements and prove them, but we shall not do so here.

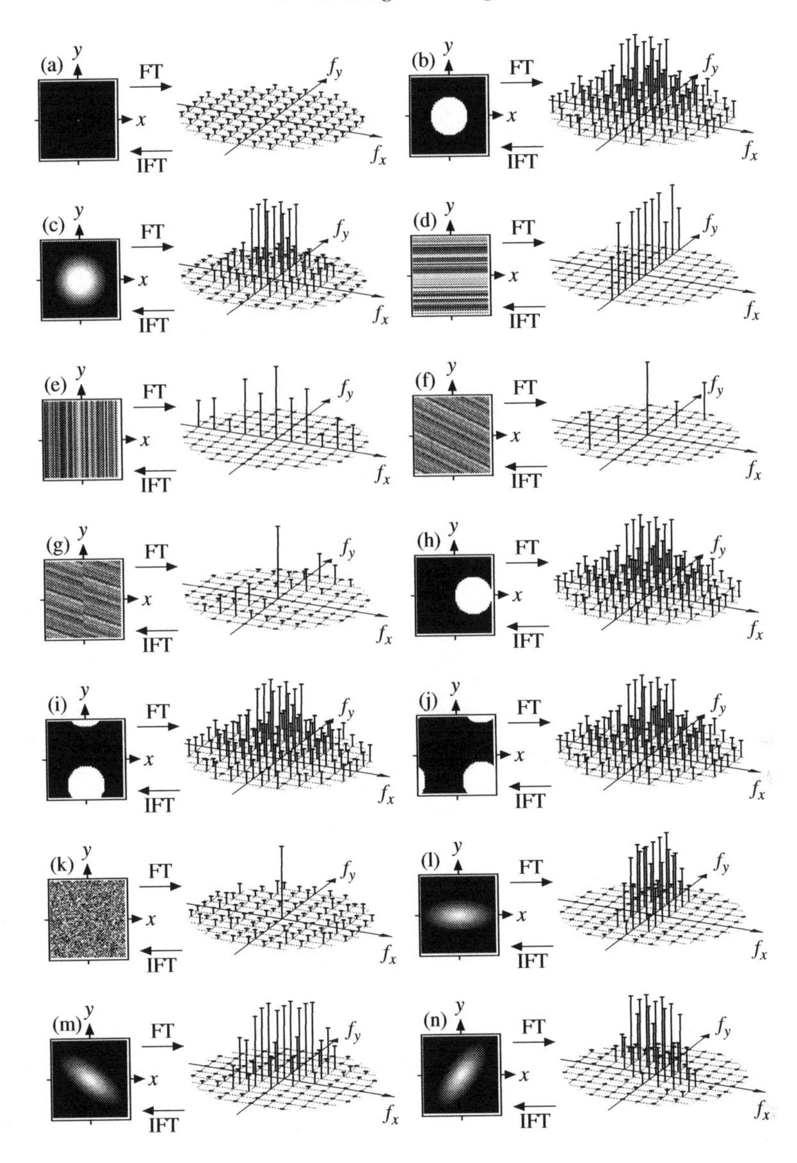

Figure 9.2 Examples of two-dimensional Fourier transforms.

Figure 9.3 shows some of the two-dimensional Fourier basis functions: these are the images whose Fourier transform has a single frequency component whose value is 1, all the others being zero. As you can see, the basis functions are sinusoidally varying patterns oriented in various directions. The frequency of the pattern depends on how far the frequency component in question is from the origin; the orientation of the pattern depends on the direction of a vector from the origin to that frequency component.

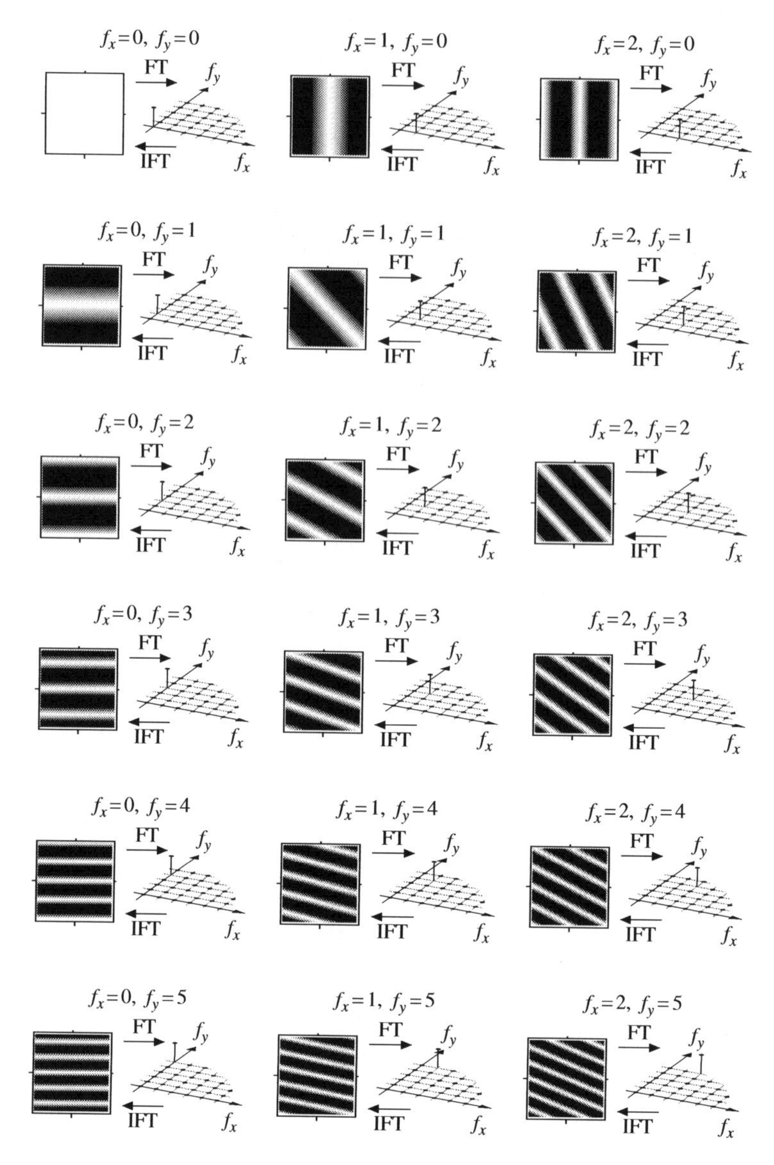

Figure 9.3 Examples of two-dimensional Fourier basis functions (real part only shown).

Aliasing in two-dimensional sampling

Aliasing occurs when sampling in two dimensions, just as it does when sampling in one dimension. The effect manifests itself (for example) as jagged edges of what should be straight lines.

By extension of the one-dimensional case, if the sample rate along the x axis is f_s, it is impossible to distinguish a frequency of f_x from a frequency of $f_x + kf_s$

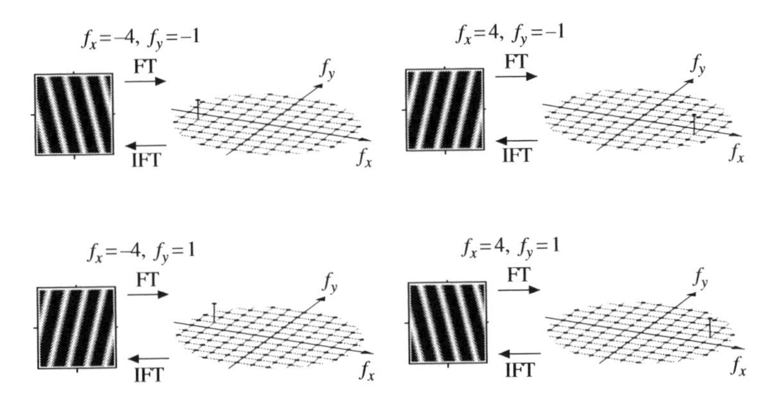

Figure 9.4 Aliasing in two dimensions.

for any integer k. The same goes for the y axis. The spectrum of the sampled image is thus periodic along both axes, and can be thought of as being tiled in just the same way as the original image.

The samples being real also gives rise to an aliasing effect. Recall that in the one-dimensional case the effect was to make the frequency f indistinguishable from the frequency $-f$; in the two-dimensional case, the effect is to make the two-dimensional frequency (f_x, f_y) indistinguishable from the two-dimensional frequency $(-f_x, -f_y)$: this accounts for the 180° rotational symmetry in the frequency components that we observed above.

This last result is perhaps slightly surprising. In one dimension the frequency $f = 4$, for example, is indistinguishable from the frequency $f = -4$; and yet, in two dimensions, the two-dimensional frequency $(f_x, f_y) = (4, 1)$ *is* distinguishable from the two-dimensional frequency $(f_x, f_y) = (-4, 1)$. Figure 9.4 shows the basis functions corresponding to the frequencies $(f_x, f_y) = (4, 1)$, $(f_x, f_y) = (-4, 1)$, $(f_x, f_y) = (4, -1)$ and $(f_x, f_y) = (-4, -1)$ for comparison: note that there are two distinct patterns.

9.4 Filtering an image

As we discussed in Chapter 5, there are two main classes of filter: FIR (finite impulse response) and IIR (infinite impulse response). Although it is possible to define IIR filters that can act on an image, they are very rarely used. The FIR filters used in image processing fall into two classes: separable and non-separable.

Separable filters

An image can be filtered by applying a one-dimensional filter first to each row, and then applying a (possibly different) one-dimensional filter to the columns of the row-filtered result. This technique is adequate for many applications, including low-pass anti-aliasing filtering prior to downsampling. It is also possible to apply the polyphase techniques described in Chapter 5 independently to the rows and columns of an image in order to scale it or resample it. Using a different filter along each axis allows the image to be scaled by different factors along each axis, for example in order to convert an image with rectangular pixels into one with square pixels.

An example of this kind of filtering is shown in Figure 9.5. At the top we have the frequency response and impulse response of a one-dimensional low-pass filter. Across the bottom we have the original image; the result after applying the filter to the rows of the image; and the final result after applying the filter to the columns of the row-filtered result. You can also construct two-dimensional band-pass and high-pass filters using this technique.

We can define the impulse response of a two-dimensional filter to be the result of applying that filter to an image consisting of a single pixel with the value 1, all other pixels having the value 0. The two-dimensional frequency response of a filter is the two-dimensional Fourier transform of its impulse response.

As Exercise 9.4 asks you to prove, if the impulse responses of the filters applied in the x and y directions are $u(x)$ and $v(y)$ respectively, the overall two-dimensional impulse response of the filter is given by $h(x, y) = u(x)v(y)$.

A two-dimensional filter which can be implemented by applying a one-dimensional filter to the rows of an image, followed by a (possibly different) one-dimensional filter to the columns of the row-filtered result is called *separable*; one which cannot be so implemented is called *non-separable*.

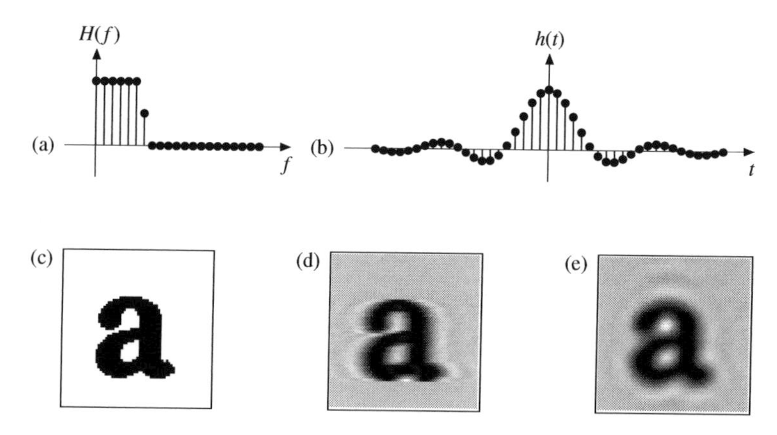

Figure 9.5 A two-dimensional separable low-pass filter: (a) one-dimensional low-pass frequency response; (b) one-dimensional impulse response; (c) original image; (d) result after row filtering; (e) result after column filtering.

Non-separable filters

A weakness of the separable filters discussed above is that they are not in general *isotropic*. 'Isotropic' means that the filter treats an image feature identically whether that feature is aligned with one of the axes of the image or not. The opposite of 'isotropic' is 'anisotropic'.

Consider the image of circles shown in Figure 9.6(a). If we filter this image using the low-pass filter shown in Figure 9.5 the result is as shown in Figure 9.6(b), which shows magnified portions where there are vertical, diagonal and horizontal lines in the image. As you can see, the diagonal lines are better preserved by the filter than the edges which are parallel to the axes. In some applications this is an undesirable artefact: for example, if the image were of text, it is clearly not ideal if the diagonal strokes of the letters appear sharper than the horizontal or vertical ones.

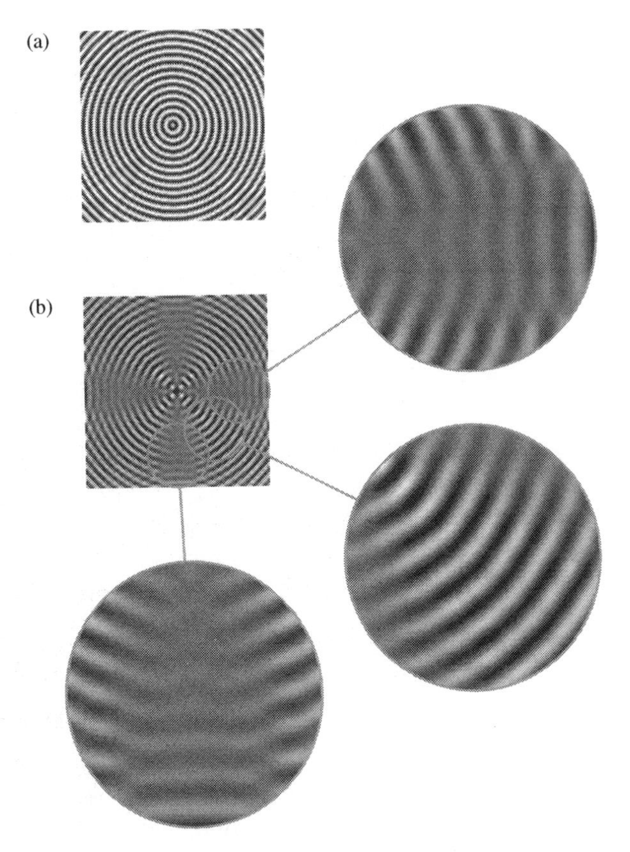

Figure 9.6 Two-dimensional separable low-pass filter acting on an image of circles: (a) original image; (b) result of filtering.

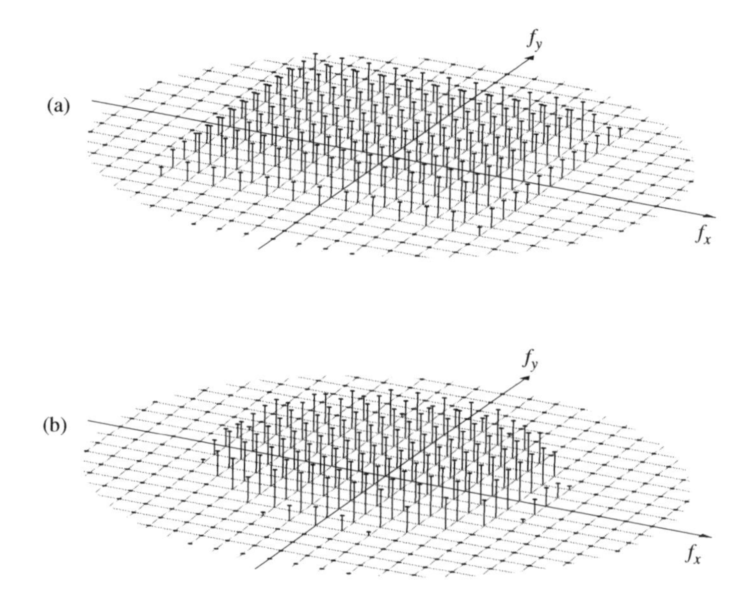

Figure 9.7 Frequency response plots of two-dimensional (a) separable and (b) isotropic low-pass filters.

If a filter is to be isotropic then its impulse response, and hence also its frequency response, must be circularly symmetric. The problem with the separable filters is, as you can see from Figure 9.7(a), that their responses are in general not circularly symmetric.

To design an isotropic filter we will therefore specify the response of the filter to a particular frequency component purely in terms of the distance of that component from the origin in the frequency domain. If a frequency component has coordinates (f_x, f_y) in the frequency domain, its distance from the origin is, by Pythagoras' theorem, $r = \sqrt{f_x^2 + f_y^2}$. For example, if we set the frequency response as a function of r to be that of a low-pass filter, the resulting two-dimensional frequency response will be as shown in Figure 9.7(b). Compare this with Figure 9.7(a), which is the frequency response of a filter like the separable one in Figure 9.5.

An inverse two-dimensional Fourier transform converts the frequency response we have designed into an impulse response. We can now apply a two-dimensional window to the impulse response, just as we did in the one-dimensional case. A popular choice is the two-dimensional separable Hamming window obtained by applying a one-dimensional Hamming window (Equation (4.3)) first to each row of the impulse response and then to each column of the result. This has the unfortunate effect of turning an isotropic filter into an anisotropic one (although it does preserve the separability of a filter). A popular circularly symmetric window, which preserves the isotropic nature of a filter (although not preserving separability), is the rotated Hamming window given by

$$H(r) = \begin{cases} 0.54 + 0.46\cos\pi\frac{r}{R-1}, & \text{if } r \leq R-1 \\ 0, & \text{otherwise,} \end{cases}$$

where $r = \sqrt{x^2 + y^2}$ and R is the radius of the window. You can check that this expression is essentially the same as Equation 4.3 with the middle of the window moved to $r = 0$.

Note that you should *not* construct a two-dimensional filter directly in the spatial domain simply by, for example, substituting $r = \sqrt{x^2 + y^2}$ for t in Equation (5.5). The number of sample points distance r from the origin in a one-dimensional signal is constant, whereas the number of points approximately a distance r from the origin in an image increases proportionally to r. Using Equation (5.5) naïvely does not correctly distribute the weight of a coefficient $h(r)$ among these points.

To apply a filter to an image we calculate a two-dimensional convolution of the image with the impulse response of the filter, exactly analogously to the case of the one-dimensional FIR filter. To illustrate the process, we will first consider the simple impulse response shown in Figure 9.8, which is a kind of two-dimensional analogue of one of the three-point filters we discussed in Chapter 5.

For each pixel in the image we take a copy of the impulse response aligned with that pixel, multiply it by the pixel value in the image, and add together the results. The process is illustrated in Figure 9.9.

In the one-dimensional case we saw (see Figure 5.9) how a convolution can be calculated by superimposing suitably scaled and shifted copies of the original signal and adding the results. We can do the same in two dimensions: Figure 9.10 shows the convolution of Figure 9.9, but calculated by superimposing five scaled and shifted copies of the original image. As you can see, the result is the same.

Finally in this section we shall return to the isotropic filter whose frequency response is shown in Figure 9.7. If we apply this filter to the test image of

(a)

0	1	0
1	2	1
0	1	0

(b)

0	$\frac{1}{6}$	0
$\frac{1}{6}$	$\frac{1}{3}$	$\frac{1}{6}$
0	$\frac{1}{6}$	0

Figure 9.8 Impulse response of a simple two-dimensional filter (a) with integer coefficients; (b) scaled so coefficients sum to 1.

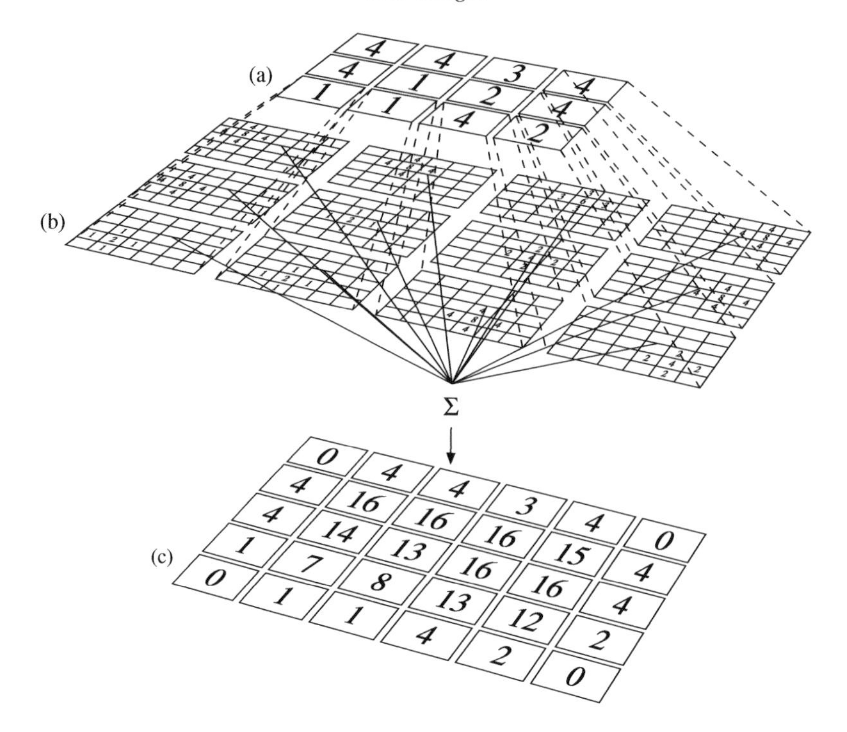

Figure 9.9 Applying a two-dimensional filter to an image: (a) original image; (b) scaled and shifted copies of impulse response; (c) sum of scaled and shifted copies of impulse response gives result.

Figure 9.11(a), the result is as shown in Figure 9.11(b). Compare this result with Figure 9.6(b): the same portions of the resulting image have been enlarged for clarity. As you can see, the isotropic filter treats horizontal, vertical and diagonal edges the same, giving a more visually satisfying result.

Filtering using Fourier transforms

Just as in the one-dimensional case, we can calculate a two-dimensional convolution using a two-dimensional Fourier transform using the following recipe: take the two-dimensional transforms of the images to be convolved, multiply corresponding frequency components, and calculate the inverse transform of this product. The procedure is illustrated in Figure 9.12, which is directly analogous to Figure 5.15. Note that in order to use this method the two images to be convolved must be the same size, and so the smaller one must be padded out with zero samples to the same size as the larger one.

Because the Fourier transform operates as if it were acting on one tile of an infinitely repeating pattern, the samples on one edge of the original image affect the samples on the opposite edge of the result. These edge effects can be avoided

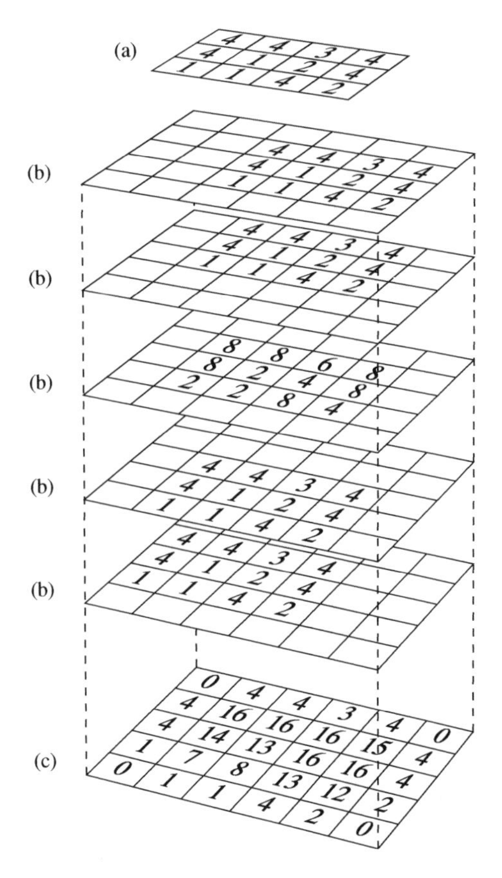

Figure 9.10 Applying a two-dimensional filter by adding shifted and scaled copies of the original image: (a) original image; (b) scaled and shifted copies of image; (c) sum of scaled and shifted copies of image gives result.

by padding the original image using samples with a value of zero (or some other suitably chosen background intensity). Using the result of Exercise 9.1 it is possible to show that if the non-zero samples of the impulse response cover a rectangle measuring u pixels in the x direction and v pixels in the y direction, the original image needs to be padded out by $u-1$ pixels in the x direction and $v-1$ pixels in the y direction to avoid edge effects.

Exercise 9.3 shows that if a fast Fourier transform algorithm is used the Fourier transform method can be considerably faster than direct calculation for large convolutions.

Two-dimensional correlation

The two-dimensional correlation of two images can be defined in an analogous way to the correlation of two one-dimensional real signals (see Section 5.4). In

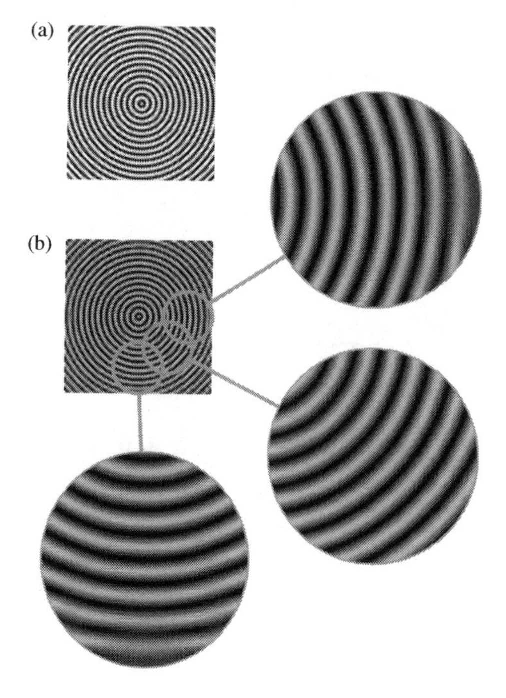

Figure 9.11 Two-dimensional isotropic low-pass filter acting on an image of circles: (a) original image; (b) result of filtering.

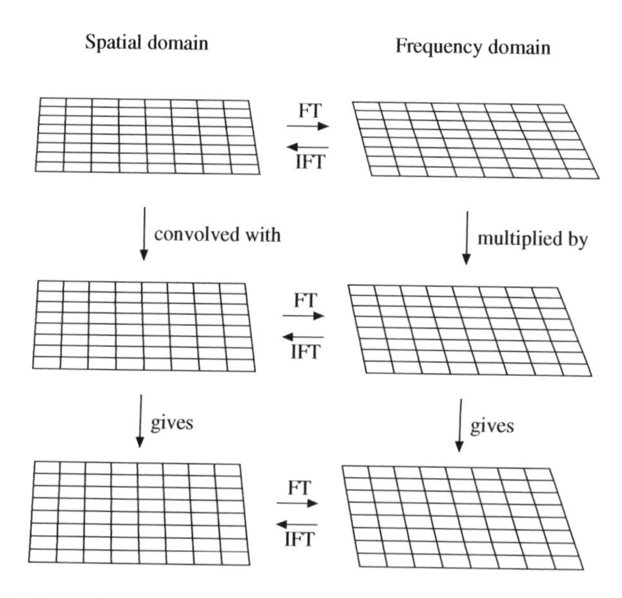

Figure 9.12 Convolution in the spatial domain is equivalent to pointwise multiplication in the frequency domain.

the one-dimensional case we calculated the correlation of two signals by convolving the reverse of one signal with the other; similarly, if $H(x, y)$ and $X(x, y)$ are two images their correlation is calculated by rotating H through 180° and calculating its convolution with X. The convolution step can be performed using two-dimensional Fourier transforms, as described above. If a fast Fourier transform algorithm is used, the result is a *fast two-dimensional correlation*.

9.5 The discrete cosine transform

The discrete cosine transform, or DCT, is a variation on the Fourier transform that is widely used in image processing, especially in image compression. The reason for its popularity is that the discrete cosine transform does not make the assumption that the signal to be transformed is periodic.

The one-dimensional discrete cosine transform

The idea behind the discrete cosine transform is to construct a periodic signal from one that is not necessarily periodic, and then take a Fourier transform, discarding any redundant parts of the result.

Imagine that the eight-sample signal shown in Figure 9.13(a) is taken from eight successive pixels in one row of a greyscale photographic image. There is a large difference between the first and the last sample values, and so when we use it as one period of a periodic signal a large discontinuity is introduced: see Figure 9.13(b).

We can construct a more promising-looking signal by following the original signal with a reflected copy of itself, as shown in Figure 9.14(a). The periodic version now has a period of 16 samples and does not suffer from any discontinuities. In the most widely used version of the discrete cosine transform zeros are

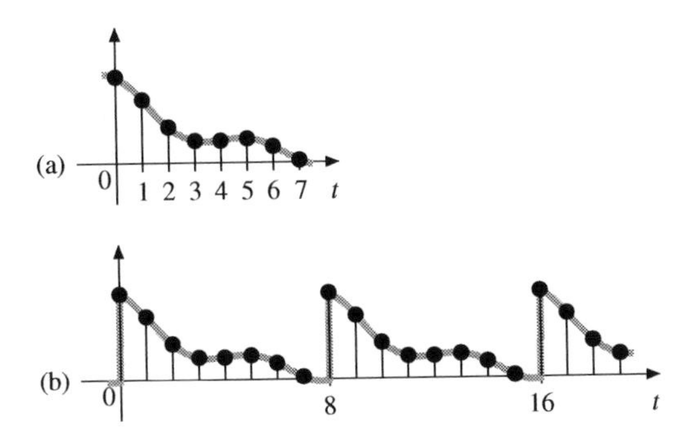

Figure 9.13 (a) Eight samples from one row of a greyscale image; (b) those samples considered as one period of a periodic signal.

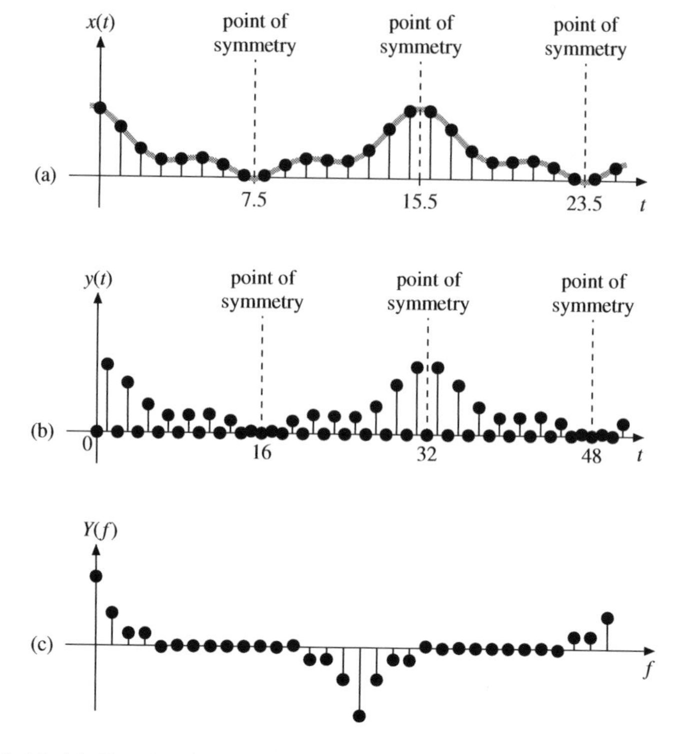

Figure 9.14 (a) Creating a sixteen-sample periodic signal from an eight-sample signal; (b) interleaving with zero samples and shifting by one place produces a signal symmetric about $t = 0$ and $t = 16$; (c) Fourier transform of (b).

inserted between adjacent samples of the newly constructed signal, and the result shifted along by one place, as shown in Figure 9.14(b): this makes the signal symmetric about times $t = 0$, $t = 16$, $t = 32$ and so on.

Now that we have constructed a continuous periodic signal, we can take a Fourier transform. The reflection has doubled the number of samples, and the interleaving with zeros has doubled the number of samples again, so we shall need a 32-point transform. The result of the transform on samples 0 to 31 of Figure 9.14(b) is shown in Figure 9.14(c).

Because the signal we transformed was symmetric about time $t = 0$ the result of the transform is purely real (see Exercise 4.4); because the signal was real, the second half of the transform is the reflection of the first half; and because of the interleaved zeros in the signal, the second half of the Fourier transform has the same sequence of magnitudes as the first half, but with opposite sign (see Exercise 4.5(c)).

The combination of all these observations means that the entire Fourier transform, and hence the entire signal, can be reconstructed from just its first eight frequency components. These frequency components form the discrete cosine transform of the original signal.

The expression normally used for the N-point discrete cosine transform of a signal $u(t)$ is

$$U(f) = C(f) \sum_{t=0}^{N-1} u(t) \cos \frac{(2t+1)\pi f}{2N} \tag{9.1}$$

where f runs from 0 to $N-1$, and where

$$C(f) = \begin{cases} \sqrt{\frac{1}{N}}, & \text{if } f=0; \\ \sqrt{\frac{2}{N}}, & \text{otherwise.} \end{cases}$$

You can check that in the case $N = 8$ this expression, apart from the values of the scaling constants $C(f)$, can be obtained by calculating the Fourier transform of the symmetric 32-point signal constructed above.

The inverse discrete cosine transform is given by

$$u(t) = \sum_{f=0}^{N-1} C(f)U(f) \cos \frac{(2t+1)\pi f}{2N}$$

where t runs from 0 to $N-1$. Note that the same scaling factors $C(f)$ appear in both the forwards and inverse transforms.

The basis functions for the discrete cosine transform, i.e., functions whose transform consists of a single non-zero value, are shown in Figure 9.15. The discrete cosine transform expresses any signal as a weighted sum of these basis functions. They are cosine waves linearly increasing in frequency; but, unlike the basis functions for the Fourier transform, the number of periods in each basis function goes up in steps of 1/2. It is the fact that not all the basis functions consist of a whole number of cycles which makes the discrete cosine transform more suitable for representing non-periodic signals such as images using a small number of coefficients.

The two-dimensional discrete cosine transform

A two-dimensional version of the discrete cosine transform can be defined for use on images. As you might expect, it is calculated by simply taking the transforms of the rows of the image, and then taking the transform of each column of the transformed result.

The expression for an 8-point by 8-point two-dimensional discrete cosine transform is thus

$$U(f_x, f_y) = C(f_x)C(f_y) \sum_{x=0}^{7} \sum_{y=0}^{7} u(x, y) \cos \frac{(2x+1)\pi f_x}{16} \cos \frac{(2y+1)\pi f_y}{16}$$

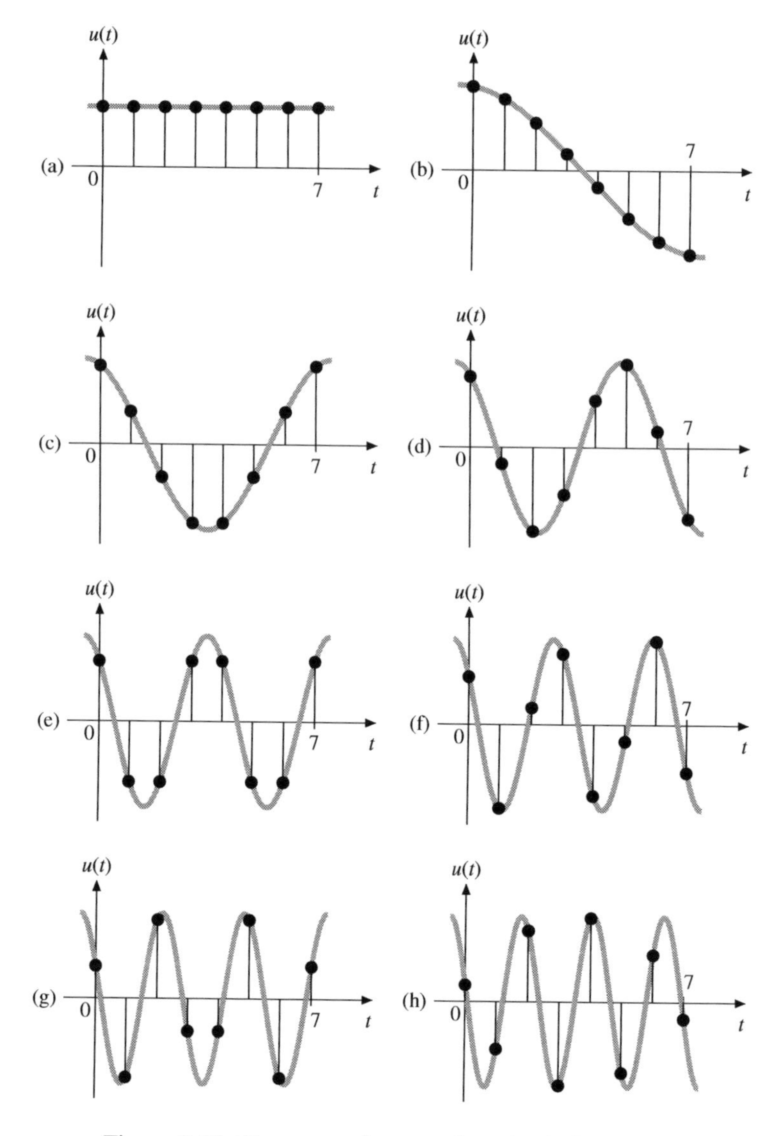

Figure 9.15 Discrete cosine transform basis functions.

with $C(f)$ defined as above. The corresponding inverse transform is

$$u(x, y) = \sum_{f_x=0}^{7} \sum_{f_y=0}^{7} C(f_x)C(f_y)U(f_x, f_y) \cos \frac{(2x+1)\pi f_x}{16} \cos \frac{(2y+1)\pi f_y}{16}.$$

The basis functions for the two-dimensional discrete cosine transform, each of which is an 8-pixel-square image, are as shown in Figure 9.16. These are the images whose transform contains a single non-zero value. The images in the figure are arranged in a grid with f_x increasing from left

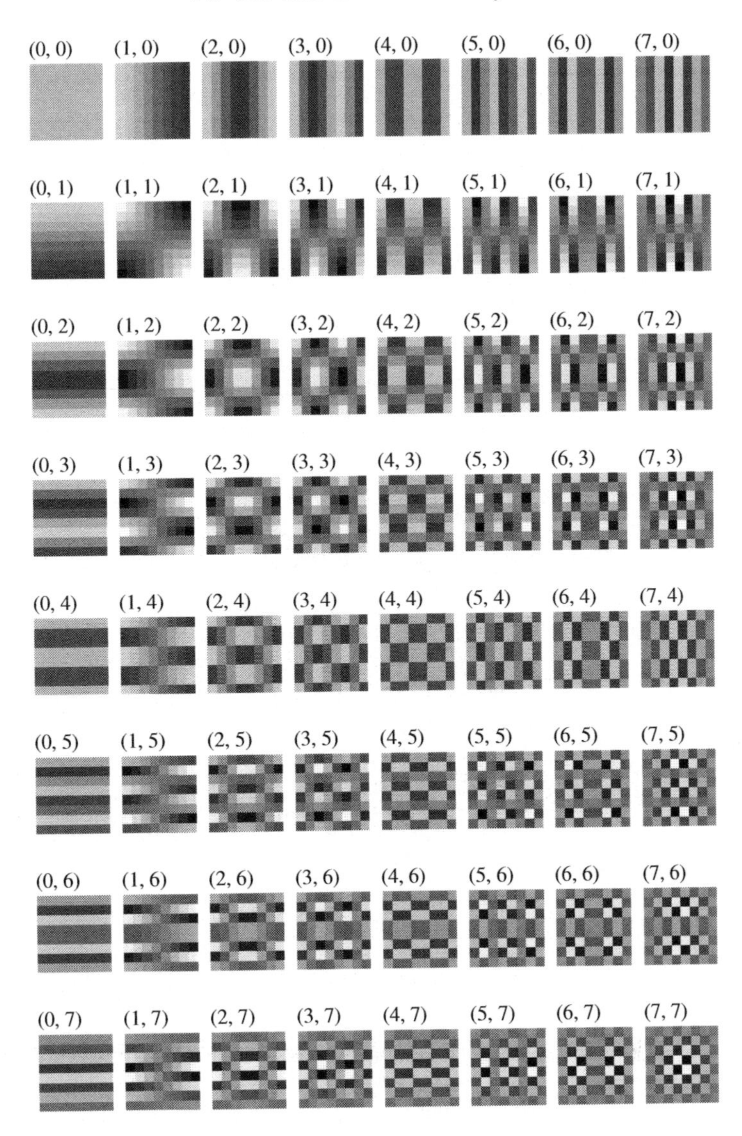

Figure 9.16 Two-dimensional discrete cosine transform basis functions: labels show (f_x, f_y).

to right and with f_y increasing from top to bottom, where $U(f_x, f_y)$ is the non-zero value. The images with higher frequency components are therefore towards the bottom and right of the grid. The two-dimensional discrete cosine transform expresses an image $u(x, y)$ as a weighted sum of these basis functions.

The discrete cosine transform component $U(0, 0)$ is called the *DC coefficient*; the other components are called the *AC coefficients*.

9.6 JPEG compression for continuous-tone images

JPEG (Joint Photographic Experts Group) is the popular name for a compression standard designed to work with greyscale or colour continuous-tone images.

We will outline one of the simpler modes of the JPEG compression algorithm, concentrating on those aspects that are most interesting from a signal processing point of view: see Figure 9.17. Decompression is simply the reverse of the encoding process.

First, if the image is colour, it is converted into luminance (Y) and chrominance (C_B and C_R) components using the equations in Section 9.1. To take advantage of the human eye's poorer resolution for chrominance information than for luminance information, the chrominance components of the image can then be downsampled

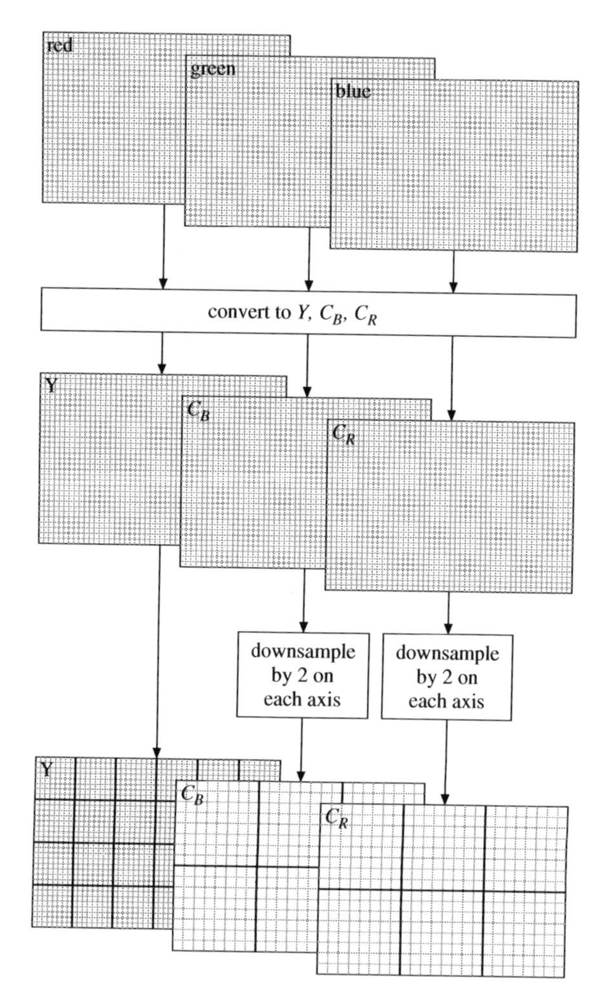

Figure 9.17 JPEG compression: initial processing.

by a factor of 2 either in just the horizontal direction or in both the horizontal and vertical directions. The downsampling can be done by a simple averaging of adjacent pixels, or one of the more sophisticated scaling techniques described in Section 9.8 can be used. The three resulting planes of image information are now processed as if each were a greyscale image.

Each image plane to be compressed is divided into square tiles, eight pixels on a side. Each tile is now processed independently: see Figure 9.18.

The two-dimensional discrete cosine transform of the pixels in each tile is calculated. Because images tend to have large constant or slowly varying areas, we would hope that most of the AC coefficients in the results of these discrete cosine transforms will be small, especially those corresponding to higher frequencies, i.e., those towards the bottom right of Figure 9.16.

The discrete cosine transform coefficients are now quantised. The quantisation levels can be chosen by the encoder, and may be different for each discrete cosine

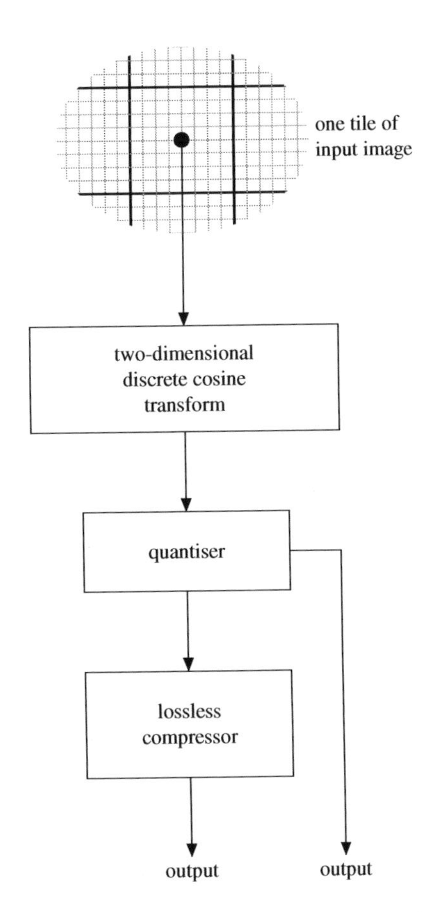

Figure 9.18 JPEG compression: processing of each eight-pixel-square tile.

transform frequency component. Some encoders allow an element of manual control over the quantisation levels: if the quantisation steps are set coarsely we will need fewer bits to represent each coefficient and so the compression ratio will be higher, but the quality of the reconstructed image will be poorer. If, on the other hand, the steps are set finely the compression ratio will be lower but the quality of the reconstructed image will be better. The quantisation levels used form part of the output of the algorithm.

The final step is to apply a lossless compressor to the quantised frequency components. The components are presented to the lossless compressor in a zig-zag order as shown in Figure 9.19: this helps to keep together the lower-frequency components, which we generally expect to be non-zero, and the higher-frequency components, which we generally expect to be zero, which in turn simplifies the design of the lossless compressor.

If an image contains a large number of sharp edges fewer of the coefficients will be quantised to zero and the lossless compression stage will not achieve such a high compression ratio. This is why JPEG compression is more appropriate for photographs than it is for line drawings.

It is found that good quality for typical photographs can be achieved using an average of about 64 bits to encode each block. This corresponds to one bit per pixel; comparing this with an original greyscale image that might have 8 bits per pixel, we see that a compression ratio of 8 : 1 can typically be achieved.

If we adjust the quantisation thresholds so that a higher compression ratio is achieved, artefacts of the compression process start to appear in the reconstructed image. The most immediately noticeable artefact is that the edges of the

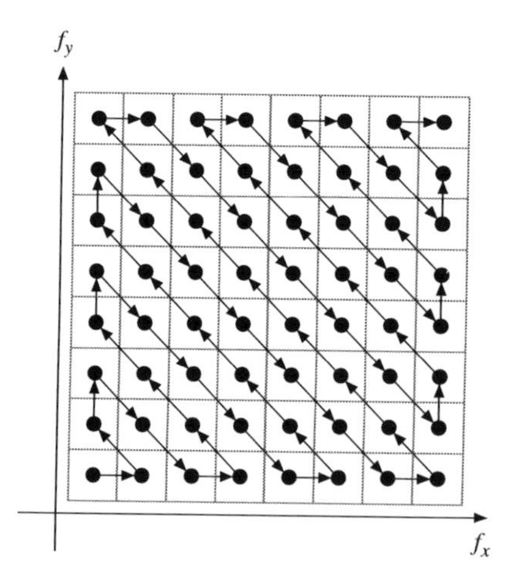

Figure 9.19 JPEG compression: order of scanning of discrete cosine transform frequency components.

8-pixel-square tiles of which we took the two-dimensional discrete cosine transform become visible, usually referred to as 'blockiness'. There are two reasons for this. First, since each tile is independently quantised, the error introduced by quantisation will change from tile to tile. In particular, the amplitude of a given frequency component may be just below the lowest quantisation threshold in one tile, and just above it in an adjacent tile: the frequency component will suddenly appear at the edge between the two tiles. The second reason for blockiness lies in the discrete cosine transform itself.

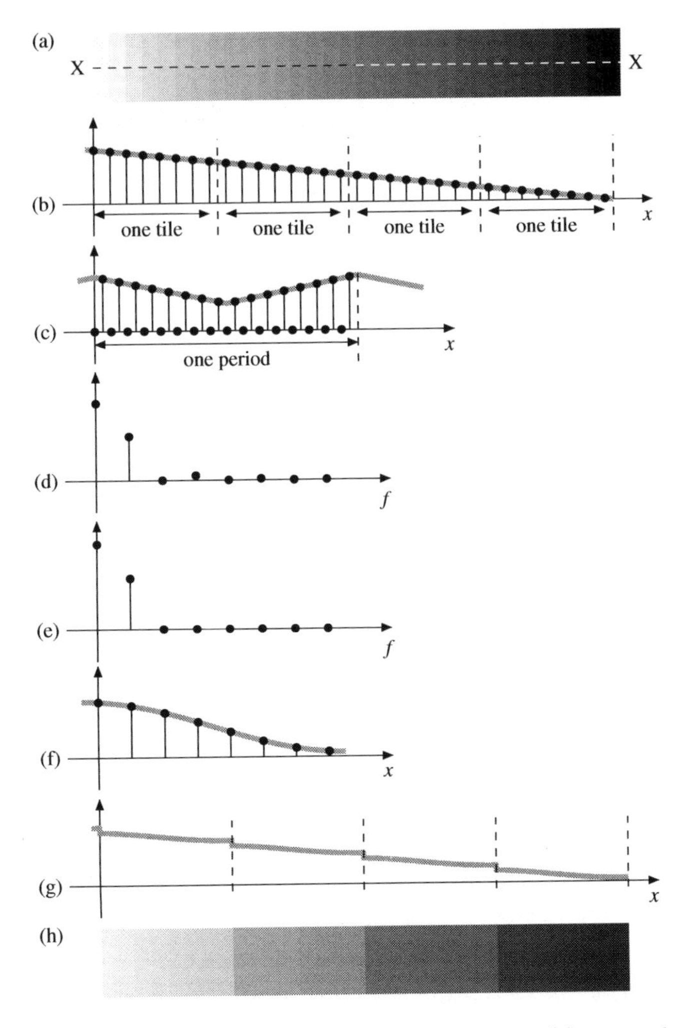

Figure 9.20 Blockiness in JPEG-compressed images: (a) example image; (b) cross-section X-X; (c) periodic signal of which Fourier transform is effectively calculated; (d) discrete cosine transform components; (e) components after quantisation; (f) inverse transform of quantised components; (g) reconstructed cross-section of image; (h) reconstructed image (effect exaggerated for clarity).

To see what can happen, consider the case of the smoothly changing greyscale image shown in Figure 9.20(a). A plot of a cross-section through the image along the line marked X-X is shown in Figure 9.20(b), and as you can see, the brightness of the image decreases linearly across it. The image is 32 pixels wide and 8 pixels high and so gets divided into tiles as shown. Consider just one of these tiles. Calculating the discrete cosine transform of one row of this tile is equivalent to calculating the Fourier transform of the periodic signal shown in Figure 9.20(c); the resulting frequency components are shown in Figure 9.20(d). The DC component and the lowest-frequency AC component are both strong, but there are also small contributions from other frequency components which arise as a result of the sudden changes in slope of the signal being transformed (see Section 4.6). These small components will probably be discarded in the quantisation process, giving rise to Figure 9.20(e); reversing the transform, we get Figure 9.20(f); if the same thing happens in the other tiles, the resulting cross-section will be as shown in Figure 9.20(g), where the tile boundaries are shown by dashed lines. The final image will then appear as in Figure 9.20(h). (The effect has been slightly exaggerated in the final image for clarity.)

As you can see from Figure 9.20(g), the maximum error introduced by the quantisation stage is small; unfortunately, however, the discontinuities introduced into the image as a side-effect are particularly distracting to the human eye. Some JPEG decompressors take measures to attempt to smooth over this type of discontinuity.

9.7 The discrete wavelet transform

The discrete wavelet transform seeks to separate the high-frequency and low-frequency components of a signal. It mainly finds application in its two-dimensional form in image compression, where this separation allows for fine control over the level of detail preserved in a compressed image. The 'JPEG 2000' image compression standard employs a discrete wavelet transform over the entire image. This gives it freedom from blocky artefacts and, on average, a slightly higher compression ratio than schemes based on the discrete cosine transform.

The one-dimensional discrete wavelet transform

Recall that in Chapter 8 we discussed how the first stage of MP3 audio compression involves a filter bank to split a signal into a number of frequency bands. The idea behind the discrete wavelet transform is similar, but implemented in a recursive fashion.

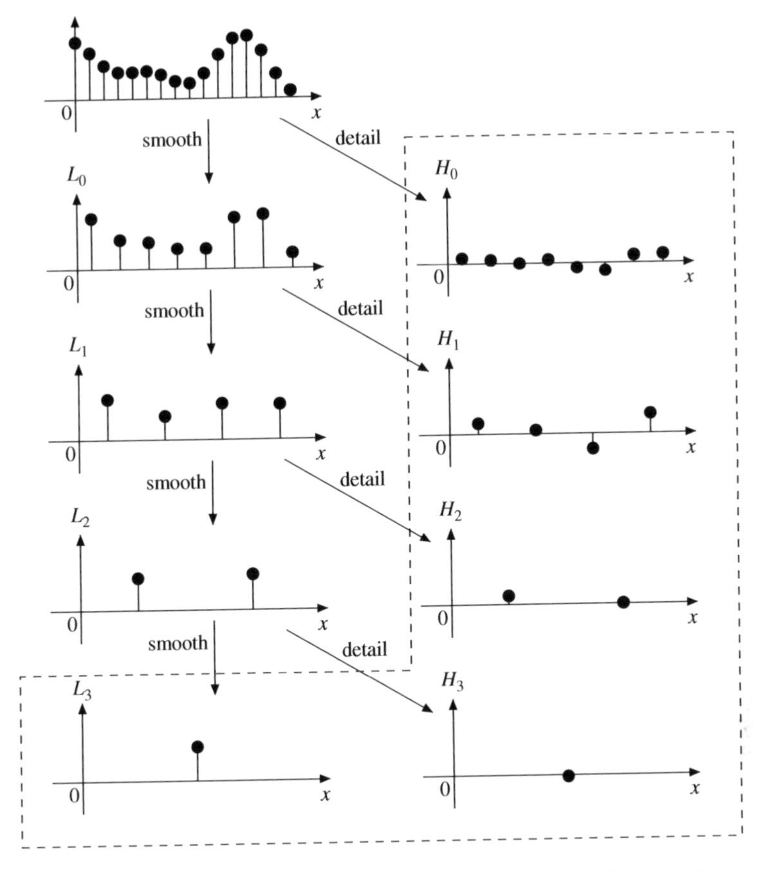

Figure 9.21 Simple example of a one-dimensional wavelet transform.

We shall start by describing a simplified version of the discrete wavelet transform; we shall assume throughout that the number of samples in the signals we are dealing with is a power of 2.

Consider the signal of 16 samples shown at the top of Figure 9.21. We can produce a crudely smoothed version of this signal by dividing it into eight pairs of adjacent samples and replacing each pair of samples (a, b) by their average $u = (a+b)/2$. Since this is a crude low-pass filter, we shall call these eight numbers the 'smooth', or L_0 part of the signal. What further information do we need to recover the original signal from the averages? It would suffice to have recorded (for example) $v = (a-b)/2$ also; we could then reconstruct a and b using $a = u+v$ and $b = u-v$. This is a crude high-pass filter, and so we call the eight v values the 'detail', or H_0 part of the signal.

We can repeat the above process on the smooth part of the signal, splitting it in turn into a 'smooth-smooth', or L_1 part and a 'smooth-detail', or H_1 part, now four samples each, as shown in Figure 9.21. (We could in principle also process the detail part of the original signal in the same way, but there would be little point as normally the detail part of a signal does not itself exhibit much smoothness.) We can repeat again on the resulting L_1 part, producing an L_2 part and an H_2 part, two samples each; and then again, producing an L_3 part and an H_3 part, each consisting of one sample.

The result of this transformation, outlined by the dotted box in the figure, consists of the values in H_0, H_1, H_2, H_3 and L_3. As you can verify this is a total of sixteen numbers representing our original sixteen samples. Since each step on the way is reversible, it is possible to recover our signal from the output. Furthermore, flat regions of the signal lead to detail values of zero, and slowly changing regions of the signal give rise to detail values that are small in magnitude. A compressor could therefore discard the smaller detail values without losing the important features of the signal.

A scheme like the one above is used in some compression algorithms simply as a way of changing the order in which the information appears in the compressed result. If the transformed version of the signal above were transmitted in the order L_3, H_3, H_2, H_1, H_0, it would be possible to reconstruct a crude version of the signal at the receiver immediately on receipt of L_3; this approximation could then be incrementally refined as the detail information arrived. When browsing an image database stored on a remote computer, for example, this would allow the user to determine quickly whether a given image was the wanted one and abort the transfer of data early if not.

The one-dimensional discrete wavelet transform in practice

The main difference between the discrete wavelet transform as used in practice and the simplified version above is in the sophistication of the filtering. The process is illustrated in Figure 9.22, which shows three levels of filtering.

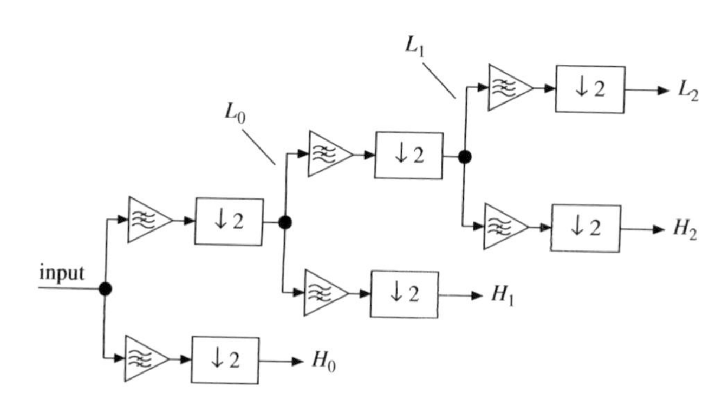

Figure 9.22 Calculating the one-dimensional wavelet transform.

First the signal, which we shall assume is real and has sample rate f_s, is split into two bands of equal width, called the 'low' and the 'high', or L_0 and H_0 bands: the L_0 band runs from 0 (or DC) up to half the Nyquist limit, or $f_s/4$, while the H_0 band runs from $f_s/4$ up to the Nyquist limit, $f_s/2$. As above, the L_0 band is now itself split into two further bands, called the L_1 and H_1 bands. The L_1 band is now split into two again, forming the L_2 and H_2 bands, and so on.

Since the output of each filter in this system only contains a narrow band of frequencies, we can use sub-Nyquist sampling (see Chapter 2) and represent its output using a reduced sample rate. In this case we can sample the output of each filter at half its input sample rate by discarding the samples with odd time indices (in the boxes marked '$\downarrow 2$' in the block diagram) without losing information. The total sample rate at the output of the pair of filters at each level in the diagram is the same as the input sample rate to that pair. In practice the system is designed so that the sample rates and filter frequency responses all scale by a factor of 2 at each level, and so the same sets of low-pass and high-pass filter coefficients can be used throughout. We shall refer to the low-pass filter as L, with impulse response $l(t)$, and the high-pass filter as H, with impulse response $h(t)$.

There is still a considerable amount of freedom left in the design of the transform. In particular, the exact response of the filters remains to be specified: the only constraint is that the low-pass and high-pass filters should between them cover the entire frequency range from 0 to $f_s/2$ without gaps, so that we can be sure that we can invert the transform. It is at this point that various formulations of the discrete wavelet transform diverge. We shall describe one of the original formulations, which is also among the simplest.

We will start by insisting, for simplicity of computation, that the bands are separated using four-point FIR filters, and that the original signal can be reconstructed from the outputs of these filters also using four-point FIR filters. We shall also insist that the filter responses are complementary (in a sense explored in Exercise 9.6) and finally we shall partially specify the high-pass filter by requiring that it have zero output when provided with an input that either is constant or is changing linearly (like the cross-section of the example image in Figure 9.20(b)).

These requirements taken together determine a pair of (non-causal) filters. The impulse response for L is given by

$$l(t) = \ldots, 0, 0, c_3, c_2, c_1, c_0, 0, 0, \ldots$$

where $l(0) = c_0$, and that for H is given by

$$h(t) = \ldots, 0, 0, -c_0, c_1, -c_2, c_3, 0, 0, \ldots$$

where $h(0) = c_3$, and where

$$c_0 = \frac{1}{4\sqrt{2}}(1+\sqrt{3}),$$

$$c_1 = \frac{1}{4\sqrt{2}}(3+\sqrt{3}),$$

$$c_2 = \frac{1}{4\sqrt{2}}(3-\sqrt{3}),$$

and

$$c_3 = \frac{1}{4\sqrt{2}}(1-\sqrt{3}).$$

The process for inverting the discrete wavelet transform is shown in Figure 9.23. At each level the signal carrying the high-frequency components is interleaved with the signal carrying the low-frequency components, with the sample from the signal carrying the low-frequency components coming before the corresponding sample from the signal carrying the high-frequency components. The result is passed through two filters, E and O. Again we discard the output samples from these filters which have odd time indices, and interleave the resulting sequences. The samples that come from filter E are placed before those from filter O so that filter E is responsible for the even-numbered samples in the final result and filter O is responsible for the odd-numbered ones.

The (non-causal) filters E and O have impulse responses given by

$$e(t) = \ldots, 0, 0, c_3, c_0, c_1, c_2, 0, 0, \ldots$$

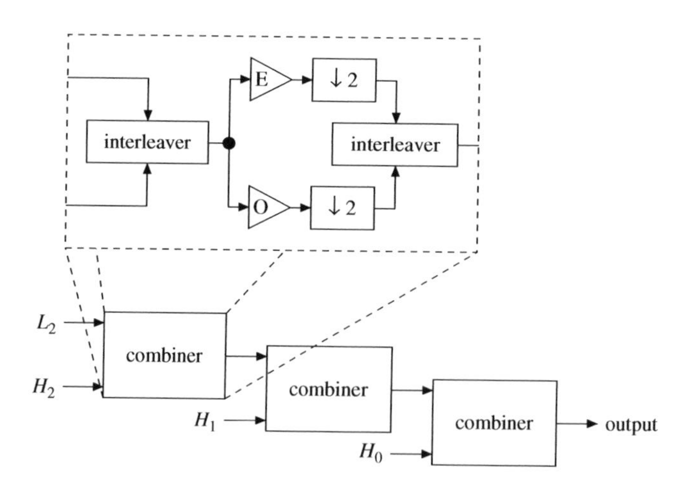

Figure 9.23 Calculating the inverse one-dimensional wavelet transform.

where $e(0) = c_0$, and

$$o(t) = \ldots, 0, 0, -c_2, c_1, -c_0, c_3, 0, 0, \ldots$$

where $o(0) = c_1$, and the c_i are as defined above.

Still more sophisticated versions of the discrete wavelet transform can be defined, using longer FIR filters to separate the high- and low-frequency components of the signal.

The two-dimensional discrete wavelet transform

A two-dimensional discrete wavelet transform can be defined on an image using the following procedure. For illustration we shall use the crude sum-and-difference wavelet transform we described first; similar-looking results would be obtained using more sophisticated filters.

First, one-dimensional discrete wavelet transforms are applied to each row of the original image (Figure 9.24(a)). A new image is formed (Figure 9.24(b)) whose left-hand half contains the L_0 results in each row and whose right-hand half contains the H_0 results in each row. (Note that the figure shows the absolute values of the filter outputs for clarity: some of the outputs of the high-pass filter are negative. The high-pass filter outputs have also been scaled up to make them easier to see.) Next, the columns of the image are transformed, one at a time, using a one-dimensional discrete wavelet transform. Again, a new image is formed, with the L_0 results in the top half of each column and the H_0 results in the bottom half of each column: see Figure 9.24(c). This result is an image divided into four squares. The top-left square contains the components with low horizontal and low vertical frequency, and is called the LL_0 quadrant. At the top right, in the HL_0 quadrant, we have the components with high horizontal and low vertical frequency such as vertical edges. Similarly, the bottom left (the LH_0 quadrant) contains the components with low horizontal and high vertical frequency such as horizontal edges. Finally, at the bottom right in the HH_0 quadrant we have the components with high horizontal and high vertical frequency, including areas where there are fine textures.

As you can see, the LL_0 quadrant contains a scaled-down version of the original image. The other quadrants contain all the extra information needed to reconstruct the full-resolution original from the scaled-down version; since the original contains large areas of uniform intensity, most of the values in the other quadrants are zero or nearly zero.

As before, the transformation step can be repeated on the LL_0 quadrant, replacing it with LL_1, HL_1, LH_1 and HH_1 quadrants: the result is as shown in

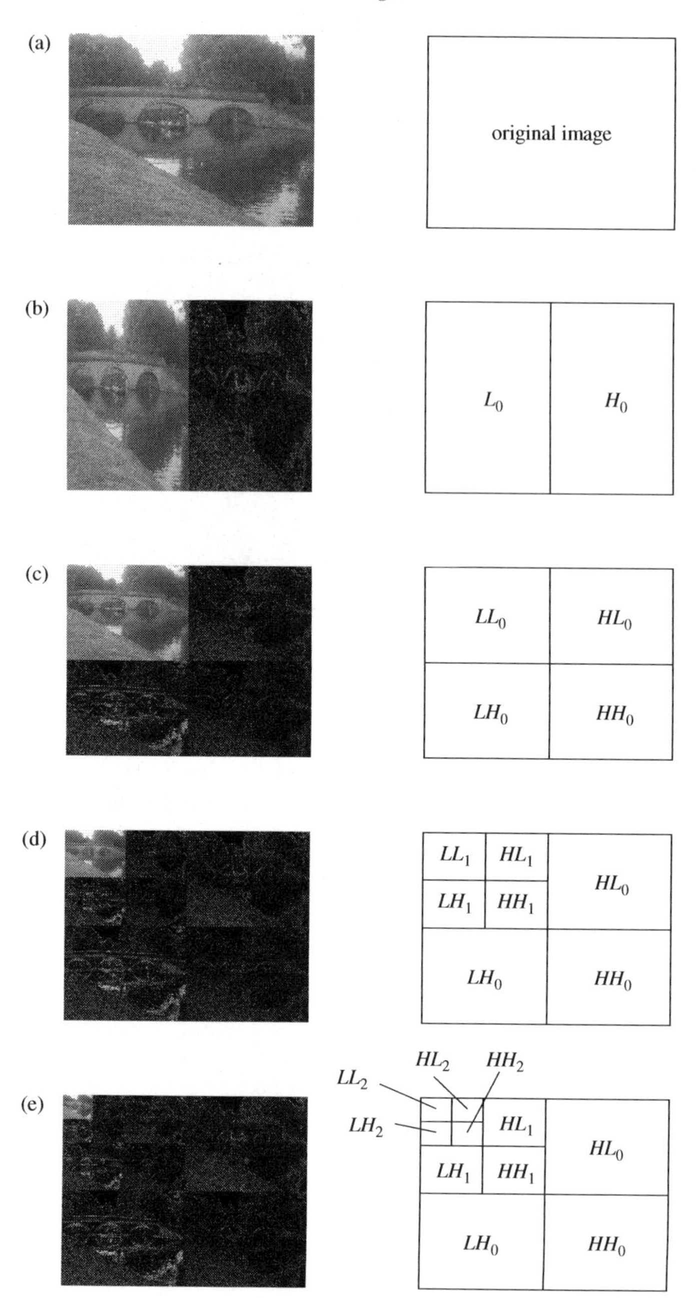

Figure 9.24 The two-dimensional discrete wavelet transform: (a) original image; (b) after horizontal low- and high-pass filtering; (c) after horizontal and vertical low- and high-pass filtering; (d) after two iterations of filtering; (e) after three iterations.

Figure 9.24(d). Iterating once more we reach the situation in Figure 9.24(e). If we wished, we could iterate further.

In its usual mode of operation the JPEG 2000 image compression algorithm uses a two-dimensional discrete wavelet transform essentially as above, followed by a quantisation stage that sets many of the small-magnitude detail components of the transform to zero. The result is then losslessly compressed. There are many further details to the standard, including dealing with images which are not square and a power of 2 pixels on a side.

9.8 Image scaling

Scaling an image is equivalent to changing its sample rate, and methods analogous to the interpolation techniques described in Chapter 5 for one-dimensional signals can also be used on images. The aliasing artefacts that are introduced when images are scaled without proper filtering are popularly called 'jaggies'.

For most applications it is sufficient to first scale the image in the x direction by changing the sample rate along each row of the image independently, and then scale the image in the y direction by changing the sample rate along each column independently. For an image measuring X pixels by Y pixels, this requires $X+Y$ applications of a one-dimensional interpolator. If the scaling factor in the x and y directions is the same, the same interpolation filter can be used for the rows and the columns; otherwise, the filters must be designed separately. Caution is required if the image is to be stretched (i.e., upsampled) along one axis and shrunk (i.e., downsampled) along the other: in this case the first interpolation filter must be designed to cut off frequencies above half the sample rate of the *original* image whereas the second interpolation filter must be designed to cut off frequencies above half the sample rate of the *new* image. In other words, the two resampling operations are entirely independent of one another.

One-dimensional polyphase interpolation filters (see Section 5.6) can be used in the above process to scale images by arbitrary factors.

The process described above is the best that we can do in the sense that it preserves as many frequency components as possible from the original image in the scaled image. However, the result does not always look perfect when judged by eye, particularly when an image is scaled up.

The reason for this is that the anti-aliasing filter we apply when scaling the image as described above is not isotropic. In fact, the frequency response of the filter might look just like the one in Figure 9.7(a), which will create artefacts resembling those shown in Figure 9.6. A result that looks better, even though it contains fewer frequency components and hence less information, can be obtained by using an isotropic low-pass filter for anti-aliasing, as shown in the example in Figure 9.11.

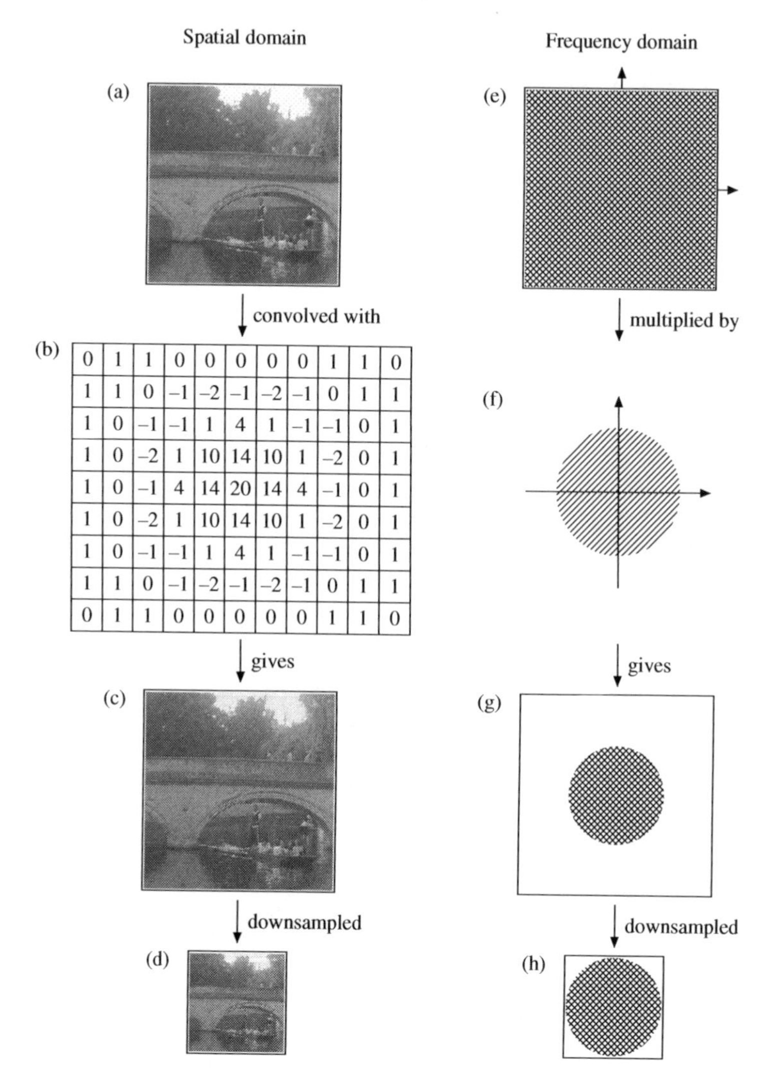

0	1	1	0	0	0	0	0	1	1	0
1	1	0	–1	–2	–1	–2	–1	0	1	1
1	0	–1	–1	1	4	1	–1	–1	0	1
1	0	–2	1	10	14	10	1	–2	0	1
1	0	–1	4	14	20	14	4	–1	0	1
1	0	–2	1	10	14	10	1	–2	0	1
1	0	–1	–1	1	4	1	–1	–1	0	1
1	1	0	–1	–2	–1	–2	–1	0	1	1
0	1	1	0	0	0	0	0	1	1	0

Figure 9.25 Downsampling an image: (a)–(d) spatial domain; (e)–(h) frequency domain, with hatched areas indicating where the coefficients may be non-zero; (a), (e) original image; (b), (f) low-pass isotropic anti-aliasing filter, shaded area in (f) showing frequencies passed; (c), (g) filtered image; (d), (h) downsampled image.

Downsampling an image

We shall first consider the case where an image (Figure 9.25(a)) is to be downsampled by the same ratio k along both axes so that the result is k times smaller on each side than the original. We shall assume that the image has square pixels, although if the pixels were not square their aspect ratio would cancel out in the calculations that follow.

We design a suitable isotropic low-pass filter using the technique described in Section 9.4, whose response will be as shown in Figure 9.25(f). The curve in the (f_x, f_y) frequency domain corresponding to the cutoff frequency of the filter is given by

$$\sqrt{f_x^2 + f_y^2} = \frac{1}{k}\frac{f_s}{2}$$

where f_s is the sample rate, the same for the two axes. The (windowed) impulse response of this filter is shown in Figure 9.25(b). As you can see, the impulse response is circularly symmetric.

After filtering, the spectrum of the image only occupies the hatched area in Figure 9.25(g). We can now safely discard all but one of every k-by-k square of pixels to reduce the sample rate by a factor of k along each axis. This leaves a final image Figure 9.25(d) whose spectrum is as shown in Figure 9.25(h).

As in the one-dimensional case, we do not in practice actually go to the bother of calculating the sample values we subsequently discard.

Upsampling an image

Next we shall look at the case where an image is to be upsampled by the same ratio k along both axes so that the result is k times bigger on each side than the original. Again, we assume for simplicity that the image has square pixels. The process is illustrated in Figure 9.26.

One way to proceed is first to expand the original image by a factor of k along both axes by inserting zero samples as necessary: see Figure 9.26(b). The spectrum of this image contains many aliases, shown in solid grey in Figure 9.26(f).

We design an isotropic filter to remove the aliases, with cutoff at

$$\sqrt{f_x^2 + f_y^2} = \frac{f_s}{2}$$

where f_s is the sample rate of the original image: see Figure 9.26(g).

The final result after filtering is as shown in Figure 9.26(d), with spectrum as shown in Figure 9.26(h). The image lacks higher frequency components, and so it will appear somewhat blurred. Note also that the spectrum shown in Figure 9.26(h) actually contains less information than the spectrum we started with in Figure 9.26(e): this is a consequence of the isotropic filtering.

Note that many of the samples in Figure 9.26(b) are zero and so many of the multiplications implied by the convolution of the signals in Figure 9.26(b) and Figure 9.26(c) are known in advance to be multiplications by zero. Practical implementations will avoid doing these multiplications in the interests of efficiency.

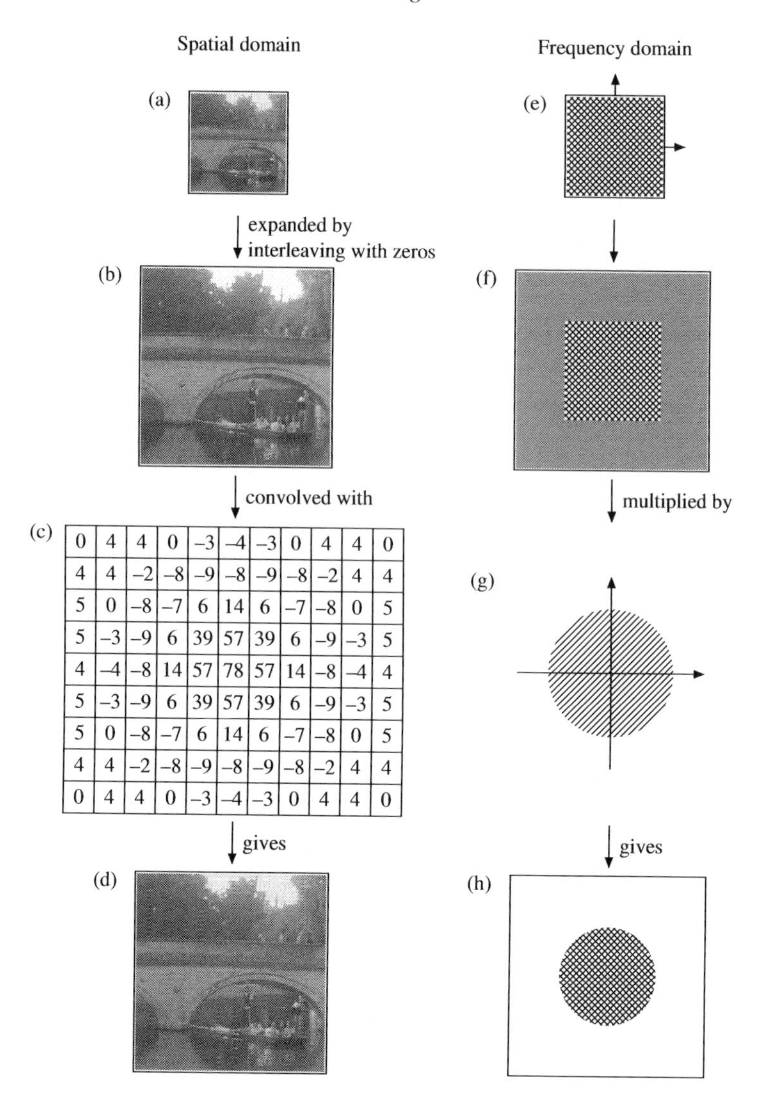

0	4	4	0	−3	−4	−3	0	4	4	0
4	4	−2	−8	−9	−8	−9	−8	−2	4	4
5	0	−8	−7	6	14	6	−7	−8	0	5
5	−3	−9	6	39	57	39	6	−9	−3	5
4	−4	−8	14	57	78	57	14	−8	−4	4
5	−3	−9	6	39	57	39	6	−9	−3	5
5	0	−8	−7	6	14	6	−7	−8	0	5
4	4	−2	−8	−9	−8	−9	−8	−2	4	4
0	4	4	0	−3	−4	−3	0	4	4	0

Figure 9.26 Upsampling an image: (a)–(d) spatial domain; (e)–(h) frequency domain, with hatched areas indicating where frequency components of original image may be non-zero and grey where aliases may appear; (a), (e) original image; (b), (f) image expanded by interleaving with zero samples; (c), (g) isotropic low-pass filter to remove aliases; (d), (h) upsampled image.

General image scaling

An image can be scaled by a simple fractional ratio by first upsampling and then downsampling by suitable integer factors. For example, to scale an image to two thirds of its original size – i.e., to resample it from sample rate f_s to rate $\frac{2}{3}f_s$ – it could be upsampled by a factor of 2 and then downsampled by a factor of 3. Exactly as in the one-dimensional case described in Section 5.6 it is possible to save work by applying an anti-aliasing filter designed for the final sample rate (which would in this case have a cutoff frequency of $\frac{1}{2} \cdot \frac{2}{3} f_s = \frac{1}{3} f_s$) during the first upsampling phase. Again, it is not necessary for the upsampling process to compute any sample values which will be discarded by the subsequent downsampling process.

The one-dimensional polyphase interpolator of Section 5.6 can also be extended to the two-dimensional case. A filter must be created for each potential fractional offset that the interpolated sample has within the original image. If interpolation to an accuracy of 1/4 pixel is required, there is a total of $4 \times 4 = 16$ possible filters, corresponding to each possible combination of fractional offsets in the x and y directions. Normally these filters will be precalculated and stored: they can occupy large amounts of storage if a high degree of interpolation accuracy is required. Exercise 9.10 indicates how these filters can be constructed.

As in the one-dimensional case, interpolation can be combined with arbitrary filtering.

9.9 Image enhancement

Image enhancement is the processing of an image to make certain features of it more clearly visible or more visually attractive. The human visual system is a complex thing and not fully understood, and so it is not always obvious why a given image enhancement technique is effective. After all, such methods can at best preserve the information in an image, and will typically remove information. It is difficult to define an objective measure of the extent to which an image has been 'enhanced'.

Contrast and brightness

A simple way to enhance a greyscale image is to expand a narrow range of shades within the image to occupy the full range displayable to the user. Shades below this range are clipped to black, and shades above the range are clipped to white. An example *transfer function*, or plot of output intensity against input intensity, is shown in Figure 9.27(a).

The effect on an image of some text is shown in Figure 9.27(b) (before processing) and Figure 9.27(c) (after processing). To the right of each image is a histogram of the shades within it: as you can see, the histogram is stretched linearly over the specified range. Notice that shades in the input outside the specified range end up in the outermost bins in the histogram. Of course, if an inappropriate input range is specified, the resulting image can be entirely black

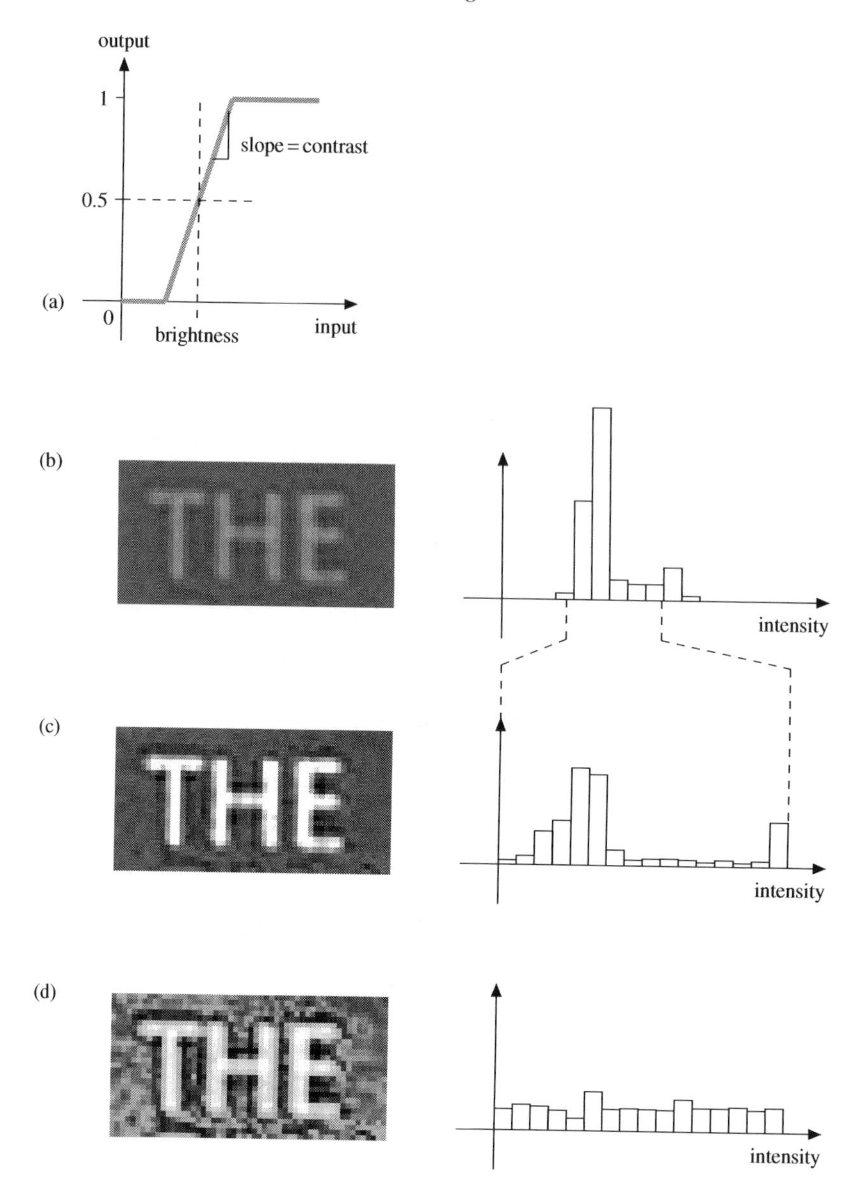

Figure 9.27 Contrast enhancement: (a) transfer function for simple brightness and contrast adjustment; (b) original image and its histogram; (c) with contrast doubled; (d) result of histogram equalisation.

or entirely white. Many image processing programs offer a contrast enhancement facility, either specified in terms of the endpoints of the input range or in terms of the slope of the line in the transfer function (the 'contrast') and the input value corresponding to an output value in the middle of the range (the 'brightness').

Non-linear transfer functions can also be used. A simple automatic choice of transfer function, suitable for use when manual control is either not practical or not desirable, is to 'equalise' the histogram: in other words, choose a transfer function that makes the histogram as flat as possible. Figure 9.27(d) shows the effect of equalising the histogram of the image of Figure 9.27(b). Exercise 9.12 asks you to design an algorithm to perform histogram equalisation.

9.10 Edge detection

A useful operation in image processing is to find the edges of objects in an image. This is in general a very hard problem, not least in that we need a definition of 'object' before we can start. There are, however, some simple techniques that we can use to attempt to find edges in an image that we can use as a first step in a pattern recognition system, for example.

Consider the one-dimensional filter with impulse response shown in Figure 9.28(a). Its frequency response is shown in Figure 9.28(b): as you can see, it is a high-pass filter. If we apply this filter to the rows of an image such as the one shown in Figure 9.29(a), the result is Figure 9.29(b). The result is zero (shown grey in the figure) over flat areas of the image because the DC component of the frequency response is zero. Each pixel is replaced by the difference between the pixels to its left and to its right, which is a measure of the gradient of the image intensity in the horizontal direction. (Note that we have chosen an impulse response of $\ldots, 0, 1, 0, -1, 0, \ldots$ rather than $\ldots, 0, 1, -1, 0, \ldots$ so that the difference pixels naturally line up with the pixels of the original image: compare this with filters A and B of Section 5.1.) Let us call the resulting image $\Delta_X(x, y)$.

The filter shows the greatest response where there are vertical edges in the image, i.e., where the image changes rapidly as we move across it. You can check that the response is positive if, scanning from left to right, the edge changes from a dark shade to a light shade, and negative if it changes from a light shade to a dark one. The sharper the edge, the greater the response. Note that the edge direction is at right angles to the direction in which the filter is applied.

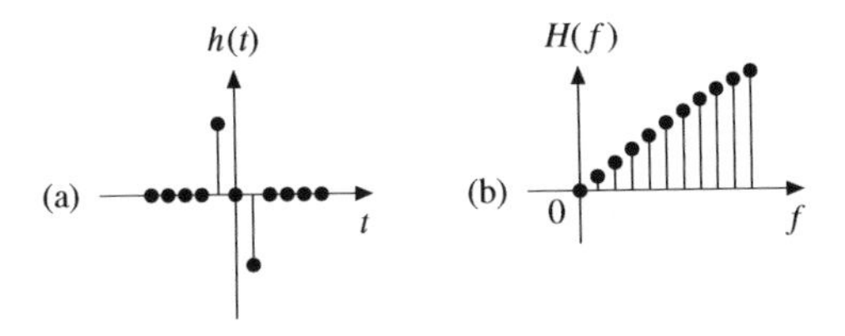

Figure 9.28 A high-pass filter: (a) impulse response; (b) frequency response.

We can also apply the same filter to the columns of the image to calculate the vertical intensity gradient $\Delta_Y(x, y)$: the result is shown in Figure 9.29(c). In this case the filter is sensitive to horizontal edges (at right angles to the direction in which the filter is applied), and again the sign of the output indicates the polarity of the edge.

A diagonal edge will give some response from both filters, again depending on the sharpness of the edge. In fact, if we consider the two numbers associated with each pixel, $\Delta_X(x, y)$ and $\Delta_Y(x, y)$, as a two-dimensional vector, we find that it points in a direction perpendicular to the edge and has a magnitude corresponding to the sharpness of the edge; you may wish to try a few examples by hand to verify this.

Figure 9.29(d) shows those pixels in the original image where the magnitude of this vector exceeds a certain threshold. This bilevel image result could form the input to further processing using the morphological methods we describe below.

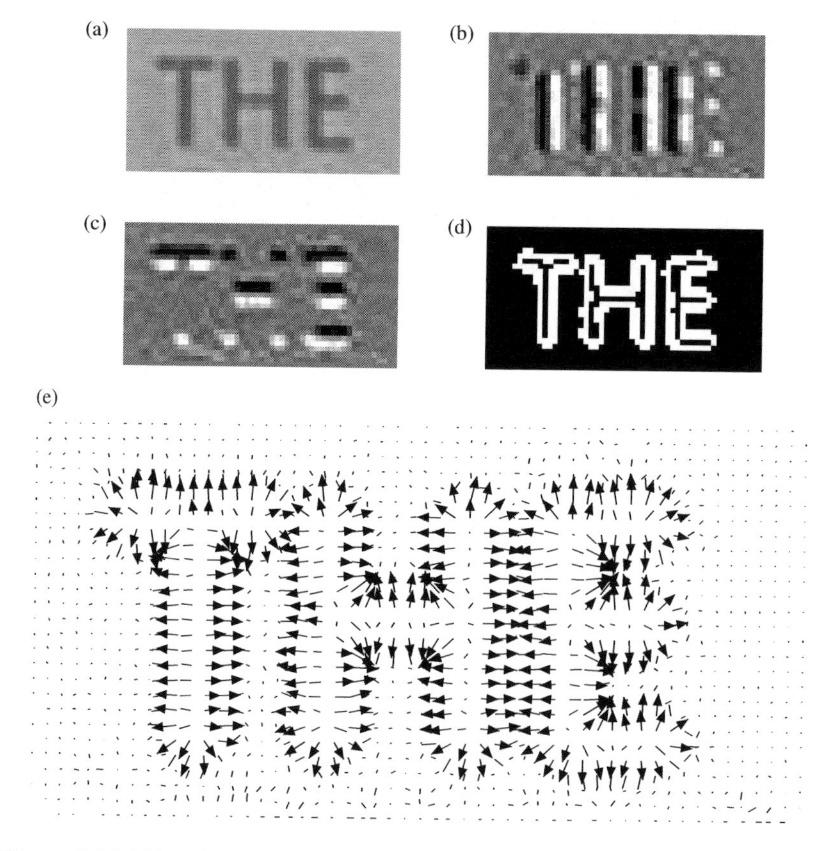

Figure 9.29 Edge detection: (a) original image; (b) result of horizontal high-pass filtering Δ_X; (c) result of vertical high-pass filtering Δ_Y; (d) thresholded magnitude of vectors (Δ_X, Δ_Y); (e) the vectors themselves.

Figure 9.29(e) shows the vectors themselves, with arrowheads on the vectors corresponding to white pixels in Figure 9.29(d). The relationships between the vector direction and magnitude and the edge direction and sharpness are clear.

There are many other methods for detecting edges in images, varying in their sensitivity to edges and robustness against noise in the original image. Many are modifications of the above technique using more complicated filters in the initial processing or starting from a circularly symmetric two-dimensional high-pass filter.

9.11 Processing of bilevel images

Bilevel images can be processed using any of the techniques we have described so far for continuous-tone images, but the results will not in general be bilevel. We shall not cover the compression of bilevel images except to say that there is a number of standards for the lossless compression of bilevel images, including 'JBIG', 'JBIG2' and 'DjVu'. Their compression performance can often be improved, especially when working on thresholded scanned images, by cleaning up the input image using some of the techniques below; of course, the system as a whole is not then lossless.

In the examples that follow, we shall refer to the two levels of the bilevel image as 0 (background) and 1 (foreground), which we shall represent in images by black and white respectively. We can define logical operations such as 'NOT', 'AND', 'OR', 'exclusive-OR' and 'difference' (A AND NOT B) between bilevel images, where the results are defined on a pixel-by-pixel basis: see Figure 9.30 for examples and for the notation we shall be using in this section.

Basic morphological operations

In the world of bilevel images the analogues of filters are *morphological operations*. A morphological operation is specified in terms of a *structuring element*, which is simply another bilevel image, usually having a simple shape such as a 3-by-3 square. The structuring element is moved over the image, and at each possible offset the set of pixels in the image under the structuring element is considered: this is analogous to the way the impulse response of a filter is moved over the image to calculate a convolution in the case of filtering greyscale images. Exactly how those pixels are 'considered' varies according to the type of operation.

It is important to note that a new image is created, pixel by pixel, as we scan the structuring element over the old image: we do *not* replace the pixels of the image one by one. A newly calculated pixel value cannot directly influence any other newly calculated values.

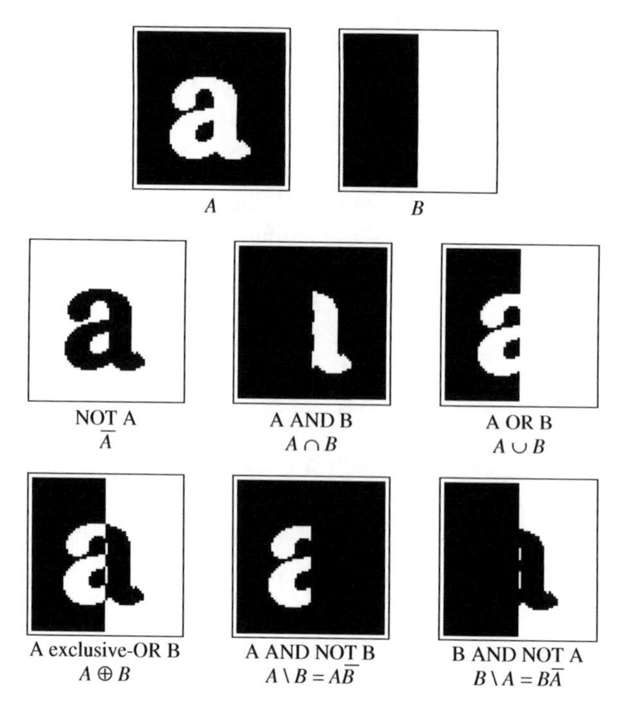

Figure 9.30 Logical operations on two example bilevel images, A and B: black represents logic zero and white logic one.

A simple operation is to take the logical 'OR' of all the pixels under the structuring element: the result is called *dilation*, and the result of dilating an image A using a structuring element B is written, perhaps slightly confusingly, $A+B$. Figure 9.31 shows the dilation of an example image using a 3-by-3 square structuring element and using a larger round structuring element. Notice how the image has been fattened by the size of the structuring element, and that the operation is similar to a convolution: see Exercise 9.13.

The complementary operation to dilation is *erosion*, where we take the logical 'AND' of all the pixels under the structuring element. Figure 9.32 shows the erosion of the example image by the same structuring elements. The result indicates those pixels where a copy of the structuring element can be fitted into the original image. The erosion of an image A using a structuring element B is written $A-B$.

Erosion operators can be used to remove isolated white speckles from noisy bilevel images ('salt noise'), and dilation operators can remove isolated black speckles ('pepper noise'). These are examples of the kind of preprocessing that can be applied to bilevel images to improve the performance of a subsequent compression step.

At the next level in complexity are the *opening* and *closing* operators.

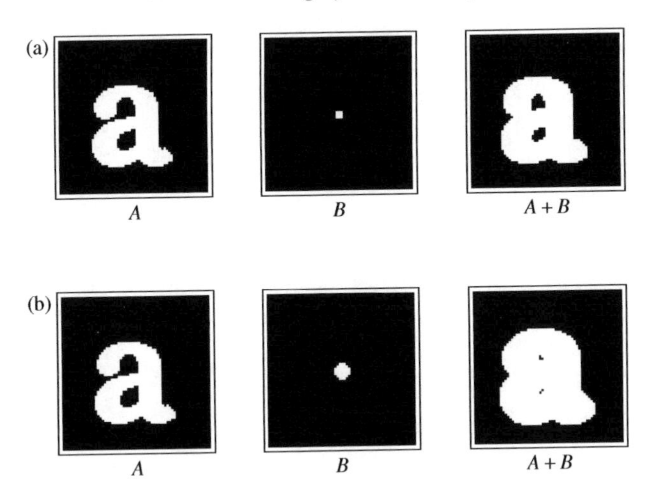

Figure 9.31 Dilation by (a) a small square structuring element; (b) a larger round structuring element.

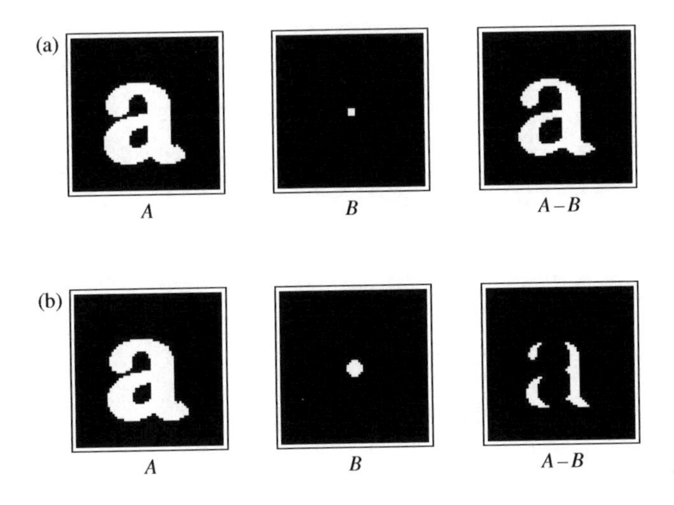

Figure 9.32 Erosion by (a) a small square structuring element; (b) a larger round structuring element.

'Opening' is an erosion followed by a dilation, using the same structural element: the opening of an image A by a structural element B is $O(A, B) = (A - B) + B$. Figure 9.33 shows an example: the rightmost image in each case is the exclusive-OR of the result with the original image, showing which pixels have been changed. The operation preserves most of the features of the original – as you might perhaps expect from the expression $(A - B) + B$ – but regions of the image into which the structural element will not fit are completely removed by the erosion step and therefore cannot be replaced by the dilation step. This means that small isolated regions in the image are removed and regions of the image

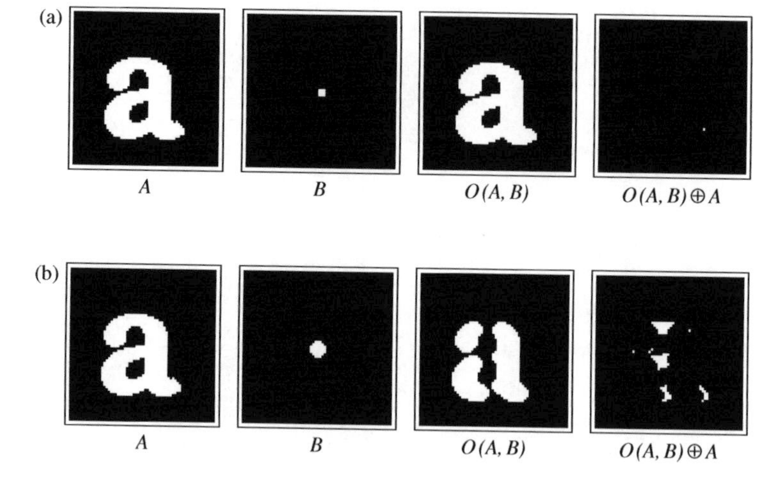

Figure 9.33 Opening (erosion followed by dilation) by (a) a small square structuring element; (b) a larger round structuring element.

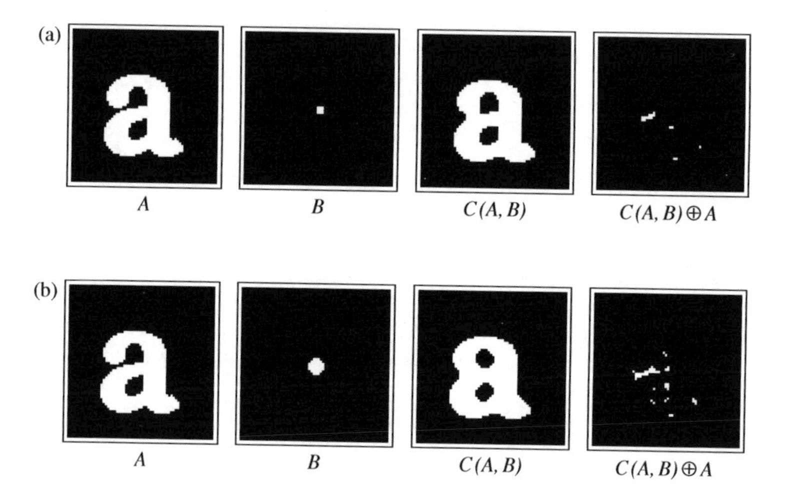

Figure 9.34 Closing (dilation followed by erosion) by (a) a small square structuring element; (b) a larger round structuring element.

that are joined by bridges thinner than the structural element become separated. In the case of the smaller structuring element, Figure 9.33(a), the image is almost unaltered: just one pixel, in the tail of the 'a', has been changed. In Figure 9.33(b), however, the larger structuring element does not fit into the narrower strokes of the letter, which have therefore been broken.

'Closing' is a dilation followed by an erosion, using the same structural element: the closing of an image A by a structural element B is $C(A, B) = (A + B) - B$. The process is illustrated in Figure 9.34. Again, most of the features are preserved, but regions which are sufficiently close together become joined by the closing operation.

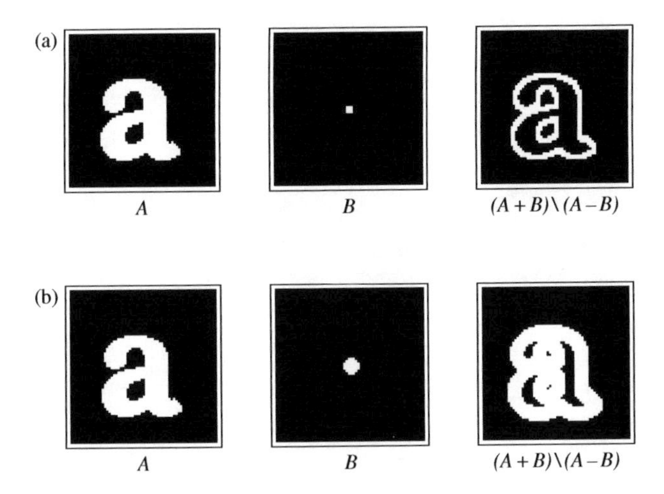

Figure 9.35 Morphological gradient operation using (a) a small square structuring element; (b) a larger round structuring element.

An opening operation followed by a closing operation (or the reverse sequence) is a better choice than a simple erosion or dilation for preprocessing a bilevel image prior to compression. It can not only remove salt and pepper noise but also will preserve the thickness of most of the strokes in the image.

The final operation we shall consider is the *morphological gradient*. This is simply the difference between the dilation of an image and its erosion, using the same structuring element: this can be written $(A+B)\backslash(A-B)$. Figure 9.35 shows an example of this operation on an image, and we can see that the morphological gradient is an edge detection operator.

Extension to greyscale images

The definitions of the morphological operators above can be extended to greyscale images. For simplicity, we shall assume here that the greyscale image is given by pixel values $u(x, y)$ in the range from 0 to 1, with 0 representing black and 1 representing white.

First we redefine the logical operations in terms of arithmetical operations. 'NOT A' is calculated as $1-A$ (where the '$-$' is a minus sign indicating subtraction, not erosion!); 'A AND B' as $\min(A, B)$; 'A OR B' as $\max(A, B)$; and '$A\backslash B$' as $A-B$ (i.e., the subtraction of pixel values, usually clipping negative result pixels to zero).

We can redefine dilation to be the operation of replacing each pixel with the maximum of the pixels covered by the structuring element centred on the target pixel: this is analogous to the logical OR operation in the bilevel case. Likewise,

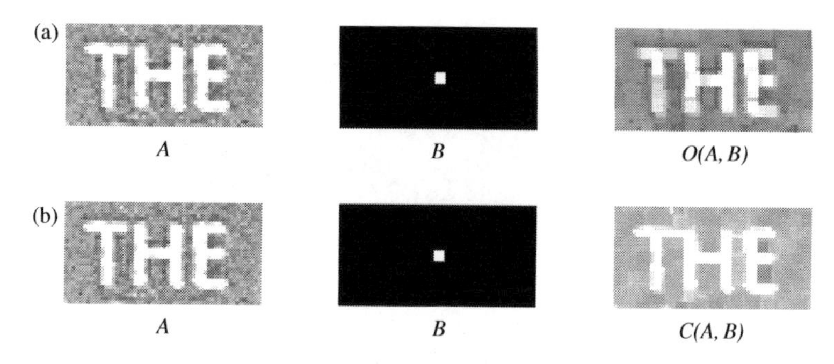

Figure 9.36 Greyscale morphological operations: (a) opening; (b) closing.

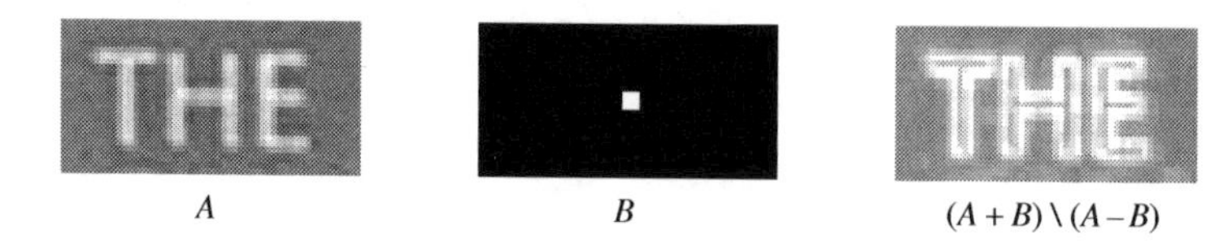

Figure 9.37 Morphological gradient operation on greyscale image.

erosion involves replacing each pixel with the minimum of the pixels covered by the structuring element. Opening and closing are defined in terms of dilation and erosion as before.

The structuring elements are still bilevel because they are only used to determine the sets of pixel values that enter into the arithmetical operation.

As in the case of bilevel images, an opening operation followed by a closing operation is a good way to remove noise from an image: Figure 9.36 shows an example. Figure 9.36(a) shows the result of the opening operation, and Figure 9.36(b) the result of the closing operation.

Figure 9.37 illustrates the greyscale analogue of the morphological gradient, which is the difference between the dilation and the erosion of an image. As you might expect, the operator acts as an edge detector for greyscale images: compare the result with Figure 9.29(d).

Other morphological operations

We can view any of the morphological operations defined above as a function of the pixel values in a neighbourhood of a given image pixel, where the shape of the neighbourhood is specified by the structuring element. We can thus create new morphological operations by simply thinking up new functions. In the case of bilevel images and a reasonably small neighbourhood it is even practical to create arbitrary functions specified by a look-up table. For a relatively large neighbourhood of 25 pixels (say a 5-by-5 square), the look-up table will have

(a) 

(b) 

Figure 9.38 Effect of a median filter on a greyscale image: (a) before; (b) after.

2^{25} entries. Each entry specifies the new pixel value, and therefore requires one bit of storage. The total look-up table size is thus only $2^{25}/8$ bytes, or about 4 megabytes.

This technique gives immense flexibility. We can construct morphological operators that detect edges at certain angles or of given curvature, lines in specified orientations and so on. Exercise 9.17 gives another example of the application of this technique.

The look-up table technique is less practical for greyscale applications. If the pixels are represented as 8-bit quantities, even a 3-by-3 square neighbourhood would require a table with $2^{9\times8}/8 = 5.9\times10^{20}$ bytes, which is prohibitively large. We can design other functions, however. A popular example is the so-called *median filter*, which again is used for reducing noise.

In the median filter, each pixel is replaced by the median of the pixel values in its neighbourhood. The median value is the one that would come halfway down the list if the list were sorted. It is usually close in value to the mean, but is less sensitive to the presence of 'outliers', extreme pixel values that can result from certain types of noise. Figure 9.38 shows a median filter with a 3-by-3 square neighbourhood applied to a noisy test image.

9.12 Pattern recognition

We will now turn to a simple example of pattern recognition based on the likelihood methods described in Chapter 6. The problem we shall examine is a simplified version of a common task in industrial machine vision applications and is as follows: given a noisy bilevel image of a washer and some nuts, find the coordinates of the centre of the washer. Some sample test images are shown on the left in Figure 9.39: the washer is circular with a circular hole, and the nuts are hexagonal with a circular hole.

Notice first that the objects in the image overlap, and second that the noise in the image is 'pepper noise', i.e., white pixels that have become black. We will suppose that the washer we are looking for has known internal radius r and known external radius $2r$.

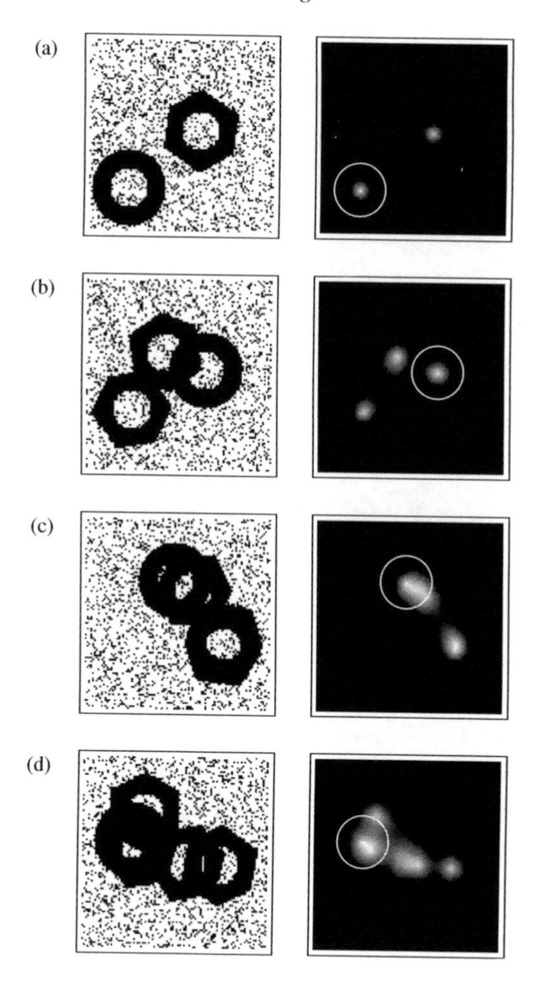

Figure 9.39 Identifying the location of a washer in a noisy image of nuts and washers: (a) one nut and one washer; (b) two nuts and one washer; (c) three nuts and one washer; (d) four nuts and one washer.

We begin by modelling the image. Suppose the washer is centred at position (x, y) in the image and consider a pixel with coordinates (u, v). Let $d = \sqrt{(x-u)^2+(y-v)^2}$ be the distance between (x, y) and (u, v). If d is between r and $2r$ then we would expect the pixel at (u, v) to be black. All we know about the other pixels in the image, including the ones inside the hole in the washer, is that they are probably white; however, they might be black owing to noise or owing to the presence of another object. We will use the following somewhat arbitrary values for the probabilities involved:

$$P(\text{black pixel at } (u, v) \mid r \leq d < 2r) = 0.999$$

$$P(\text{black pixel at } (u, v) \mid d < r \text{ or } d \geq 2r) = 0.1.$$

In practice we would determine reasonable values for these probabilities by examining a number of example images and comparing the noise models in the way described in Chapter 6.

We can now work out the probability of any particular image arising given that there is a washer centred at (x, y). The equations above tell us the probability associated with each pixel being black; and, since the image is bilevel, we have

$$P(\text{white pixel at } (u, v)) = 1 - P(\text{black pixel at } (u, v)).$$

Now, Bayes' theorem (Equation (6.2)) tells us that

$$P(\text{washer at}(x, y) \mid \text{image}) = \frac{P(\text{image} \mid \text{washer at}(x, y))P(\text{washer at}(x, y))}{P(\text{image})}.$$

We shall assume that $P(\text{washer at } (x, y))$, our prior distribution on the location of the washer, is flat; and if all we are interested in is comparing this quantity for various points (x, y) for a fixed image, the term $P(\text{image})$ is constant. This means that we can write

$$P(\text{washer at } (x, y) \mid \text{image}) \propto P(\text{image} \mid \text{washer at } (x, y)).$$

For brevity we shall write $P_{(x,y)} = P(\text{image} \mid \text{washer at } (x, y))$. To evaluate $P_{(x,y)}$ for a given (x, y) we scan over all the pixels (u, v) in the image. For each we calculate d as above and thus either $P(\text{black pixel at } (u, v) \mid d)$ or $P(\text{white pixel at } (u, v) \mid d)$, according to the colour of the pixel at (u, v) in the image. Finally, we multiply all these probabilities together to get the final result $P_{(x,y)}$. In practice we would represent the probabilities involved by their logarithms as suggested in Chapter 6.

Figure 9.39 shows the results obtained: the plot on the right in each case shows the value of $P(\text{image} \mid \text{washer at } (x, y))$ for each possible value of (x, y) by the intensity of the image at that point. The intensities are on a logarithmic scale and normalised for clarity. The point (x, y) giving rise to the maximum probability in each case is circled in white: as you can see, the algorithm correctly finds the washer in each case. You may need to inspect closely the original image in case (d), where several objects overlap to a significant extent, to convince yourself that its answer here is right.

Exercise 9.18 asks you to show that the calculation above can be thought of in terms of a two-dimensional correlation. An advantage of the approach discussed here, however, is that it is easy to generalise to cases where the sampling of the image is not on a regular grid or where parts of the image are unknown, for example because they are obscured by another object.

Exercises

9.1 The non-zero samples of the impulse response of a two-dimensional filter cover a rectangle u pixels wide and v pixels high. This impulse response is convolved with an image p pixels wide and q pixels high. Show that the size of the maximum rectangular region occupied by the non-zero samples in the result is $p+u-1$ pixels by $q+v-1$ pixels. (This question is the two-dimensional analogue of Exercise 5.1.)

9.2 You wish to take the Fourier transform of a 1024 pixel by 1024 pixel square greyscale image using a library function to perform the individual row and column transforms. Sadly, your Fourier transform library only offers a 1024-point one-dimensional transform with complex input and output values. Show how to calculate the two-dimensional transform using a minimum number of calls to this function. (**Hint:** Use the result of Exercise 4.9.)

9.3 Using your results from Exercises 4.9, 4.11 and 9.2, calculate the total number of complex operations (multiplications and additions) involved in working out the two-dimensional Fourier transform of a 1024 pixel by 1024 pixel square greyscale image. Ignore the operations required to convert from real to complex data to allow the complex transform to be used.

How many complex operations are needed to work out the circular convolution of two 1024 pixel by 1024 pixel square greyscale images directly, treating them as complex-valued signals? How many are needed if we transform to the frequency domain (using a fast Fourier transform), multiply corresponding frequency components, and then transform back again?

9.4 A separable two-dimensional filter is implemented by applying filters in the x and y directions in turn, with impulse responses $u(x)$ and $v(y)$ respectively. Show that the overall two-dimensional impulse response of the filter is given by $h(x, y) = u(x)v(y)$.

9.5 Show that the filter whose impulse response is shown in Figure 9.8 is not separable.

9.6 Suppose you have a one-dimensional filter whose frequency response is given by $H(f)$. Show that, if you negate every second sample going into the filter, the overall frequency response of the new system is $H(f + f_s/2)$. (**Hint:** Negating every other sample is equivalent to multiplying the signal by $e^{2\pi j f t}$ where $f = f_s/2$.)

Hence show that if the filter with impulse response (c_0, c_1, c_2, c_3) is a low-pass filter for the band from 0 to $f_s/4$, the filter with impulse response $(-c_0, c_1, -c_2, c_3)$ is a high-pass filter for the band from $f_s/4$ to the Nyquist limit $f_s/2$.

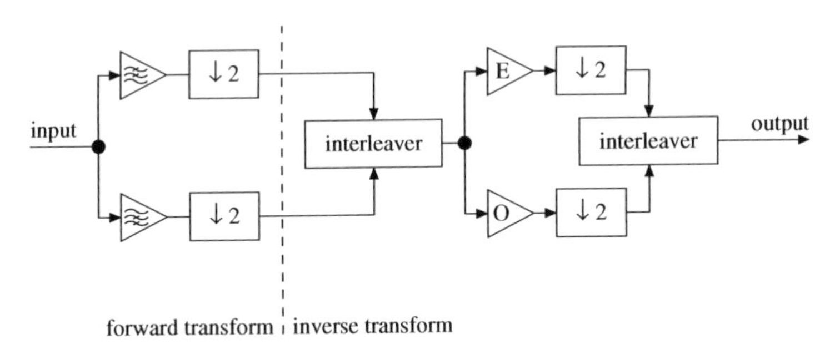

Figure 9.40 One-stage forward and inverse wavelet transform.

9.7 Show that, for the values of c_0, c_1, c_2 and c_3 given in the description of the discrete wavelet transform,

$$c_0^2 + c_1^2 + c_2^2 + c_3^2 = 1$$

and

$$c_0 c_2 + c_1 c_3 = 0.$$

Show, by following the effect of an impulse through the one-level discrete wavelet transform and inverse discrete wavelet transform shown in Figure 9.40, that the multi-level inverse discrete wavelet transform shown in Figure 9.23 does indeed correctly reconstruct the input signal to the multi-level discrete wavelet transform shown in Figure 9.22. Because samples with even time indices are treated differently from samples with odd time indices, you will need to check separately the cases where the impulse occurs at an even or an odd time.

9.8 Digital cameras often have a resolution specified in megapixels, giving the total number of elements in the sensor. A camera specified as having, for example, '3 megapixels' might have a sensor that is 2000 pixels wide by 1500 pixels high. However, a colour image sensor claimed to have n pixels might have $n/4$ pixels sensitive to red light, $n/4$ pixels sensitive to blue light and $n/2$ pixels sensitive to green light, and so the chrominance resolution of the sensor is not as high as you might perhaps expect.

The pixels of a camera sensor are usually arranged in the pattern shown in Figure 9.41, called the 'Bayer pattern'. In the figure, 'R' represents a pixel sensitive to red light, 'G' a pixel sensitive to green light, and 'B' a pixel sensitive to blue light. Notice that the patterns of red and blue pixels are the same as one another (apart from an offset), but different from the pattern of green pixels.

Describe how you would design a system to calculate a red intensity value at a pixel position where there is (a) a green or (b) a blue pixel.

B G B G B G
R G R G R G R G
B G B G B G B G B G
G R G R G R G R G R
B G B G B G B G B G
G R G R G R G R G R
G B G B G B G B
G R G R G R

Figure 9.41 The Bayer pattern.

(a)

(b)

Figure 9.42 Shearing an image: (a) before; (b) after.

How would you calculate a green intensity value where there is a red or blue pixel? (**Hint:** Turn the page through 45° and compare the situation with (b) above.)

9.9 Show how to apply a one-dimensional polyphase interpolator to the rows of an image to *shear* it: for example, given the image of Figure 9.42(a), produce the image of Figure 9.42(b). Note that the final image has the same height as the original one.

Show that

$$\begin{pmatrix} 1 & -\alpha \\ 0 & 1 \end{pmatrix} \begin{pmatrix} 1 & 0 \\ \beta & 1 \end{pmatrix} \begin{pmatrix} 1 & -\alpha \\ 0 & 1 \end{pmatrix} = \begin{pmatrix} \cos\theta & -\sin\theta \\ \sin\theta & \cos\theta \end{pmatrix}$$

if

$$\alpha = \tan\frac{\theta}{2}$$

and

$$\beta = \sin\theta.$$

(**Hint**: $\sin\theta = 2\alpha/(1+\alpha^2)$ and $\cos\theta = (1-\alpha^2)/(1+\alpha^2)$; work in terms of α.)

Hence show how to rotate an image using three shear operations.

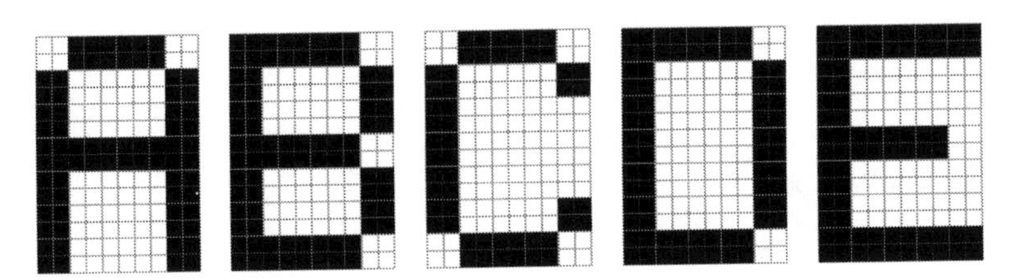

Figure 9.43 The original blocky characters.

9.10 Suppose you have an image N pixels wide and N pixels high. Show that moving the image one pixel to the right (with pixels falling off the right-hand end reappearing on the left) multiplies its two-dimensional Fourier transform by $e^{-2\pi j f_x/N}$. Show that moving the image one pixel in the y direction (again with pixels wrapping round from one edge to the opposite one) multiplies its two-dimensional Fourier transform by $e^{-2\pi j f_y/N}$. Show that moving the image by a vector (d_x, d_y), where d_x and d_y are integers and where the image wraps around as above, multiplies its two-dimensional Fourier transform by $e^{-2\pi j(d_x f_x + d_y f_y)/N}$.

This last relation holds for non-integer values of d_x and d_y as well. By analogy with the one-dimensional polyphase filter design procedure given in Section 5.6, describe how to construct a set of n^2 filters suitable for polyphase interpolation in two dimensions to a resolution of $1/n$ pixel along each axis. Assume that the resolution of the output image will be at least as high as that of the input image, so you do not need to worry about aliasing.

9.11 Extending the method of the previous exercise, describe how to construct a set of n^2 filters for polyphase interpolation in two dimensions where the resolution of the output image will be lower than that of the input image by a known factor that is not a simple fraction.

9.12 Design and implement an algorithm to equalise the histogram of an image.

9.13 Suppose you have a library function that performs very fast convolutions of greyscale images with two-dimensional filters. Show how to use this function to calculate the dilation of a bilevel image by a given structuring element. Can you extend your method to calculate the erosion of a bilevel image by a given structuring element?

9.14 Show that if A is a bilevel image and B is a structuring element, $A - B = \overline{\overline{A} + B}$.

9.15 Is it possible to construct an example of an image A and a structuring element B such that $(((A+B)-B)+B)-B$ is not the same as $(A+B)-B$? (In other words, can closing an image for a second time using the same structuring element have any effect?)

9.16 Show that $C(A, B) = \overline{O(\overline{A}, B)}$ and that $O(A, B) = \overline{C(\overline{A}, B)}$.

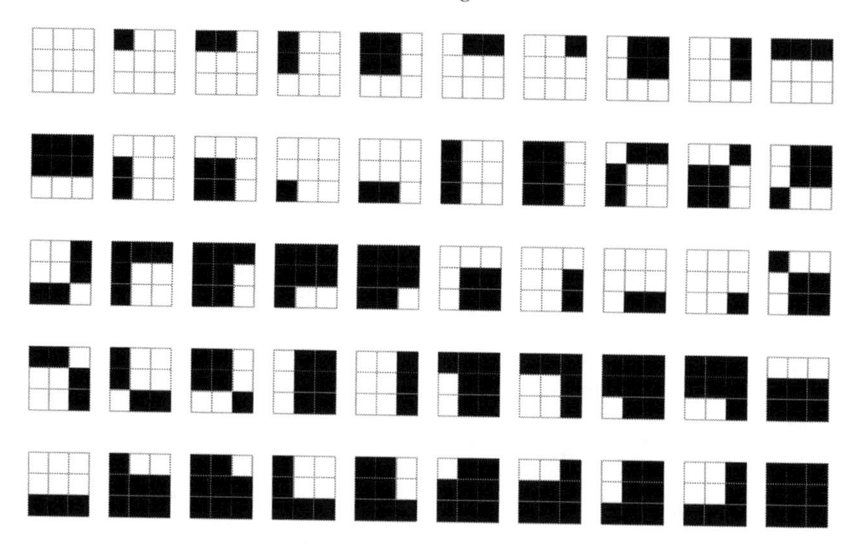

Figure 9.44 The fifty 3-by-3 patterns that can occur.

9.17 A font designed on a crude 5-pixel by 7-pixel matrix has been enlarged by a factor of 2 along each axis for use on a computer display. The resulting characters have the blocky appearance shown in Figure 9.43. Design a morphological operator with a square 3-pixel by 3-pixel structuring element which you can apply to these characters to improve their appearance. The only 3-by-3 patterns that can occur are those shown in Figure 9.44, and so you only need to specify the output of the operator in each of those 50 cases. Many of the cases are reflections or rotations of one another, which you probably will want to treat alike. You may find it easiest just to work out in which cases you would want to *change* the state of the central pixel in the square.

Draw the results of applying your algorithm to the characters shown in Figure 9.43.

9.18 Consider the pattern recognition example of Section 9.12. Let $I(x, y)$ be the image to be processed and $W(x, y)$ be a reference image of a washer on a blank background. Show how to use the two-dimensional correlation of $W(x, y)$ with $I(x, y)$ to calculate the value of $P_{(x,y)}$ for any desired (x, y). (**Hint:** Imagine moving the reference image over the image to be processed and let $N_b(x, y)$ and $N_w(x, y)$ be the number of black pixels and white pixels respectively covered by the reference. Work out $P_{(x,y)}$ in terms of $N_b(x, y)$ and $N_w(x, y)$.)

10
Moving images

This chapter is about the application of digital signal processing techniques to video signals.

A video stream is composed of a sequence of images normally sampled at regular intervals in time, as shown in Figure 10.1. A video signal therefore has a three-dimensional domain: two of the dimensions are the x and y axes of the image, and the third dimension is time. A video camera usually samples the individual images that make up the video stream more coarsely than a typical still camera. For example, an ordinary 'broadcast quality' video signal, depending on the exact format, consists of images sampled to a resolution of less than half a million pixels in total; even the high-definition television standards require at most about four times that figure. By contrast, even a low-cost digital still camera has a resolution of several millions of pixels.

10.1 Standard video formats

Video is typically sampled in time at a rate between 24 Hz and 60 Hz, although lower rates are sometimes used in highly constrained applications such as video telephony.

For ordinary television applications video is usually sampled either at 50 Hz, with a resolution of 720 (horizontal) by 576 (vertical) visible pixels, or at 60 Hz, with a resolution of 720 by 480 visible pixels. (Notice that these alternatives correspond to the same total number of pixels per second.) For analogue transmission the images are padded with black space on all sides, which is why the two standards are also called '625-line' and '525-line' respectively. The latter standard is used in the United States and Japan, while the former is widely used in Europe and elsewhere. There are machines that can convert video signals between these sample rates, and we shall see below how they work.

Figure 10.1 A video signal consists of a number of two-dimensional images sampled in time.

You will sometimes see the terms 'NTSC' (National Television Standards Committee) and 'PAL' (Phase Alternation Line) applied to video standards. Strictly speaking, these terms denote the method used for encoding chrominance information in an analogue transmitted television signal, but they are used loosely to refer to 525-line (NTSC) and 625-line (PAL) systems in general.

Many other different resolutions are used for representing video signals, including some designed to match the resolutions of typical computer monitors, but we shall not go into them in detail.

Sometimes video is captured at a higher rate than 50 Hz or 60 Hz so that the result can be easily and smoothly played back in slow motion. This is often done for television coverage of sporting events and in scientific applications.

Most of the image processing methods we discussed in Chapter 9 can be applied to video signals by simply treating each captured image independently. The rest of this chapter concerns what we can do by comparing the images captured over a number of samples in time.

Aliasing

Sampled video signals can suffer from aliasing in all three dimensions, sometimes all at once! Aliasing occurs in either of the two spatial dimensions in a similar way to ordinary images. Although most broadcast television signals are adequately filtered to remove aliases, it is sometimes possible to see aliasing effects in finely detailed parts of images such as those of brickwork or of Venetian blinds. Sometimes diagonal sharp edges of text used for titles can appear jagged, which is another example of aliasing.

Aliasing in the time dimension is much more common for television images. Suppose a presenter is wearing a shirt with a square chequered pattern: if he moves on the screen by more than the width of one of the squares in the pattern in the time between the capture of consecutive images, aliasing will be introduced.

As you might imagine, this kind of event is extremely common, especially in scenes with rapid camera movement. For various reasons, however, the human eye is very good at not noticing this kind of alias, and so we can get away with a relatively low temporal sample rate. Unfortunately, any simple digital processing we could do on a signal will inevitably work on the assumption that the sampling process has not introduced aliases. Where this assumption is false the processing might easily produce visually unacceptable artefacts. As we shall see, much effort in video signal processing is devoted to modifying the processing we do in order to minimise the visual effect of these artefacts.

Interlace

A video signal is converted into a one-dimensional signal for analogue transmission over a cable or over the air by scanning each sampled image in turn. Each image is scanned in reading order from left to right and from top to bottom.

When analogue television formats were originally designed a compromise was made between the spatial resolution and the temporal resolution of the images. Rather than display every line of every image, the odd-numbered lines appear in one image, and the even-numbered lines in the next: so-called *interlaced scan*. The lines are traditionally numbered from top to bottom of the image, starting from 1; when processing video images it is usually more convenient to start numbering from zero. Each image, containing only half the total number of lines, is called a *field*; a pair of images adjacent in time, together forming a complete set of lines, is called a *frame*. Figure 10.2 shows which lines of the image are involved in each field. (A real analogue television signal also includes half-length scan lines at the bottom of each odd field and the top of each even field, but we have not shown these here.)

Looking at Figure 10.2 edge-on, we can see that interlace corresponds to a non-rectangular sampling pattern in the y-t axes: see Figure 10.3. In fact, the pattern is similar to the hexagonal sampling pattern we saw in Figure 2.8.

If the video signal is not interlaced, containing a complete set of lines in every field, the y-t sample pattern is rectangular and the scan is said to be *progressive*. In this case there is only one field per frame. 'Interlaced' and 'progressive' are sometimes abbreviated to just 'i' and 'p'. For example, a '1080i' signal is one that is interlaced with 1080 lines per frame, or 540 lines per field. A '720p' signal is not interlaced and has 720 lines per field (or, equivalently, per frame).

Interlace was originally a solution to a problem in analogue television broadcasting to cathode ray tube (CRT) displays. It was designed to keep the width of the frequency band occupied by the signal narrow while still offering high vertical resolution and a reasonable field update rate; the latter is necessary with

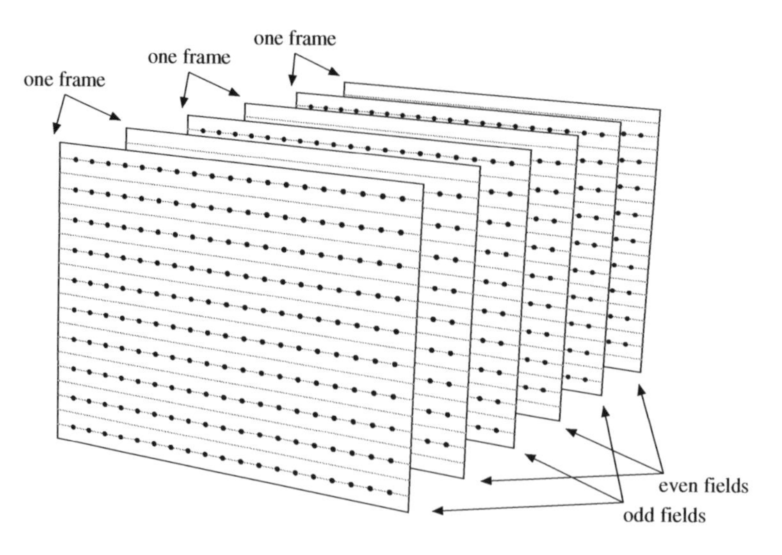

Figure 10.2 Scan lines involved in an interlaced video signal.

CRT displays to avoid visible flicker. The compromise interlace offers between vertical resolution and temporal resolution is fixed. With modern digital signal processing and compression methods, however, we are in a position to make a dynamic tradeoff between these resolutions: where an image is relatively still, we can achieve greater spatial resolution, and where an image is moving quickly, we can sacrifice spatial resolution for greater temporal resolution. Nevertheless, mostly for reasons of backward compatibility, interlace is still a feature of more modern digital television standards.

10.2 Deinterlacing

Deinterlacing is the process of converting an interlaced signal to a progressive one at the same field rate and total vertical resolution: in other words, we have to reconstruct the missing samples in Figure 10.3. This processing is required, for example, if we wish to show video signals designed for interlaced displays on computer monitors, which almost invariably use progressive scan.

The problem is similar to the one explored in Exercise 9.8, where we suggested the idea of rotating the page through 45° and treating the samples as if they were on a regular square grid. In that example, there was no particular reason to treat the axes of the sampled signal differently from one another since they were just the x and y axes of an image; here, however, the axes of the sampled signal (y and t) are rather different in nature and so we need to analyse the problem a little more carefully.

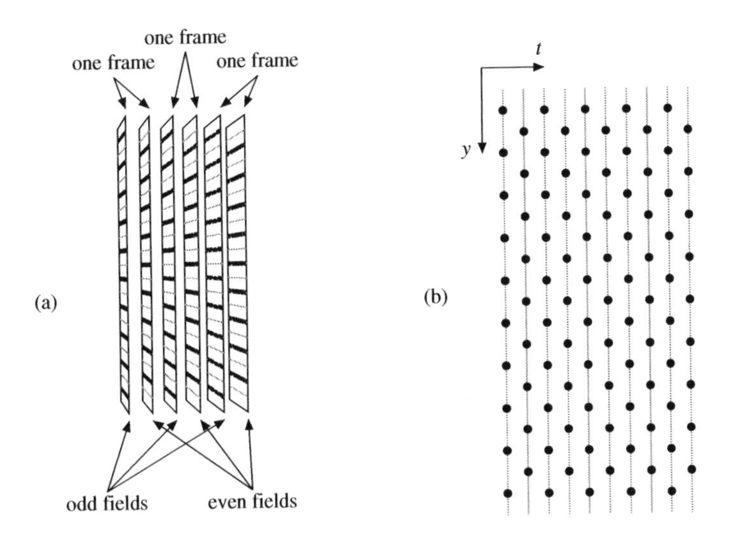

Figure 10.3 The interlace pattern seen (a) edge-on; (b) in the y-t plane.

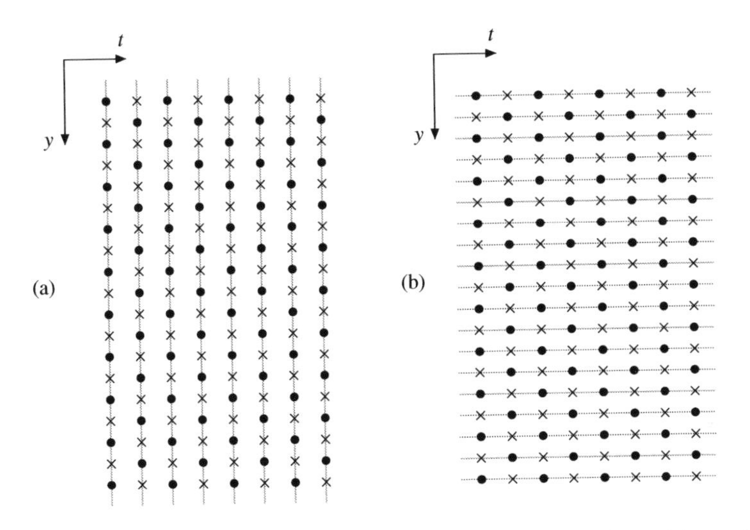

Figure 10.4 Deinterlacing by interpolation (along grey lines) in (a) y; (b) t; newly generated samples are marked by crosses.

From Chapter 5 we know how to construct interpolation filters that can produce samples of a signal positioned between the samples we already have. We could apply such a filter to the columns of Figure 10.3 as shown in Figure 10.4(a). This would construct the missing lines in each field from its (existing) neighbours within that field by *spatial interpolation*. An alternative would be to interpolate horizontally in Figure 10.3, constructing a missing line from the same-numbered lines in fields adjacent in time where that line does exist, as shown in Figure 10.4(b): this is called *temporal interpolation*. The two processes give different answers. The first produces a result lacking vertical detail, but with good response to rapidly changing scenes, while the second produces better vertical detail, but gives poor results from rapidly changing scenes. Neither process gives ideal results under all conditions.

Low-cost deinterlacers interpolate only spatially, within each field. When operating on an analogue video input this method has the advantage that only a few lines of the image need to be stored in the deinterlacer, and the output need only be delayed by a few lines with respect to the input.

More sophisticated deinterlacers use a combination of spatial and temporal interpolation. One method is to calculate both the spatially interpolated and temporally interpolated versions of the missing line, and then select between them by analysing how quickly the brightness is changing in the image between corresponding fields. If the brightness is changing quickly, spatial interpolation is used so that fast-moving scenes are handled well; if the brightness is not changing quickly, temporal interpolation is used to improve resolution on stationary objects.

We can use yet another interpolation to make the switchover between the two types of interpolation smooth, helping to avoid artefacts introduced by a sudden switchover. The brightness analysis can be carried out on the image as a whole or it can be done more locally, so that different regions of the image are interpolated using different algorithms. Because the eye tends not to perceive detail well in moving objects, this hybrid method gives good results.

Any deinterlacer which performs temporal interpolation necessarily stores a number of fields of the image. It will also introduce a delay of a number of fields between its input and its output since the temporal interpolation filter needs access to samples on either side of the desired output point.

Frequency-domain analysis of deinterlacing

To understand how to perform deinterlacing in a more sophisticated fashion, we need to study the interlaced sampling process in more detail.

Figure 10.5 shows Figure 10.3 rotated by 45°: the t axis now runs down and to the right, while the y axis runs down and to the left. With this rotation we now have a regular square sampling grid. In the frequency domain this sampling will give rise to a square-tiled patterns of aliases, as we saw in Chapter 10; we also saw there that rotation in the frequency domain corresponds to the same rotation of our original signal. Interlaced sampling therefore produces the diamond-shaped pattern of aliases as shown in Figure 10.6. In the figure we have shown the typical

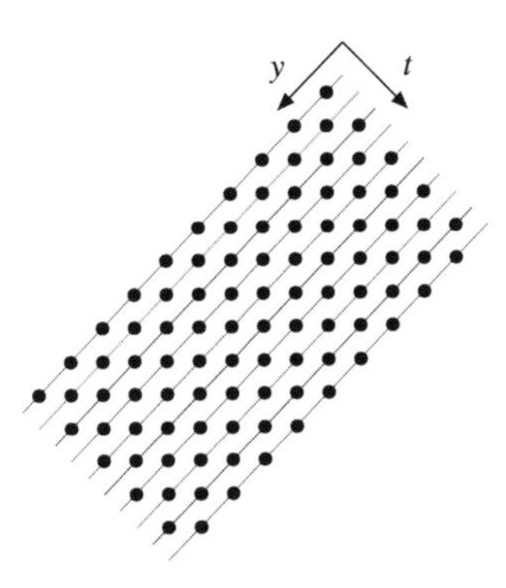

Figure 10.5 The interlace sampling pattern rotated through 45°.

frequency components that will be present as two shaded ellipses superimposed: the horizontal ellipse corresponds to video material that has low vertical resolution but which is moving rapidly and the vertical ellipse corresponds to stationary video material with high vertical resolution.

It is clear from Figure 10.6 that, if we were constrained to use a fixed anti-aliasing filter, the one we require to reconstruct the missing samples should only pass the frequency components within the dotted box in the figure; in other words, its ideal frequency response should look like Figure 10.7. The shaded area shows the frequencies that the filter should pass, and the blank area the frequencies that it should block.

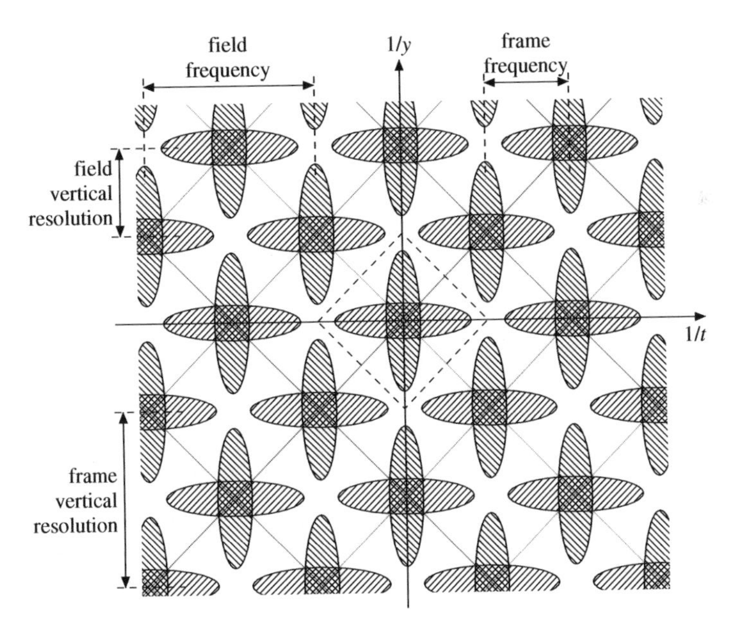

Figure 10.6 Frequency-domain pattern of aliases for an interlaced video signal.

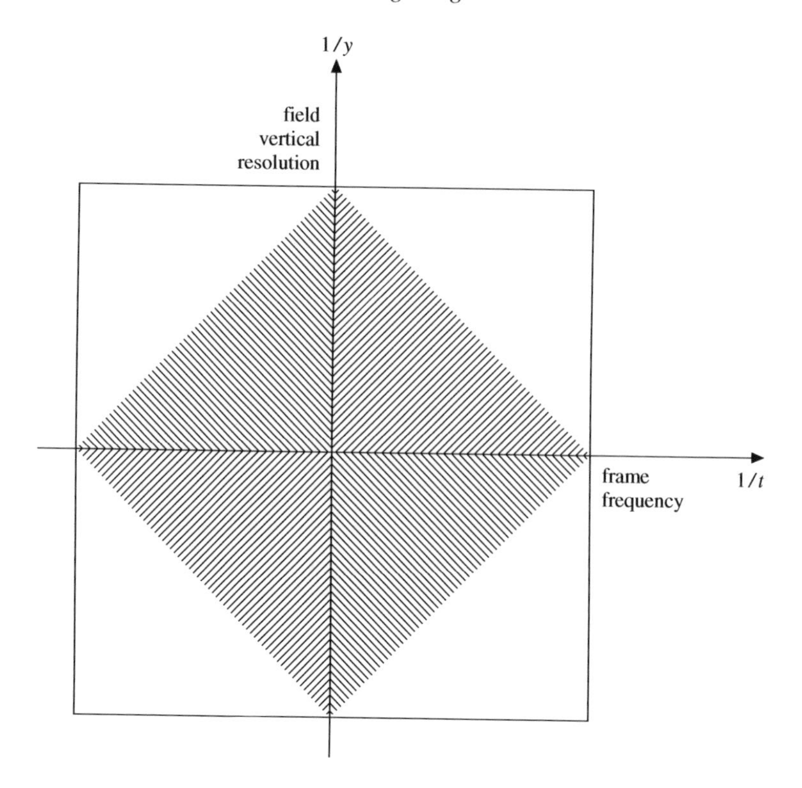

Figure 10.7 Ideal two-dimensional frequency response of deinterlacing filter passes frequencies in shaded area and blocks other frequencies.

Digital signal processing systems working in real time on video signals often require a great deal of processing power. For this reason, a practical deinterlacer of this type might use a filter with an impulse response covering just four consecutive fields and eight consecutive scan lines (i.e., four scan lines from each field). Having such a short impulse response necessarily implies some compromises in the frequency response of the filter, and the same filter might not be ideal for all situations. Some deinterlacers offer a choice of filters: in 'action mode' the deinterlacer might have a more nearly ideal anti-aliasing performance along the time axis to reduce the artefacts produced during rapid camera pans (although at the cost of overall vertical resolution), while in 'drama mode' it might offer the reverse compromise.

Other approaches to deinterlacing

Other image processing algorithms discussed in Chapter 9 can be applied to the problem of deinterlacing. A popular choice is to apply a median filter along the y and t axes: one method involves calculating the median of three input samples to

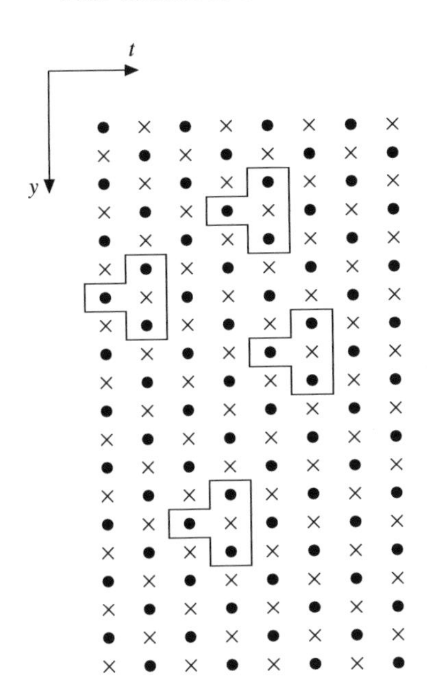

Figure 10.8 Deinterlacing using a three-point median filter: newly generated samples are marked by crosses.

produce each output sample as illustrated in Figure 10.8, where the boxes outline a few examples showing which three samples are used.

The next level of sophistication in deinterlacing is to analyse the motion of objects in a scene by attempting to match areas of image in one field to nearby areas in the previous one. The missing lines in the image of the moving object can then be interpolated between the versions of an object that appear in successive fields, shifted so that they are spatially aligned. The details of this process are outside the scope of this book, although we shall look below in more detail at methods that can be used for estimating the direction and speed of motion of objects in a scene.

10.3 Standards conversion

Standards conversion is the process of converting a video signal from one resolution and frame or field rate to another. The commonest applications are in converting from 525-line, 60 Hz video made for the United States market into 625-line, 50 Hz video for Europe and the rest of the world, and in converting video to, from and between high-definition formats. We will look at the first of these examples in more detail.

The horizontal resolutions of the two systems are for practical purposes identical, and so no resampling is required along the x axis. We can thus confine ourselves to thinking about the y and t axes, which makes the process very similar to deinterlacing as described above. Figure 10.9 shows the regular sampling pattern of the 525-line, 60 Hz interlaced signal (dots) with the positions of the samples that need to be produced overlaid (crosses). As you can see, the relative positions of the samples repeat every twelve input fields. (Strictly speaking this is not quite correct, as a '60 Hz' video signal according to the United States NTSC television standard actually has a field rate very slightly slower than 60 Hz; the discrepancy is often ignored.)

Reconstructing the samples we need is just a case of applying a two-dimensional polyphase filter, as discussed in Chapter 9. The frequency response of the filter needs to have the diamond shape as discussed in the case of deinterlacing above, but a little care is needed. Since we are changing the temporal sample rate

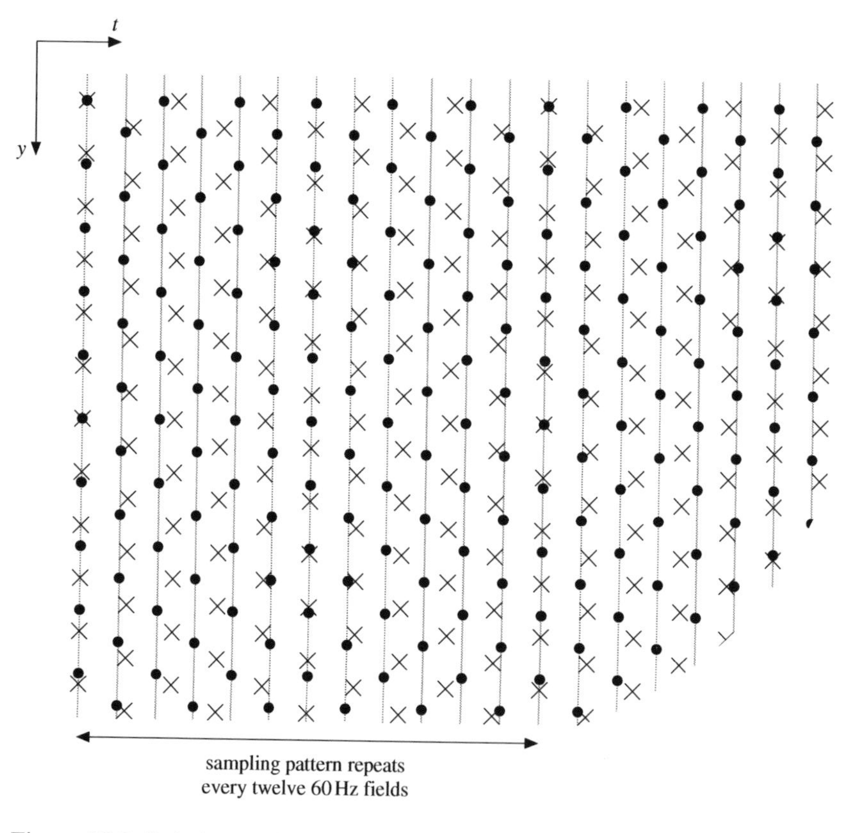

Figure 10.9 Relative positions of samples in 525-line, 60 Hz interlaced video with 480 visible lines per frame (dots) and 625-line, 50 Hz interlaced video with 576 visible lines per frame (crosses) in the y-t plane.

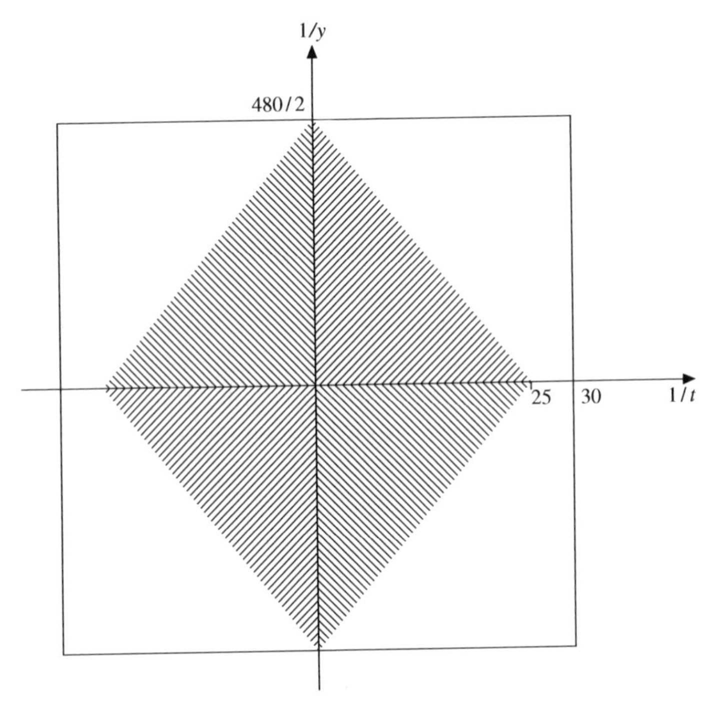

Figure 10.10 Ideal two-dimensional frequency response of filter for standards conversion between 525-line, 60 Hz interlaced video with 480 visible lines per frame and 625-line, 50 Hz interlaced video with 576 visible lines per frame passes frequencies in shaded area and blocks other frequencies.

downwards, the filter needs to cut off at a slightly lower temporal frequency to avoid aliasing in time: the response is shown in Figure 10.10. Note that the frequency on the vertical axis is expressed in terms of the resolution of the original image, which has 480 visible lines.

10.4 Motion estimation

We mentioned above that video signals are often aliased in time. This means that simple frequency-domain techniques and linear processing can fail to produce satisfactory results.

On the bright side, however, there are some properties of video signals that we can exploit to try to avoid the problems of aliasing. In particular, video signals tend to depict scenes containing a small number of objects which are mostly stationary or moving slowly. We would expect these objects to be repeated from frame to frame, appearing in a slightly different position each time. There might also be overall camera motion, contributing extra movement to all the objects in the scene. Identifying the objects in a scene and their motions is a step used in several video processing and compression algorithms. The more sophisticated compression algorithms, such as MPEG-4, allow arbitrarily shaped foreground and background objects in a scene to be encoded explicitly. A full survey of the methods an encoder might use to find such objects in a scene is outside the scope of this book, but we shall look at the simpler methods used by encoders for the MPEG-2 compression standard (see below) or by the more advanced deinterlacers and standards converters.

Motion estimation is usually carried out on the luminance part of the video signal only, although most of the methods could in principle be extended to use the chrominance information as well. This would help a motion estimator give accurate results if a scene contained two equally bright objects of different colours moving in different directions.

In some applications it is sufficient to work out only the general direction of motion in an entire scene, which can be done by calculating the two-dimensional correlation between consecutive frames. The location of the maximum value of the correlation function will give an indication of the overall direction of motion. This technique is used in some standards converters and deinterlacers, as well as forming a part of some of the more sophisticated versions of the motion estimation algorithms we describe next.

Block matching

Suppose we have a pair of consecutive frames from a video sequence as shown in Figure 10.11: call the frames 'Frame 0' and 'Frame 1' as indicated. We shall assume that the video is not interlaced; the ideas presented below can be extended to the interlaced case, although the details are fiddly.

First we divide Frame 1 into square tiles or *blocks*. Typically these blocks will be 16 pixels on a side, but other sizes (including non-square shapes) are sometimes used. For each block we try to find a matching block in a nearby position in Frame 0. Note that this matching block need not be aligned to the 16-pixel grid shown by crosses in the figure: it can be at any offset in the frame.

If the block represents a part of the image that is stationary, like the one towards the top right in the example, it will find a match at zero horizontal and vertical offset: we say that the *motion vector* for that block is $(0, 0)$.

If a block in Frame 1 is in part of the image that is moving to the right, such as the lower one in the example which covers part of the moving car, it will be to the right of its match in Frame 0. The motion vector for that block will then have positive x component. If the motion is at the rate of three pixels per frame, the offset of the block from its match, i.e., the motion vector, will be $(3, 0)$.

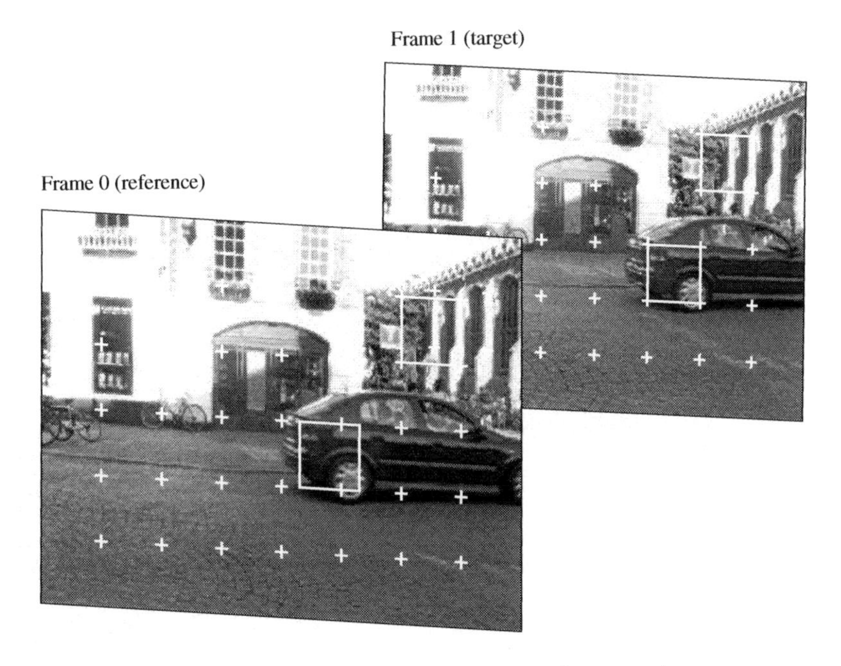

Figure 10.11 Motion of a block from one frame to the next.

If the motion were downwards and slightly to the left, the motion vector might be $(-1, 4)$.

Figure 10.12 shows the two frames side-by-side with a set of motion vectors, obtained using an algorithm described below, superimposed on Frame 1. The dots show blocks where the motion vector is zero. The vectors in the region of the car are following its motion.

Figure 10.12(d) shows an attempt to reconstruct Frame 1 from Frame 0 by copying blocks of pixels from it as directed by the motion vectors shown. As you can see, it is a reasonable approximation to the original except in areas where regions of background detail have been exposed by a motion. For example, the back of the car has exposed part of the doorway of the building. This part has been replaced by a copy of part of the image from slightly higher up on the building's frontage, as directed by the downward-pointing motion vectors. Here we have been lucky because of the appearance of the building and the result is surprisingly good: we might not always have such a convenient region of image to copy from.

How can we calculate these motion vectors? The simplest approach is just to try every possibility: that is, for each target block in Frame 1 we iterate over all the possible motion vectors, considering each candidate reference block in turn. A comparison function is used to calculate the difference between the content of the candidate reference block and content of the target block, giving a measure of

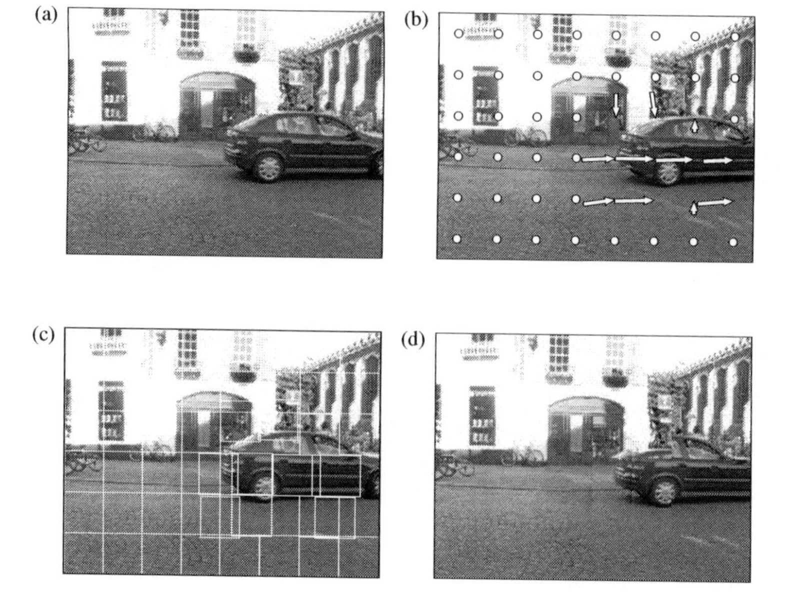

Figure 10.12 Motion vectors: (a) reference frame; (b) target frame; (c) blocks selected from reference frame; (d) approximation to target constructed from blocks copied from the reference frame.

the error between them. The candidate reference block with least error determines the motion vector.

The difference between two blocks is normally defined as the *mean absolute error*, or *MAE*, which is the sum of the absolute values of the differences between corresponding pixel values in the blocks divided by the block size:

$$\mathrm{MAE}(B_0, B_1) = \frac{1}{N}\sum_{x,y} |B_0(x, y) - B_1(x, y)|.$$

Here the $B_0(x, y)$ are the pixels in the reference block, $B_1(x, y)$ are the pixels in the target block, and N is the total number of pixels in the block: for 16-pixel-square blocks, $N = 256$. This measure (without the constant $1/N$ scaling factor) is also called the *sum of absolute differences*, or *SAD*.

An alternative measure is the *mean square error*, or *MSE*, defined by

$$\mathrm{MSE}(B_0, B_1) = \frac{1}{N}\sum_{x,y} \left(B_0(x, y) - B_1(x, y)\right)^2$$

where the symbols have the same meaning as before. When used in compression applications, the two measures give similar results. Mean absolute error is generally favoured as it does not require any multiplication operations and thus can be quicker and more straightforward to compute, especially in hardware.

This exhaustive search for motion vectors requires an enormous amount of computation. For real-time applications a number of optimisations are made to speed up the search.

The most obvious optimisation is to restrict the size of the search space by constraining the values the motion vector is allowed to take on to a *search window*. For example, if the components of the motion vector are only allowed to range from -15 to $+15$, then we only need to examine candidate reference blocks in this range. Often the allowable range for the x component of the motion vector will be set wider than the range for the y component to reflect the fact that horizontal camera motion ('pan') is more common than vertical camera motion ('tilt'). The vectors shown in Figure 10.12 were calculated using the MSE measure and a search window given by $-15 \leq x \leq +15$, $-15 \leq y \leq +15$.

Another optimisation, which speeds up the search considerably at the cost of sometimes giving poorer results, is called *logarithmic search*. The idea is to find a crude approximation to the motion vector, which we then iteratively refine.

For example, suppose we wish to find a motion vector (x, y) within the search area given by $-15 \leq x \leq +15$, $-15 \leq y \leq +15$. We first try a grid of nine candidate motion vectors $(8, 8)$, $(0, 8)$, $(-8, 8)$, $(8, 0)$, $(0, 0)$, $(-8, 0)$, $(8, -8)$, $(0, -8)$, $(-8, -8)$, as shown in Figure 10.13(a).

Suppose that, of these candidates, $(0, 8)$ is found to be the best. We now narrow down our search around this candidate by trying a grid of motion vectors around this value obtained by adding $+4$, 0 or -4 to the components of the vector. The grid of vectors we try at this stage is thus $(4, 12)$, $(0, 12)$, $(-4, 12)$, $(4, 8)$, $(0, 8)$, $(-4, 8)$, $(4, 4)$, $(0, 4)$, $(-4, 4)$: see Figure 10.13(b). Notice that the error corresponding to the fifth of these nine candidate motion vectors has already been calculated at the previous stage, and so does not need to be calculated again.

Suppose the best motion vector from this set is $(-4, 8)$. In the same fashion as before, we now inspect a set of nine motion vectors in an even smaller grid around this point, by adding $+2$, 0 or -2 to it: see Figure 10.13(c). Again, the error corresponding to the middle motion vector has already been calculated at the previous stage, and so only eight further error calculations are required.

Finally, we take the best of these matches and perform an even finer search around it, adding $+1$, 0 or -1 to the motion vector as shown in Figure 10.13(d) – another eight error calculations.

In total we have done $9+8+8+8=33$ block match operations to arrive at our final result. An exhaustive search would have required $(15+1+15)^2=961$ block matches, and so the logarithmic search is potentially about thirty times faster than exhaustive search in this case.

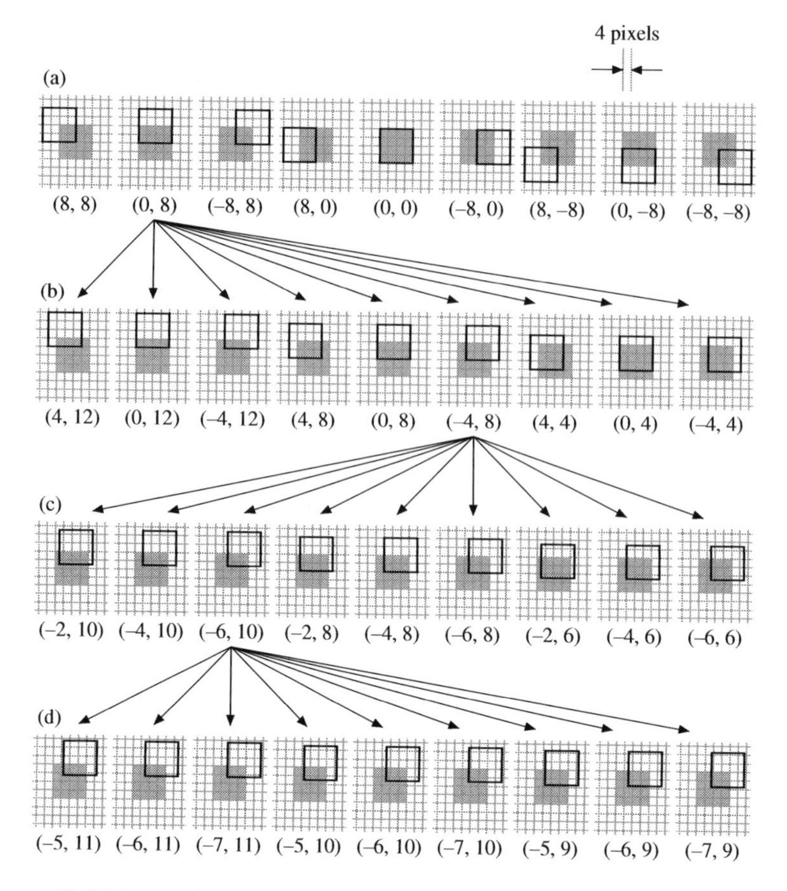

Figure 10.13 Logarithmic search for motion vectors between blocks in target frame (solid grey) and reference frame (outlined black box): (a) in steps of 8 pixels; (b) in steps of 4 pixels; (c) in steps of 2 pixels; (d) in steps of 1 pixel.

Figure 10.14 shows the result of using logarithmic search to find motion vectors for the frames shown in Figure 10.11. Compare these results with those of Figure 10.12, where exhaustive search was used. The results here are slightly less good, and in particular the algorithm has not managed to reconstruct the part of the frontage of the building exposed by the motion of the car.

As you might imagine, there is a host of variations on the logarithmic search algorithm that offer different tradeoffs between search speed and result accuracy.

10.5 MPEG-2 video compression

MPEG-2 is a compression standard used for DVD (digital versatile disc) video and for broadcast digital television, including high-definition television, in many countries. The full standard is rather complicated and includes many details that

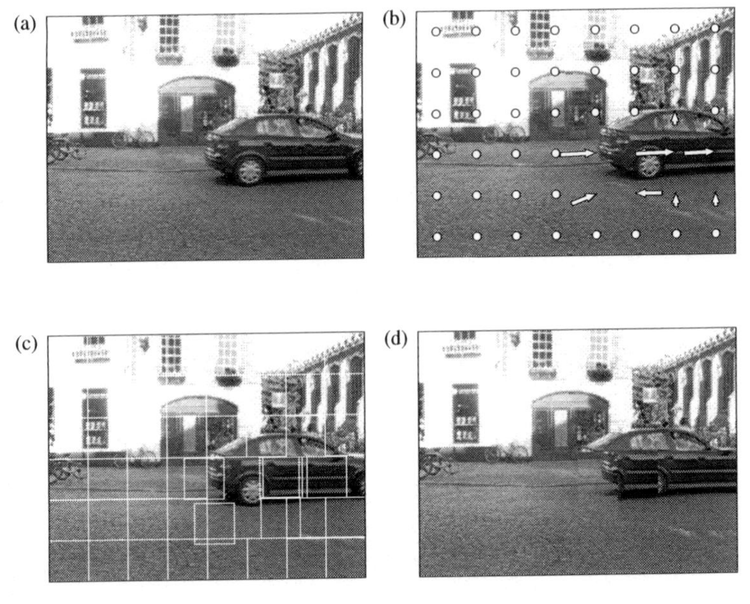

Figure 10.14 Motion vectors: (a) reference frame; (b) target frame; (c) blocks selected from reference frame; (d) approximation to target constructed from blocks copied from the reference frame.

are of no interest from the point of view of signal processing; we shall concentrate below on the more relevant parts of how the scheme works, with simplifications where necessary to make the description clearer.

As with MP3 audio compression, the standard does not specify how an encoder should work: it only specifies what the decoder should do. Implementers can make whatever tradeoffs they wish between the speed of the encoder and the quality of its output. For example, MPEG-2 includes a refinement to the concept of a motion vector discussed above whereby vectors are stored to a resolution of half a pixel. These motion vectors are calculated using the algorithms described above, but operating on interpolated versions of the images. Not all encoders can produce these sub-pixel motion vectors; those that do not produce them typically offer poorer compression ratios for the same resulting quality.

In outline, a typical MPEG-2 encoder operates as follows.

The sequence of video images (which are referred to as *pictures* to avoid the potentially confusing use of the term 'frame' or 'field'), is divided into groups. A *group of pictures*, or *GOP*, will typically contain 12 or 15 pictures: the example in Figure 10.15 shows a sequence encoded using 12-picture groups. Three different compression schemes are used on the pictures in the sequence. Which scheme is used for a particular picture depends on its position within its group.

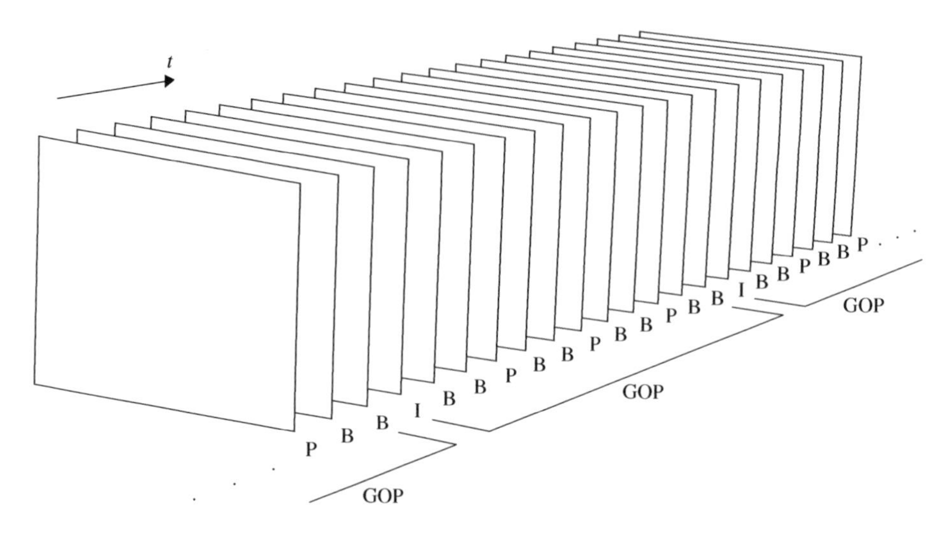

Figure 10.15 Groups of pictures (GOPs) in MPEG-2 compression.

Picture types

The first picture in each group is called an *intra picture* or *I-picture*, and is compressed using a scheme essentially the same as JPEG still image compression (see Section 9.6). I-pictures are compressed entirely independently of any other pictures in the video sequence, which allows an MPEG-2 stream to be decompressed starting from the beginning of any group. This feature is used to allow random access to video stored on DVDs, for example, and limits the propagation of any errors that might be present in the compressed data stream.

Every third picture in the group after the I-picture is called a *predictive picture* or *P-picture*. The first P-picture in the group is compressed with reference to the preceding I-picture; the second and subsequent P-pictures are compressed with reference to the previous P-picture, so picture 6 in the group is compressed with reference to picture 3, and picture 9 with reference to picture 6.

In each case the P-picture is divided into square blocks 16 pixels on a side. A motion vector is found for each block by comparison with the corresponding reference picture, and a new temporary picture is formed by extracting blocks from the reference picture as directed by the motion vectors: the idea is that if the motion vectors are accurate, the temporary picture will approximate the P-picture we are trying to encode. Figure 10.12(d) shows the kind of picture this process will generate.

The temporary picture is now subtracted from the actual P-picture, leaving a difference picture like the one shown in Figure 10.16 (which is actually the difference between Figure 10.12(b) and Figure 10.12(d)). The difference can be negative, so we have shown zero as grey in the figure; black indicates a large

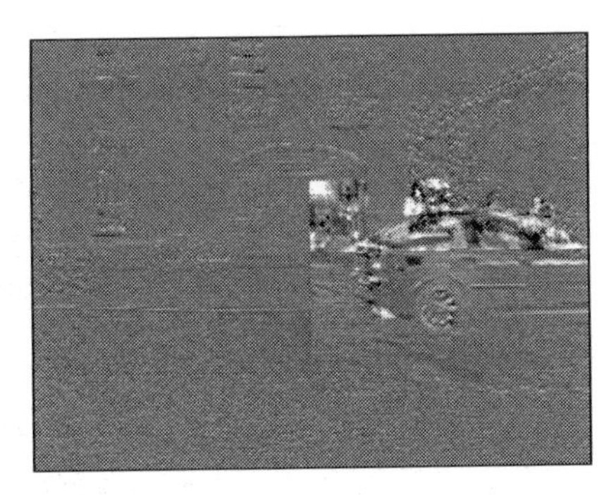

Figure 10.16 Error in approximation to P-picture.

negative difference, and white a large positive difference. As you can see, the difference is mostly zero or near zero in the stationary parts of the picture; there are some errors in the moving parts. The difference picture is now compressed using a JPEG-like scheme with minor modifications to allow it to deal with the negative pixel values. Since the image is mostly zero a very high compression ratio can be achieved. Typically, a P-picture might be compressed to one third of the size of a compressed I-picture.

If no good motion vector can be found for a particular block, the encoder has the option of encoding that block independently as if it were a block in an I-picture: such blocks are called *intra-coded* blocks.

The result of the above is that a P-picture has been converted into a compressed difference image, a set of compressed blocks where no good motion vector could be found, and a list of motion vectors. The motion vectors are themselves then compressed using a lossless scheme which takes advantage of the fact that motion vectors for adjacent blocks tend to be similar, and more generally that motion vectors tend to be small in magnitude.

Given a copy of the reference picture, a decoder can recreate the P-picture. First it reconstructs the temporary picture by copying blocks from the reference picture as directed by the motion vectors. It then decompresses the difference image and adds the result to the temporary picture. Finally it decompresses the intra-coded blocks and inserts them into the result.

The remaining pictures in a group are called *bidirectionally predictive pictures*, or *B-pictures*. B-pictures are encoded in just the same way as P-pictures, except that each motion vector may refer either to the *previous* I-picture or P-picture, or to the *next* I-picture or P-picture. So, for example, a block in picture 11 could have a motion vector which directs the decompressor to copy a block from picture 9, while a neighbouring block in the same picture could refer to picture 12, the next I-picture. Allowing the B-pictures to look forwards or backwards in time like this helps to improve compression in the case where there is a sudden change of scene

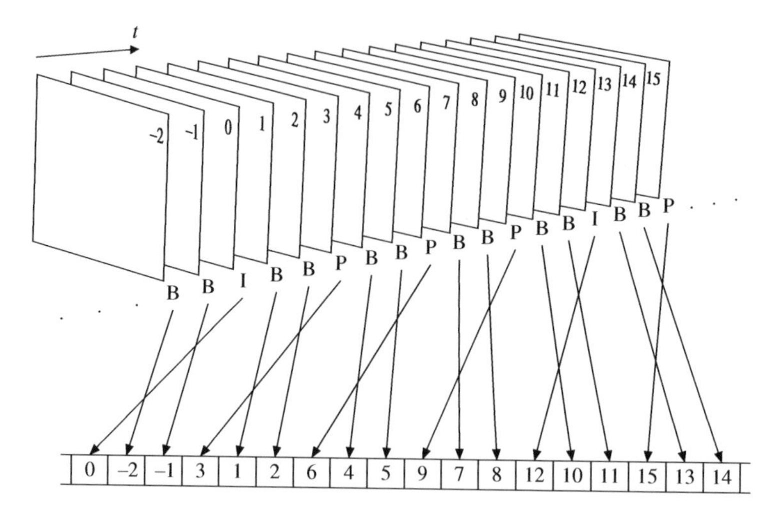

Figure 10.17 Order of picture transmission in MPEG-2 compression.

within a group or where the motion of an object exposes detail in the background. Blocks in B-pictures may also be intra-coded.

A B-picture might be compressed to one half of the size of a compressed P-picture, or about one sixth of the size of a compressed I-picture.

Order of transmission

In a real-time decoder decompressing a transmitted MPEG-2 data stream it will not necessarily be possible to reconstruct a B-picture until the content of the next P-picture or I-picture is known. For this reason, the pictures are stored in the data stream in the order shown in Figure 10.17: you can verify that each compressed picture appears in the data stream after the pictures on which it depends, so that, for example, picture 11 appears after both picture 9 and picture 12.

Exercises

10.1 European viewers, when first exposed to analogue television in the United States, find the images slightly blurry. What complaint do you think United States viewers have when they first see European analogue television?

10.2 How often does the sampling pattern in Figure 10.9 repeat vertically?

10.3 Sketch a filter response, analogous to that shown in Figure 10.10, suitable for use in a video standards converter whose input is a 625-line, 50 Hz signal and whose output is a 525-line, 60 Hz signal.

10.4 A system is being designed to compute motion vectors by block matching for a 625-line video system with a resolution of 720 pixels by 576 pixels. Blocks 16 pixels square are to be used.

(a) How many blocks are there in a frame?

(b) If exhaustive search is done with a maximum offset of ± 15 pixels in the x and y directions, how many block match operations have to be carried out per frame (ignore the effects of the edge of the frame and the possibility of early search termination)?

(c) What is the total number of pixel match operations that must be carried out per frame?

(d) If the frame rate is 25 Hz, what is the total number of pixel comparison operations that must be carried out per second?

10.5 You are watching a football game on television where the video has been compressed using the MPEG-2 standard. The camera pans rapidly across the pitch, following the action. The spectators form a blurred sea of faces in the background: looking closely at this part of the image, you notice that it sometimes seems to stop moving or break up into blocks, even though the camera is still panning smoothly. Suggest a reason for this phenomenon.

10.6 Impulsive interference garbles some of the data stream representing a video signal encoded using the MPEG-2 standard and the picture pattern shown in Figure 10.15, with the result that the receiver incorrectly decodes a single I-picture. How many other pictures in the output of the decoder might be affected?

10.7 A long film is compressed using an MPEG-2 encoder using the grouping pattern illustrated in Figure 10.15 and the picture ordering shown in Figure 10.17. There are 25 pictures per second. Each compressed I-picture is exactly 140 000 bytes long, each compressed P-picture is exactly 40 000 bytes long and each compressed B-picture is exactly 20 000 bytes long. The encoder outputs compressed data continuously at a constant rate. Ignoring other information in the compressed stream, how many bytes per second does the encoder produce? If the pictures are 720 pixels by 576 pixels, how many bits per pixel of the original signal is this? Compare this with the value you would obtain with 'motion JPEG' compression, where every picture is encoded as an I-picture.

10.8 A decoder is decompressing the data stream produced by the encoder of Exercise 10.7 and is displaying the results at a constant rate of 25 pictures per second. Assuming that the decoder cannot display a decompressed picture until it has received all the data for that picture, what is the minimum possible time between the decoder receiving the first byte of a compressed I-picture and displaying that picture?

10.9 Some video sensors capture all the pixels in one frame simultaneously, while others scan the image in a continuously repeating pattern from left to right and top to bottom in reading order, like a CRT monitor. Design a (moving) test image that would let you distinguish between the two types of sensor.

11
Communications

The techniques described in this chapter are used to convert data into and out of a form suitable for transmission between two devices via some medium. We will usually assume that we are dealing with wireless communication using radio waves, but the ideas are more generally applicable, for example to analogue communication over telephone lines.

Two aspects of the problem make it challenging: first, any given region of space is shared by many simultaneous communications, and so we need a way to keep them separate so that they do not interfere with one another; and second, there are practical difficulties associated with making a single transmitter that can efficiently generate radio waves over a wide range of frequencies.

We can see how to meet these two constraints if we think in terms of the frequency domain. If the transmitter can only operate over a narrow range of frequencies, that simply means that the signal we use for communication should only occupy a narrow band in the frequency domain. This conveniently also gives us a way to allow multiple simultaneous communications: we simply divide a range of frequencies into a number of smaller ranges and allocate them to the various transmitters. A receiver can pick out the transmission it wishes to listen to by applying a suitably designed band-pass filter, as shown in Figure 11.1.

Normally a relatively wide range of frequencies, called a *band*, is allocated to a particular purpose: for example, in some countries the frequencies between approximately 470 MHz and 862 MHz are used for television broadcasting, while the range between 88 MHz and 108 MHz is used for radio. The band is further divided into narrower frequency ranges called *channels*, usually regularly spaced over the band. The frequency range covered by each channel is called the *channel width*. In the example of broadcast television above the channel width is typically 8 MHz and so $(862-470)/8 = 49$ channels are available.

A scheme like this depends upon cooperation between the operators of the transmitters so that they do not accidentally transmit a signal containing frequency

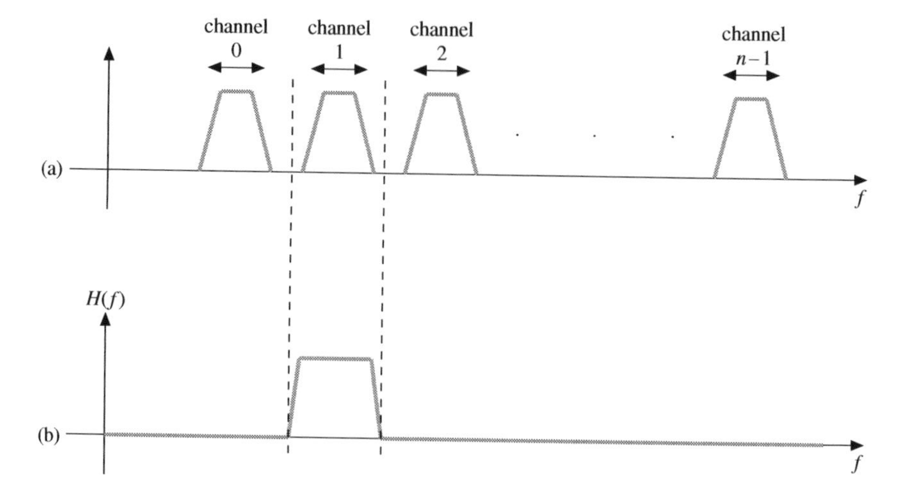

Figure 11.1 Dividing a frequency band into channels: (a) example channel allocation; (b) response of band-pass filter to select channel 1.

components in someone else's allocated channel, creating *crosstalk* or *interference*. It is for this reason that radio transmitters generally require regulatory approval before they can be used.

Dividing a band into channels as illustrated in Figure 11.1 is one way to avoid conflict between various transmitters working in that band: it is not the only possible method, as we shall see later.

The process of converting an ordinary analogue or digital signal, such as an audio waveform or a compressed stream of bits from an MPEG encoder, into a narrow-bandwidth signal suitable for transmission within a channel is called *modulation*; the reverse process, carried out at the receiver, is called *demodulation*.

11.1 Amplitude modulation

Amplitude modulation, or AM, is a simple scheme for transmitting information within a guaranteed channel width. The information to be transmitted must be represented by a positive real signal with limited bandwidth, such as that shown in Figure 11.2(a). Here the maximum frequency component of the signal is at f. Note that the magnitude spectrum is symmetric about zero frequency because the signal is real: more precisely, each negative frequency component is the complex conjugate of its positive partner.

This real signal is simply multiplied by a cosine wave called the *carrier* whose frequency f_c is at the centre of the desired channel: the result of this operation,

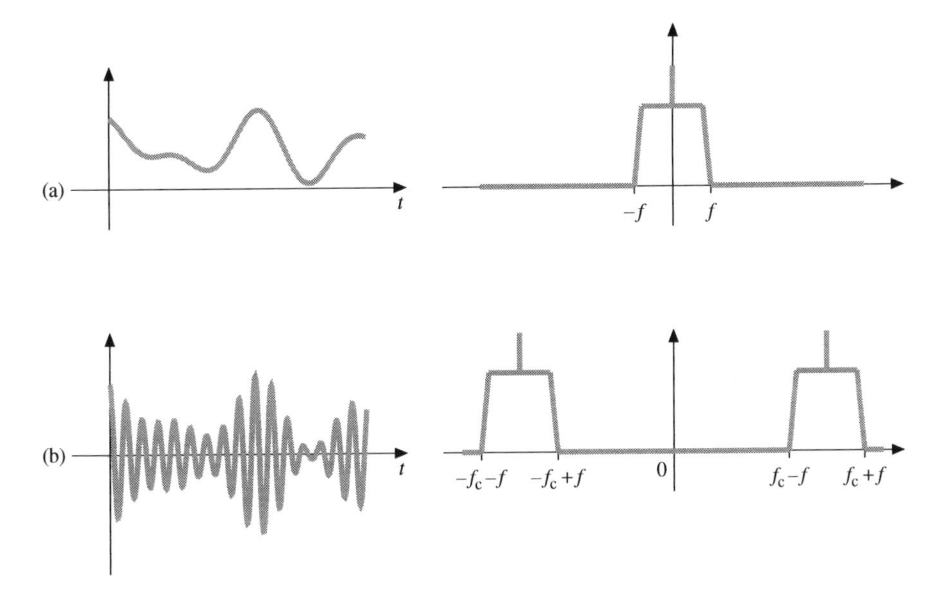

Figure 11.2 Amplitude modulation: (a) original signal and its spectrum; (b) modulated signal and its spectrum.

Figure 11.2(b), is clearly another real signal. Since

$$\cos 2\pi f_c t = \frac{1}{2}(e^{2\pi j f_c t} + e^{-2\pi j f_c t})$$

and, as we saw in Chapter 4, multiplying a signal by $e^{2\pi j f_c t}$ shifts its spectrum up by f_c, the resulting signal will be the sum of two shifted copies of the original signal. The first copy will be shifted up by f_c and the second copy will be shifted down by f_c, as shown in Figure 11.2(b). The result is again symmetric about zero frequency, which is consistent with the signal being real. (In this chapter, all the modulated signals will be real and from now on we shall therefore not show the negative frequency components of modulated signals.)

In radio terminology, the multiplication operation (or the approximation to it in analogue electronic implementations) is called *mixing*, and the multiplier a *mixer*; the process of mixing a signal with a fixed-frequency sine or cosine wave to shift its spectrum is called *heterodyning*. Shifting the spectrum upwards to higher frequencies is called *upconversion*; shifting the spectrum downwards is called *downconversion*.

A receiver for AM signals can be very simple: Figure 11.3 shows a design which, in its most basic implementation in analogue electronics, is called a 'crystal set'. First we apply a band-pass filter designed to pass frequencies in the wanted channel and reject signals we may have received in other channels. The centre

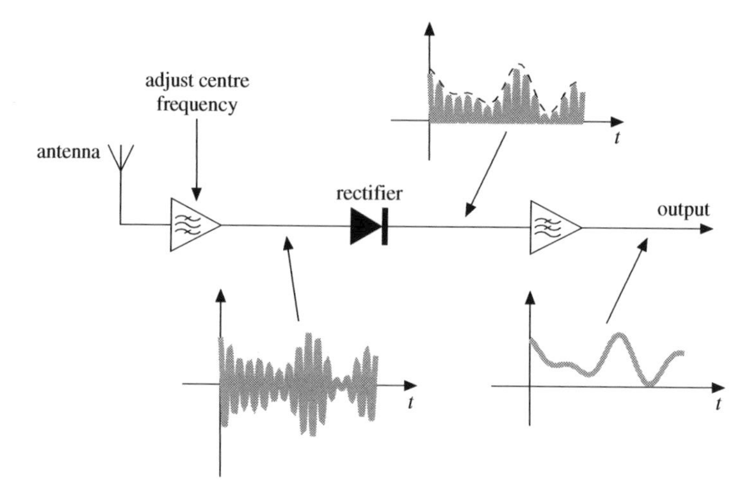

Figure 11.3 Simple AM demodulator.

frequency of the band-pass filter can be adjusted or *tuned* to select different channels.

In ideal conditions, the output of this filter will be the modulated signal shown in Figure 11.2(b). The *envelope* of this output, i.e., the curve joining its peaks, has the same shape as the original signal. Unfortunately, if we simply apply a low-pass filter to try to extract this envelope the result will be zero because the signal has no low-frequency components. Instead we first pass this signal through a rectifier (the 'crystal' in a crystal set) represented in the figure by a diode symbol. This sets to zero any negative parts of the signal, a process called *rectification*. We can now use a low-pass filter to smooth the rectified signal and recover its envelope. This is the original signal.

The disadvantage of this receiver design is that it is difficult, using analogue electronics at least, to make a band-pass filter both which has a sharp cut-off and whose centre frequency can be adjusted so that different channels can be received.

An approach that avoids this problem is shown in Figure 11.4: this is called a *superheterodyne* receiver. Suppose that we wish to receive channels in a band running from 3 MHz to 4 MHz, and that the channel width is 100 kHz: there are ten channels, as shown in Figure 11.5(a). The shaded area in the figure indicates the frequencies outside our band, and the arrow indicates the channel we are interested in: its centre frequency is at 3.65 MHz. We first apply a band-pass filter to the signal which covers the entire band of interest, not just one specific channel, as shown in Figure 11.5(b). The filter need not have a very sharp response. The result is shown in Figure 11.5(c).

The output of the band-pass filter is now multiplied by a cosine wave produced by an oscillator (the *local oscillator* or *LO*) in the receiver whose frequency can

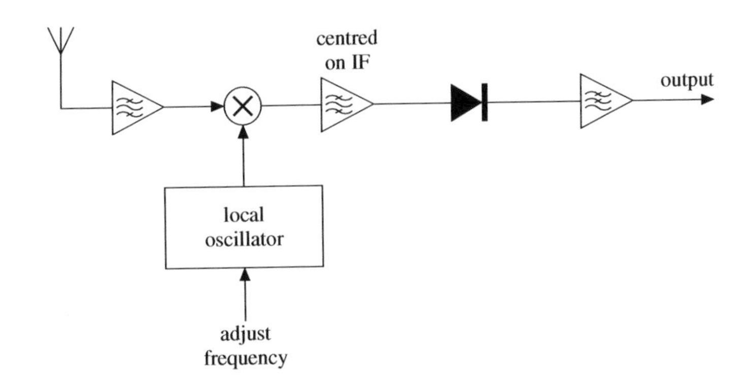

Figure 11.4 Simple superheterodyne AM demodulator.

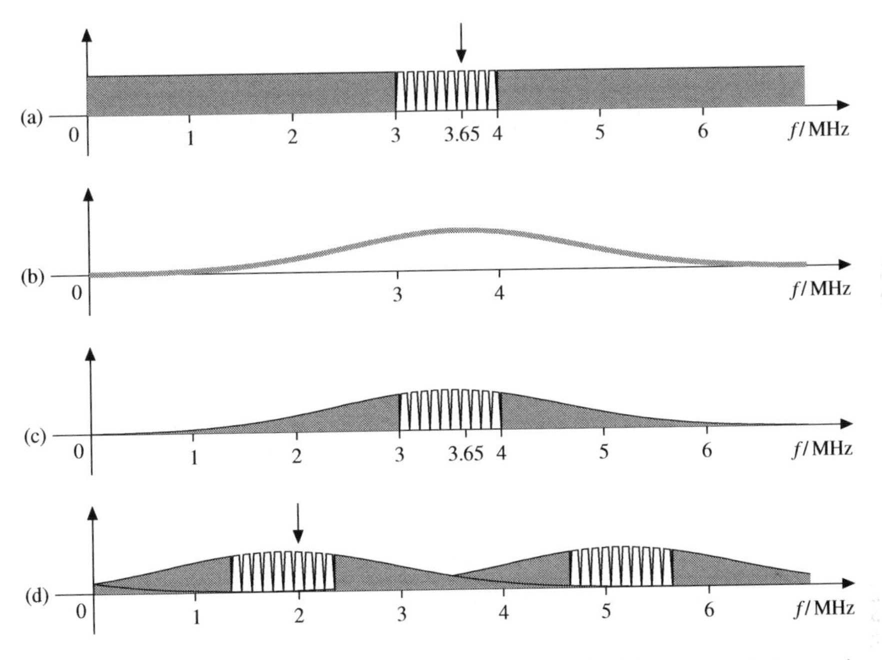

Figure 11.5 Superheterodyning: (a) channels in band of interest and (in grey) out-of-band frequencies; (b) first band-pass filter response; (c) result of band-pass filtering; (d) output of mixer (IF = 2 MHz).

be adjusted. As we described above, this produces two copies of the spectrum of the signal, shifted up and down by an amount equal to the frequency of the oscillator. In this example we adjust the frequency of the oscillator to bring the centre frequency of the wanted channel to exactly 2 MHz, which is the centre frequency of the second band-pass filter. This is called the *intermediate frequency*, or *IF*. Most superheterodyne receivers use a fixed intermediate frequency.

The frequency shift required is 3.65 MHz − 2 MHz = 1.65 MHz. However, the mixing process produces *two* copies of the original spectrum: the other will be located exactly 1.65 MHz higher than the original. The situation is illustrated in Figure 11.5(d). The resulting spectrum may look a mess, but the important point is that at 2 MHz we have a copy of the channel we are interested in, with no interference at that frequency from other channels, either originally inside our band or originally outside it. Notice that a small amount of the lower tail of the shifted-down copy of the spectrum extends into negative frequencies. Since the band-pass filter output is real, these frequencies are indistinguishable from their positive counterparts, and so they appear in Figure 11.5(d) as a reflection extending upwards from zero frequency.

The rest of the superheterodyne receiver is exactly the same as the simple design of Figure 11.3, except that the band-pass filter only has to work at a fixed frequency, and so a sharp response is much easier to obtain in practice.

When building a superheterodyne receiver care is needed in the design of the first band-pass filter and in the selection of the intermediate frequency. Figure 11.6 shows the same situation as Figure 11.5, but with an intermediate frequency of 3 MHz: as you can see, the lower tail of the shifted-up copy of the spectrum in Figure 11.6(d) interferes with the signal around the intermediate frequency.

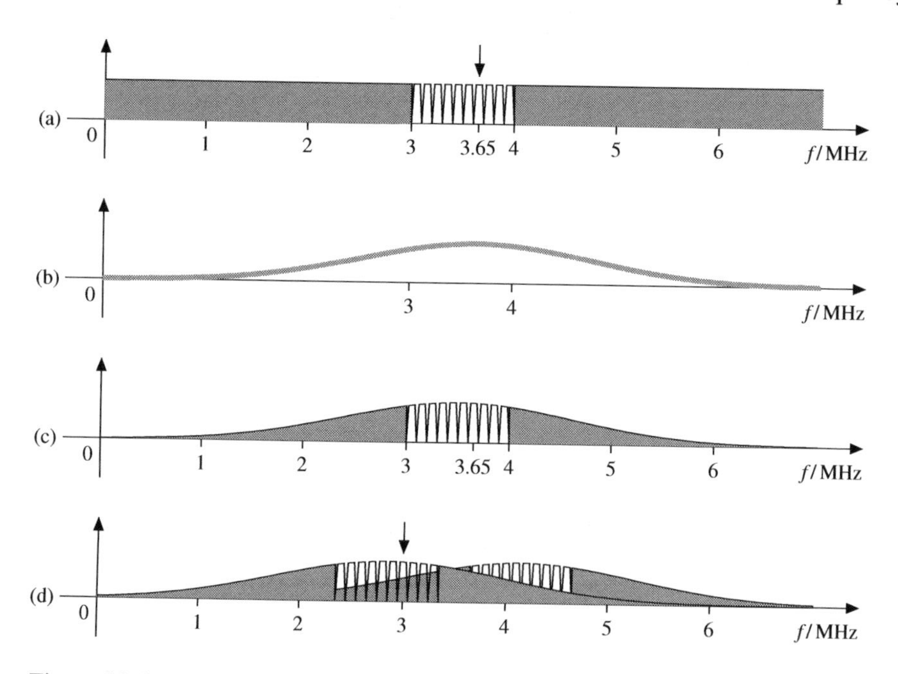

Figure 11.6 Superheterodyning with an unsuitably high choice of intermediate frequency: (a) channels in band of interest and (in grey) out-of-band frequencies; (b) first band-pass filter response; (c) result of band-pass filtering; (d) output of mixer (IF = 3 MHz).

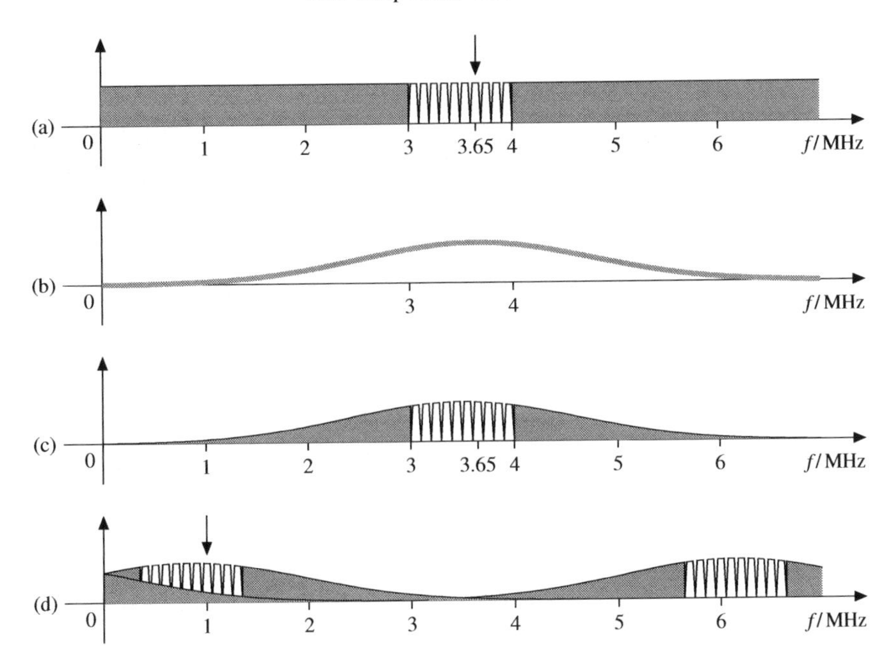

Figure 11.7 Superheterodyning with an unsuitably low choice of intermediate frequency: (a) channels in band of interest and (in grey) out-of-band frequencies; (b) first band-pass filter response; (c) result of band-pass filtering; (d) output of mixer (IF = 1 MHz).

Figure 11.7 shows the situation of Figure 11.5, but with an intermediate frequency of 1 MHz. This time it is the lower tail of the shifted-down copy of the spectrum that causes the problem: it extends sufficiently far into the negative frequencies that its reflection produces interference around the intermediate frequency.

The commonest arrangement in digital communications systems using the more sophisticated modulation schemes we will describe below is to implement the parts of Figure 11.4 up to and including the second band-pass filter using analogue electronics. The output of this filter is then fed to a digital-to-analogue converter and subsequent processing done digitally. If the second band-pass filter has a sharp response then its output only occupies a narrow band of frequencies. It is then possible to use sub-Nyquist sampling (see Chapter 2), which allows the digital-to-analogue converter to operate at a lower sample rate than would otherwise be necessary.

Applications of amplitude modulation

Amplitude modulation, with slight variations, is used for analogue radio communications on ‘long wave’, ‘medium wave’ and ‘short wave’, with carrier frequencies up to around 30 MHz.

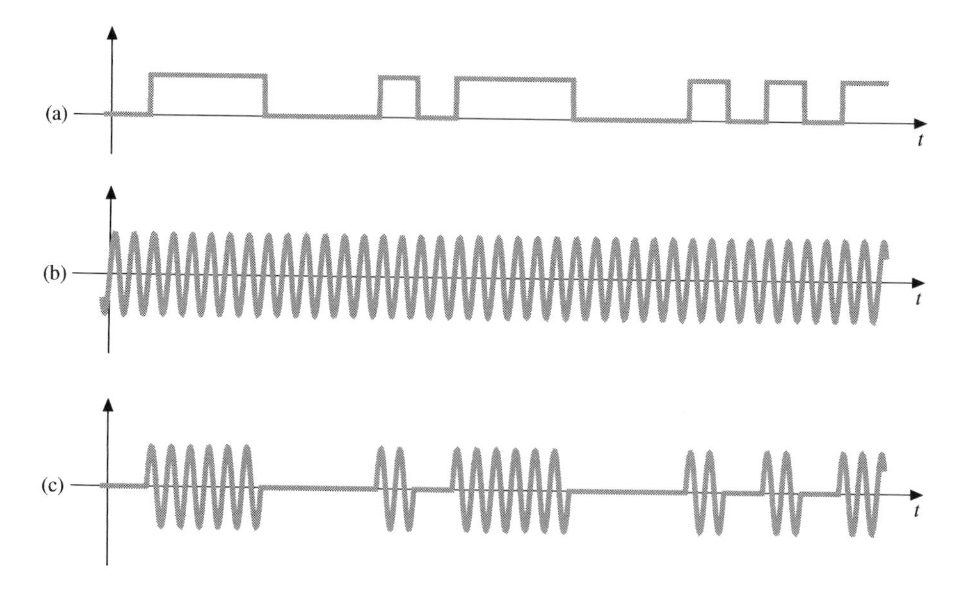

Figure 11.8 On–off keying: (a) digital signal to be sent; (b) carrier wave; (c) modulated result.

It is also used for digital transmissions in simple applications, often using *on–off keying*, where the carrier wave is simply switched on and off by a series of pulses, much as in Morse code: see Figure 11.8. In this example a '0' bit is represented by a short pulse followed by a short gap, a '1' bit by a long pulse followed by a long gap. The receiver simply has to detect the presence or absence of the carrier wave, and then measure the lengths of the pulses. This and other simple codes like it are used in low-cost short-range radio devices such as garage door openers, remote locking devices for cars, wireless outdoor thermometers and wireless doorbells. Modulation and demodulation is almost invariably done with analogue electronics, while the encoding and decoding of messages is done by digital means, often using a microcontroller.

A disadvantage of amplitude modulation, and especially on–off keying, is that the amplitude of the received signal depends not only on the data being sent, but also on how far away the transmitter is. An automatic gain control system, perhaps along the lines explored in Exercise 7.7, is usually required.

11.2 Frequency modulation

In a frequency modulation, or FM, transmitter we again have a carrier oscillator. This time, however, it is the frequency of its output, rather than the amplitude of its output, that changes in sympathy with the signal being transmitted. The

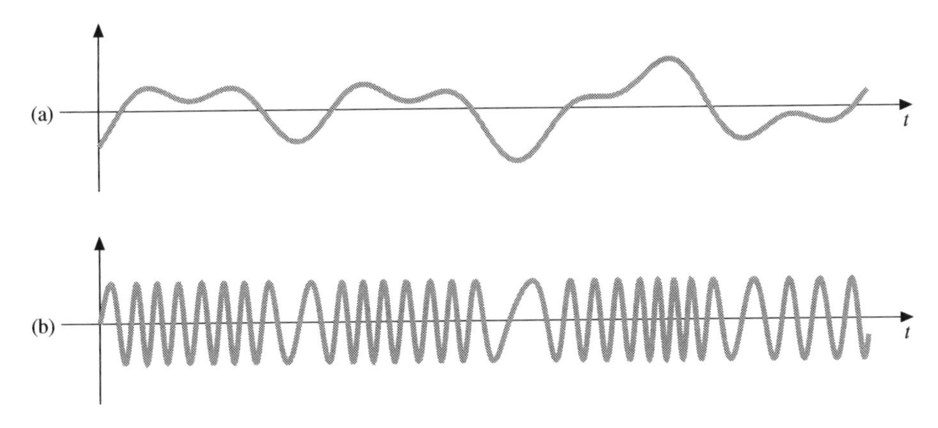

Figure 11.9 Frequency modulation: (a) original signal; (b) modulated result.

resulting signal is shown in Figure 11.9: as you can see, the oscillations are faster when the signal is positive and slower when the signal is negative.

The spectrum of a frequency-modulated signal is complicated to describe mathematically, and we shall not derive the formulae here. We shall just say that the spectrum spreads out to a considerable extent on either side of the carrier frequency, and so a transmitter usually includes a band-pass filter to ensure that no stray frequency components are generated outside the allocated channel. The ease with which the signal can be generated, coupled with the complexity of its spectrum, has led to the widespread use of FM synthesis in low-cost electronic musical instruments to produce 'interesting-sounding' waveforms.

FM receivers normally use heterodyne techniques to shift the carrier frequency to a known fixed IF value and then filter out other channels, in just the same way as for AM reception above. There are then several ways to proceed in order to convert the varying frequency back into the original signal.

A simple method is to pass the signal through a filter whose response rises linearly in the region around the IF as shown in Figure 11.10. The signal at the output of the filter will still be modulated in frequency, but will also now be modulated in amplitude as a result of the filter characteristic. This signal can now be treated as if it were just an AM signal using a rectifier and low-pass filter, as above.

The phase-locked loop

A more sophisticated approach is to try to track the changes in the signal frequency directly using a device called a *phase-locked loop*, or *PLL*: see Figure 11.11. The input to the phase-locked loop is a sine wave varying in frequency, such as a (possibly heterodyned) FM signal. The phase-locked loop contains an oscillator

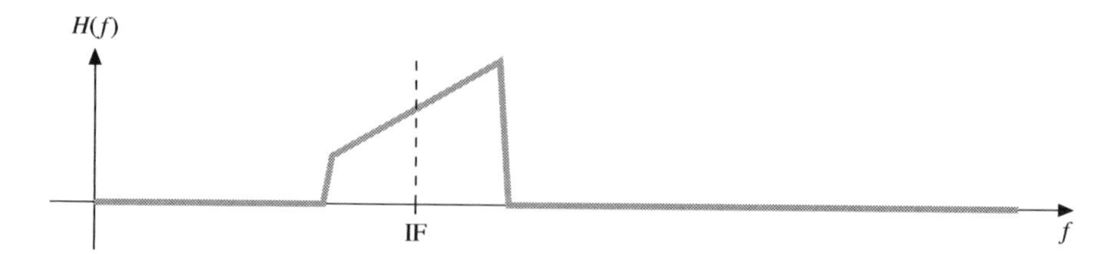

Figure 11.10 Response of a filter for demodulating FM.

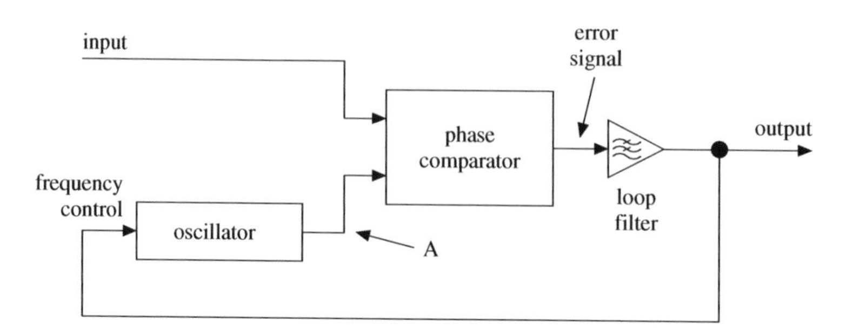

Figure 11.11 FM demodulator using a phase-locked loop.

and a control system whose objective is to make the output of the oscillator, marked A in the diagram, exactly match the input to the phase-locked loop. When this is the case, the phase-locked loop is said to be *locked*.

The control system works by comparing the oscillator output with the phase-locked loop input in a *phase comparator*, which is a device that produces a signal that depends on the phase relationship between its inputs. If the input signal is ahead of the signal coming from the oscillator, the output of the phase comparator will be positive; if the input signal lags behind the oscillator, the phase comparator's output will be negative. The output of the phase comparator is passed through a low-pass filter to produce the *error signal*, a smoothed estimate of the phase error between input and oscillator. This is then fed back to the frequency control input of the oscillator to correct its frequency until it is in phase with the input.

Great care is needed in setting the parameters of the loop filter and of the phase comparator, for which there are several different designs to choose from. A compromise has to be made among the range of frequencies to which the loop can lock, its speed of response to sudden changes in input frequency, its accuracy at tracking smooth changes in input frequency, and its sensitivity to input noise. A discussion of these issues is outside the scope of this book: you are recommended to investigate the various software packages that are available to help with phased-locked loop design.

11.3 Quadrature amplitude modulation

As we said above, the spectrum of the AM signal shown in Figure 11.2(b) is symmetric about the carrier frequency because the modulating signal is real: the component at a frequency $f_c + f$ is the complex conjugate of the component at frequency $f_c - f$. There is no particular reason why the modulated signal should have this symmetry property: we can set the frequency components to whatever values we like as long as we keep within our allocated channel.

We can generate an asymmetric signal by a method similar to the amplitude modulation scheme above. Suppose that the signal we wish to generate has the spectrum shown in Figure 11.12(a). First we construct a version of the desired signal centred around zero frequency rather than around the carrier frequency, as shown in Figure 11.12(b): this is called the *baseband* signal and is analogous to the modulating signal in the AM case. We then shift it up in frequency to the desired point.

Since the spectrum in Figure 11.12(b) is not symmetrical about zero frequency, it will correspond to a complex, rather than a real, signal. We can write the baseband signal as $z(t) = x(t) + \mathrm{j}y(t)$ where $x(t)$, the real part of the baseband signal, is called the *in-phase*, or *I*, component, and $y(t)$, the imaginary part of the baseband signal, is called the *quadrature*, or *Q*, component.

In the situation where the baseband signal was real we shifted its spectrum by multiplying it by a cosine wave. The analogous process to shift a complex baseband signal up to the desired channel is to multiply it by a complex exponential:

$$u(t) = z(t)\mathrm{e}^{2\pi \mathrm{j} f_c t}.$$

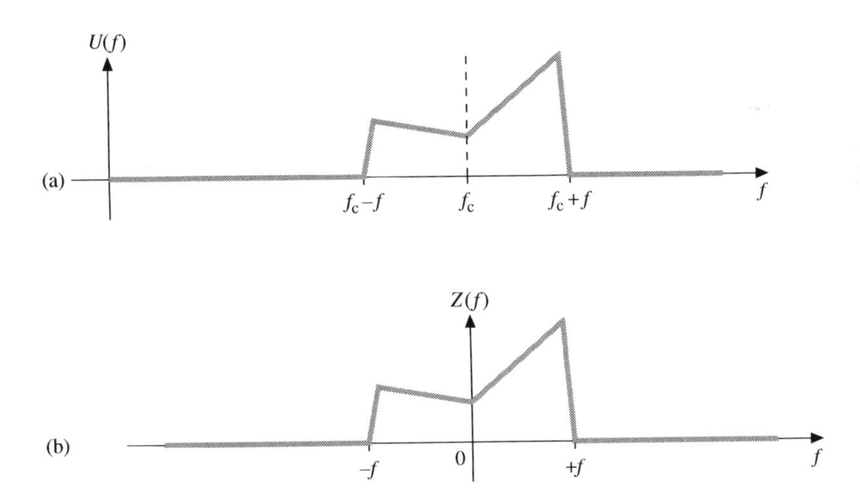

Figure 11.12 AM with asymmetric spectrum: (a) transmitted signal; (b) corresponding baseband signal.

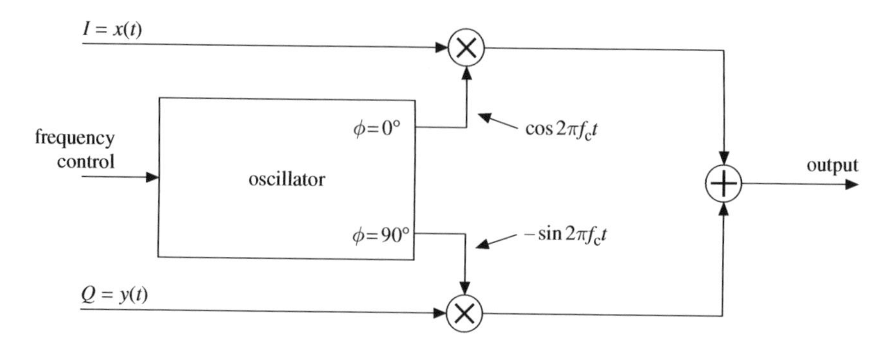

Figure 11.13 QAM modulator.

The complex signal $u(t)$ has frequency components only between $f_c - f$ and $f_c + f$ and has the spectrum shown in Figure 11.12(a). Taking the real part of $u(t)$ leaves the shape of this part of the spectrum unchanged (the effect is just to add aliases at the corresponding negative frequencies) and gives us a signal we can transmit.

Now,

$$
\begin{aligned}
\operatorname{Re} u(t) &= \operatorname{Re} z(t) e^{2\pi j f_c t} \\
&= \operatorname{Re} (x(t) + jy(t))(\cos 2\pi j f_c t + j \sin 2\pi j f_c t) \\
&= x(t) \cos 2\pi j f_c t - y(t) \sin 2\pi j f_c t
\end{aligned}
$$

and so the modulator looks like the diagram in Figure 11.13. The modulator produces the sum of two modulated carriers, $\cos 2\pi j f_c t$ and $-\sin 2\pi j f_c t$. These carriers are 90°, or one quarter of a revolution, out of phase with respect to one another: we say that the carriers are *in quadrature*. This type of modulator is hence called a *quadrature amplitude modulator*, and the modulation scheme *quadrature amplitude modulation*, or *QAM*.

Analogue broadcast television in most countries uses a QAM system to transmit chrominance information in a separate channel alongside the luminance information. The two colour difference, or chrominance, signals (see Section 9.1) are used to modulate the sine and cosine carriers: the schemes used in different countries vary in their details.

Digital communications using QAM

The (complex) baseband signal and the carrier frequency together completely specify the signal within the channel. If the channel width is f, so that the channel extends from $f_c - f/2$ to $f_c + f/2$, the frequency components of the baseband signal will extend from $-f/2$ to $f/2$. We know from Chapter 2 that this baseband signal can be exactly represented by a series of complex samples taken

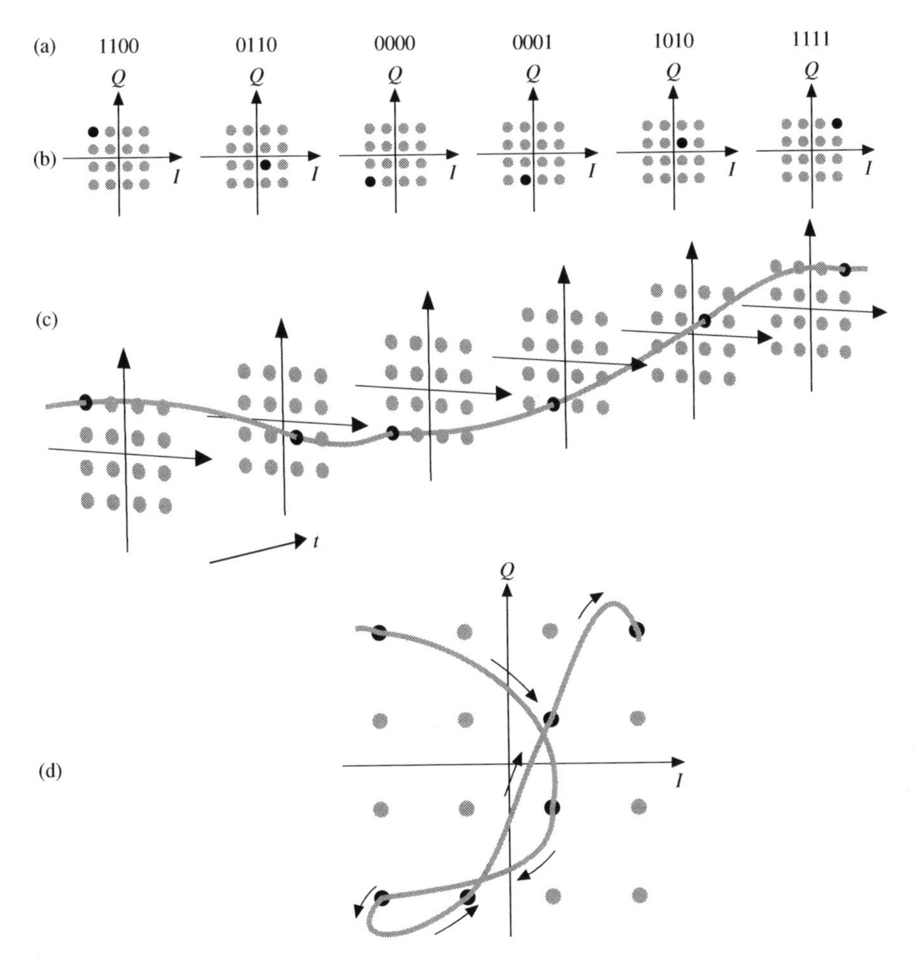

Figure 11.14 Transmitting a stream of bits using QAM: (a) original bitstream, divided into symbols of four bits each; (b) sequence of symbols mapped to constellation points in the complex plane; (c) continuous complex baseband signal reconstructed by interpolation; (d) trajectory of point in complex plane.

at frequency f. Thus the transmitted signal is completely specified by the carrier frequency along with the set of complex baseband samples.

This suggests a simple procedure for transmitting a stream of bits using QAM: see Figure 11.14. We first divide the stream of bits into short groups, say of four bits each. Each group is called a *symbol*. In this example, each symbol can take on one of $2^4 = 16$ different possible values. Each symbol is now mapped to a complex number (perhaps using a look-up table), and these complex numbers form the samples of our baseband signal. Figure 11.14(c) and (d) also show the reconstructed continuous baseband signal represented by these samples as a smooth grey curve.

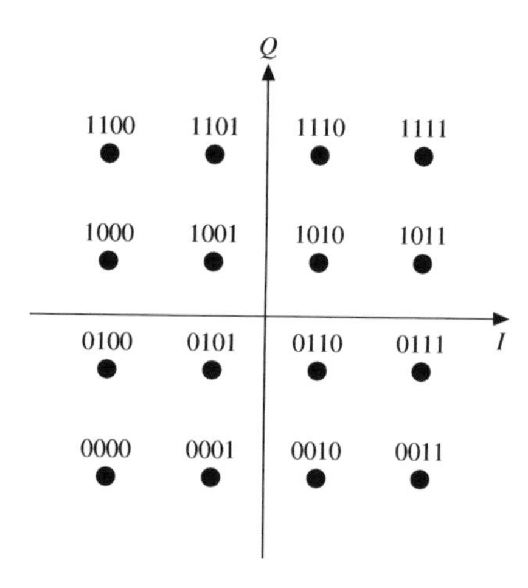

Figure 11.15 A constellation diagram for 16-QAM.

An example mapping between symbol values and complex numbers for the case where each symbol contains four bits is shown in Figure 11.15. A chart like this, showing the complex plane with points labelled by symbol values, is called a *constellation diagram*, presumably hinting at a resemblance to the stars of the night sky. For reasons we shall see later, the constellation points are typically laid out on a regular, usually square, grid.

QAM schemes are sometimes named after the number of constellation points, so this example would be called 16-QAM.

A digital QAM modulator

Figure 11.16 shows a block diagram of a digital QAM modulator, which simply carries out the steps described above in order. The bitstream to be transmitted enters at the left. It is first split into symbols, in this example of four bits each. Each symbol is then mapped to a constellation point in the complex plane, producing a complex signal sampled at the *symbol rate*, here one quarter of the input bit rate. This complex signal is upsampled to the desired final output sample rate for the system. Upsampling complex signals simply involves separately upsampling the real and imaginary parts using the methods described in Section 5.6. Notice that the real and imaginary upsamplers are identical and so the coefficients for the interpolation filters required in the two cases are the same.

Imperfections in the filters used for upsampling, usually arising as a result of constraints on the design, can lead to aliases in the resulting baseband signal with frequencies greater than half the symbol rate. In the next stage, these frequency components are shifted to frequencies outside the allocated channel. The

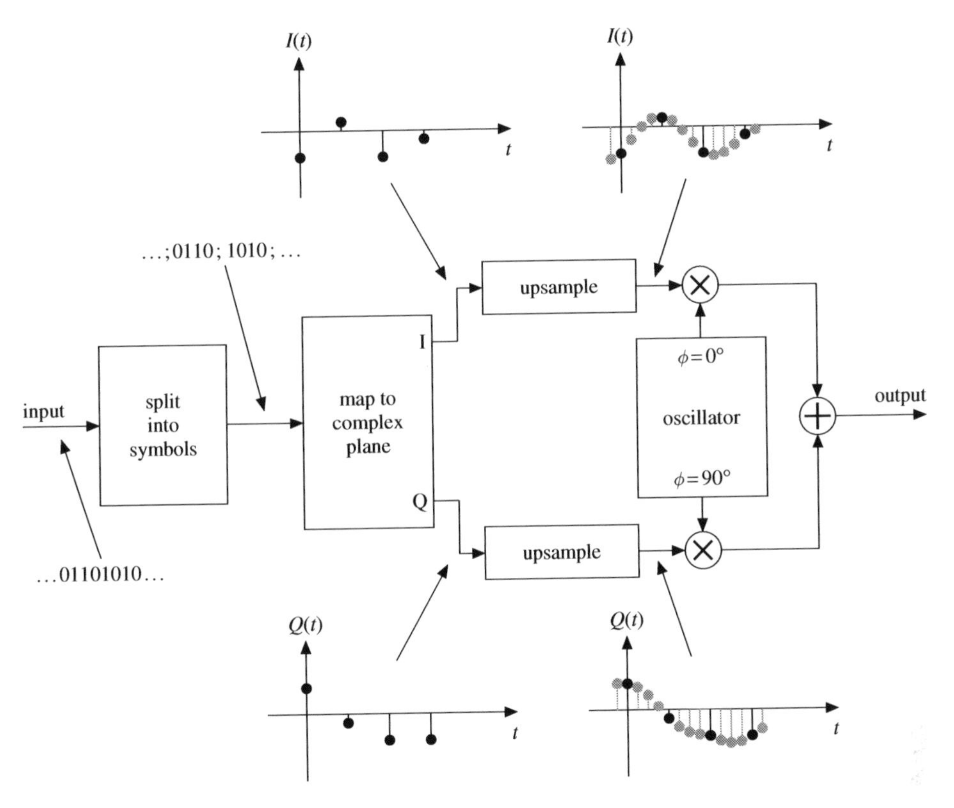

Figure 11.16 Digital transmission using QAM modulator.

interpolation filters must be carefully designed to keep the amplitude of these aliases to an acceptably low level to avoid causing interference on neighbouring channels. The impulse response of the interpolation filter is sometimes called the *pulse shape* and the filters themselves *pulse shaping filters*.

Certain designs of upsampling filters, in particular where the output sample rate is not an integer multiple of the input sample rate, can produce an output signal that does not go exactly through the input sample points: the value of one symbol has an effect on the value of the output at the time corresponding to a neighbouring symbol. This is called *inter-symbol interference*.

The final stage in the QAM modulator is to shift the spectrum of the upsampled complex signal so that it is centred on the desired carrier frequency using the method shown in Figure 11.13. Depending on the exact details of the design, it may be possible to make an optimisation by combining the multiplications by the outputs of the oscillator into the upsampling filters.

We have shown some detail of the implementation of the oscillator in Figure 11.17. At its heart is a register (or, in software implementations, a variable)

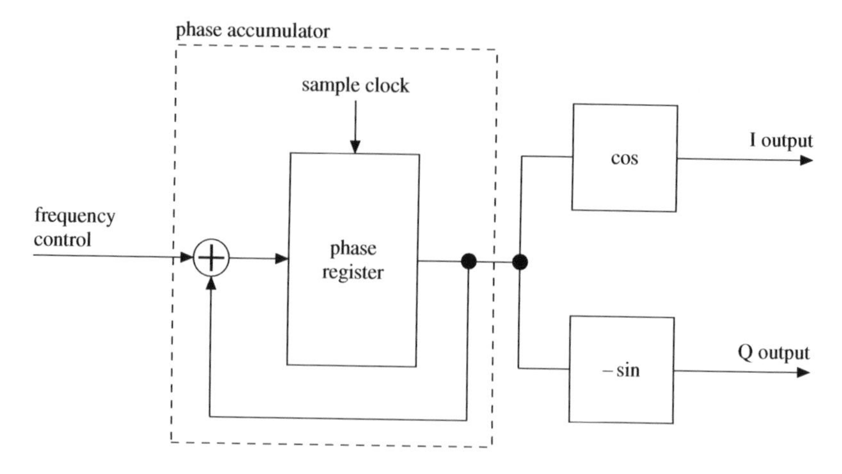

Figure 11.17 Digital oscillator implementation using a phase accumulator.

that stores the current phase ϕ of the oscillator output, which is a number in the range from 0 to 2π. At any time the outputs of the oscillator are $\cos\phi$ and $-\sin\phi$.

For each output sample to be generated the phase register is incremented by an amount that corresponds to the desired output frequency of the oscillator. For example, if it is desired that the output of the oscillator make a complete cycle every six samples, the amount added to the phase register for each sample would be $2\pi/6$. The phase register and adder together form what is called a *phase accumulator*.

It is usual, especially in fixed-point implementations, to store the phase variable as a fraction of a revolution rather than in radians. The 'modulo 2π' operation then becomes a simpler 'modulo 1' operation (which corresponds to allowing the phase accumulator to 'wrap around') and the phase increment value expresses the oscillator frequency as a fraction of its output sample rate.

The number of bits used to store the phase variable depends on the precision required in the output frequency. If the cosine and sine functions are calculated using look-up tables, it is often only necessary to drive them from a few of the most significant bits of the phase accumulator output: the error introduced by discarding the less significant bits at this stage is comparable to the effect of jitter described in Chapter 3.

The value of the phase increment can be changed dynamically in order to switch the modulator between different channels. If this function is not required, and if the output sample rate is an integer multiple of the channel centre frequency, the phase accumulator can be replaced with a simple counter.

Variations of QAM

There are many different modulation methods based on the general QAM scheme or which can be explained simply as variations upon it.

Binary phase-shift keying

The simplest possible version of QAM is where the constellation diagram has just two points, as shown in Figure 11.18. Each symbol consists of just one bit. If the bit is a zero, the corresponding constellation point is $+1$; if the bit is a one, the corresponding constellation point is -1. These two points correspond to transmitting the carrier signal or its negative; negating a cosine wave is equivalent to a phase shift (of 180°), and so this modulation scheme is called *binary phase-shift keying*, or *BPSK*. Since the symbol values are real, the baseband signal and its upsampled version are both also real. When making a BPSK modulator we can therefore dispense with the lower branch on the right of Figure 11.16, which deals with the imaginary part of the baseband signal.

Figure 11.19 shows an example stream of bits to be transmitted, the corresponding sequence of baseband samples, the interpolated baseband signal and the final modulated signal in two cases: on the left the interpolating filters are 'zero-order interpolators' (see Section 2.6), while on the right more accurate FIR interpolation filters have been used.

The final signal on the left shows the sudden phase inversions clearly. As we remarked in Chapter 4, signals that show discontinuities like this have spectra that extend over a wide range of frequencies. The modulated signal could thus cause interference in neighbouring channels.

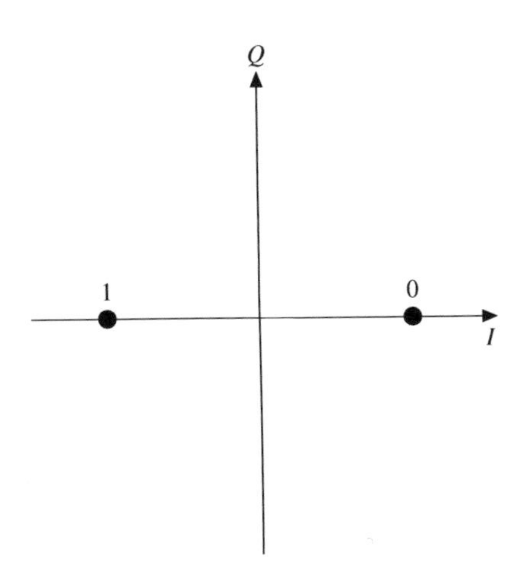

Figure 11.18 Constellation diagram for BPSK.

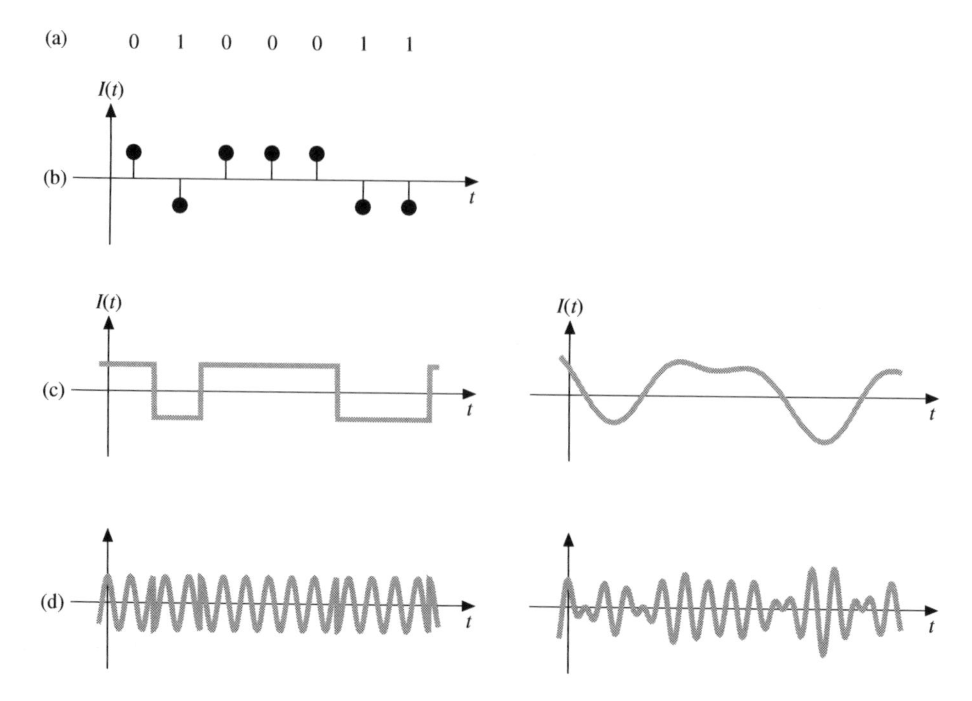

Figure 11.19 Transmitting a stream of bits using BPSK: (a) original bitstream; (b) (real) baseband signal; (c) interpolated baseband signal and (d) modulated signal using zero-order interpolation (left) and accurate interpolation (right).

The final signal on the right manages the phase inversions more smoothly, reducing the signal amplitude to zero at the point where the phase changes. If you imagine a point in the constellation diagram in Figure 11.18 moving smoothly from the symbol position at $+1$ to the symbol position at -1, you will see that it passes through the origin, which corresponds to a carrier amplitude of zero.

Quadrature phase-shift keying

If we wish to transmit two bits in each symbol, the constellation diagram will have $2^2 = 4$ points. The most obvious regular arrangement is as shown in Figure 11.20 (although some people prefer a version rotated through 45°), and QAM using such a constellation diagram is called *quadrature* (or *quaternary* or *quadri-*) *phase-shift keying*, or *QPSK*.

Figure 11.21 shows an example of QPSK modulation. The arrangement of the figure is similar to Figure 11.19: at the top is an example stream of bits to be transmitted with the corresponding sequence of symbols and baseband samples below it. On the left are the interpolated baseband signal and the final modulated signal where the interpolation is done using zero-order interpolators, while on the

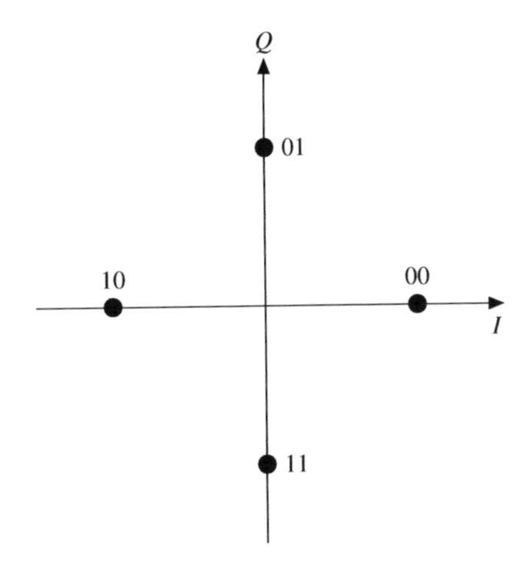

Figure 11.20 Constellation diagram for QPSK.

right we have used accurate FIR interpolation. Again, there is the potential for interference caused by imperfect interpolation, similar to the situation with BPSK.

Notice that it is again possible for the interpolated baseband signal to go through zero, and therefore for the carrier amplitude to become zero. Figure 11.22 shows the trajectory of the baseband signal from the right-hand example in Figure 11.21: the point in the complex plane corresponding to the signal at a given instant in time moves from constellation point to constellation point, tracing out smooth paths between them.

Continuous-phase frequency-shift keying

The trajectories shown in Figure 11.22 suggest an alternative method of generating an interpolated baseband signal: rather than allowing it to wander freely over the complex plane, we could consider constraining it to keep to the circle joining the four constellation points shown in Figure 11.23. The resulting modulation scheme is called *continuous-phase frequency-shift keying*, or *CPFSK*.

In CPFSK modulation the amplitude of the modulated signal remains constant, which helps certain types of radio transmitter, including those used on some satellites, produce less distortion in the spectrum of the transmitted signal.

Under this constraint we can represent the complex baseband signal at a given point in time by a single number, the angle $\phi(t)$ which is the argument of the signal. When we interpolate between two baseband samples in a general QAM system, the result usually approximates a straight-line path in the complex plane. For example, in moving from constellation point 01 to point 11 (see Figure 11.20) the trajectory would pass through (or at least near to) the origin. In the CPFSK case we have to decide on the route we wish to take around the circle from one

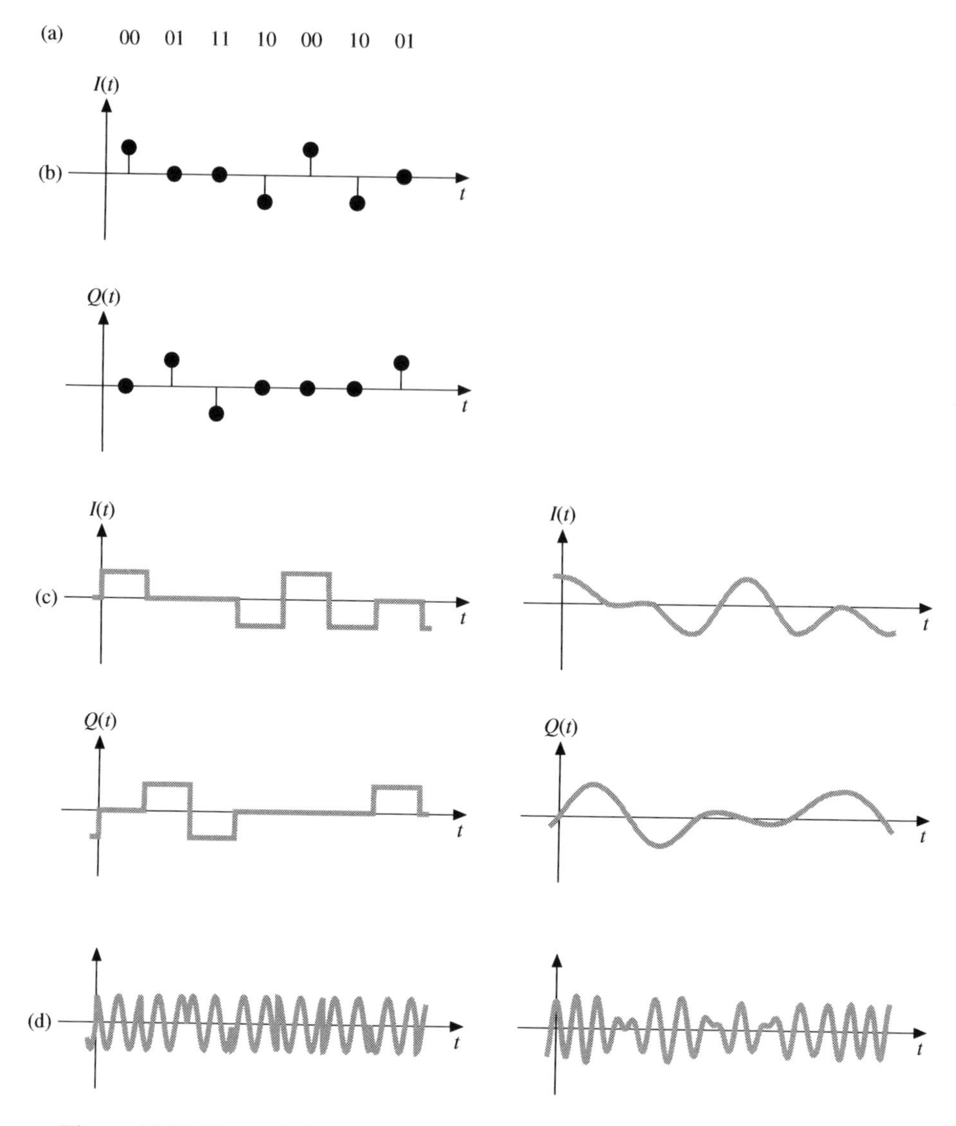

Figure 11.21 Transmitting a stream of bits using QPSK: (a) original bitstream divided into two-bit symbols; (b) complex baseband signal; (c) interpolated baseband signal and (d) modulated signal using zero-order interpolation (left) and accurate interpolation (right).

constellation point to another: we can go clockwise or anticlockwise. Different algorithms for making this choice as a function of the input symbol stream give rise to a range of CPFSK variants.

A further variant, called *minimum shift keying* or *MSK*, uses a constellation diagram with four points like the one shown in Figure 11.23 but only allows two types of transition: from a given constellation point the signal is only allowed

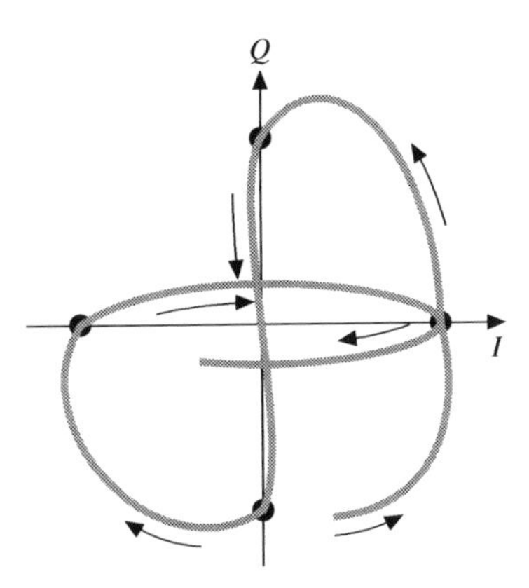

Figure 11.22 Trajectory of QPSK $I(t)$ and $Q(t)$ baseband signals with accurate interpolation.

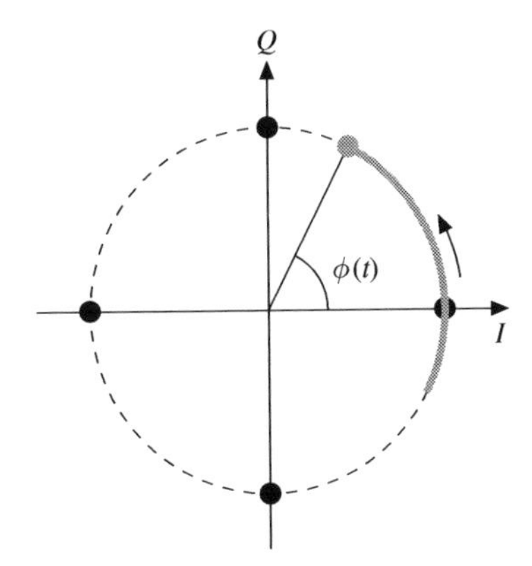

Figure 11.23 Trajectory of CPFSK $I(t)$ and $Q(t)$ baseband signals.

to move either to the clockwise neighbour or to the anticlockwise neighbour. In other words, the signal must undergo a phase shift of either $+90°$ or $-90°$ for each symbol. There are thus two possible symbols, which can encode one bit of information. Figure 11.24 shows the *phase trellis* for MSK, which is a plot of $\phi(t)$ against time showing all the possible trajectories starting from $\phi(0) = 0$. Movement up and to the right in the trellis corresponds to increasing $\phi(t)$ and thus to anticlockwise motion in Figure 11.23; movement down and to the right in the trellis corresponds to clockwise motion in Figure 11.23. Note that the phase is 'unwrapped' in the phase trellis so that we have shown $\phi(t) = 270°$ separately from $\phi(t) = -90°$ even though they represent the same point in the constellation diagram. The bold line shows the trajectory corresponding to the binary sequence 0010111.

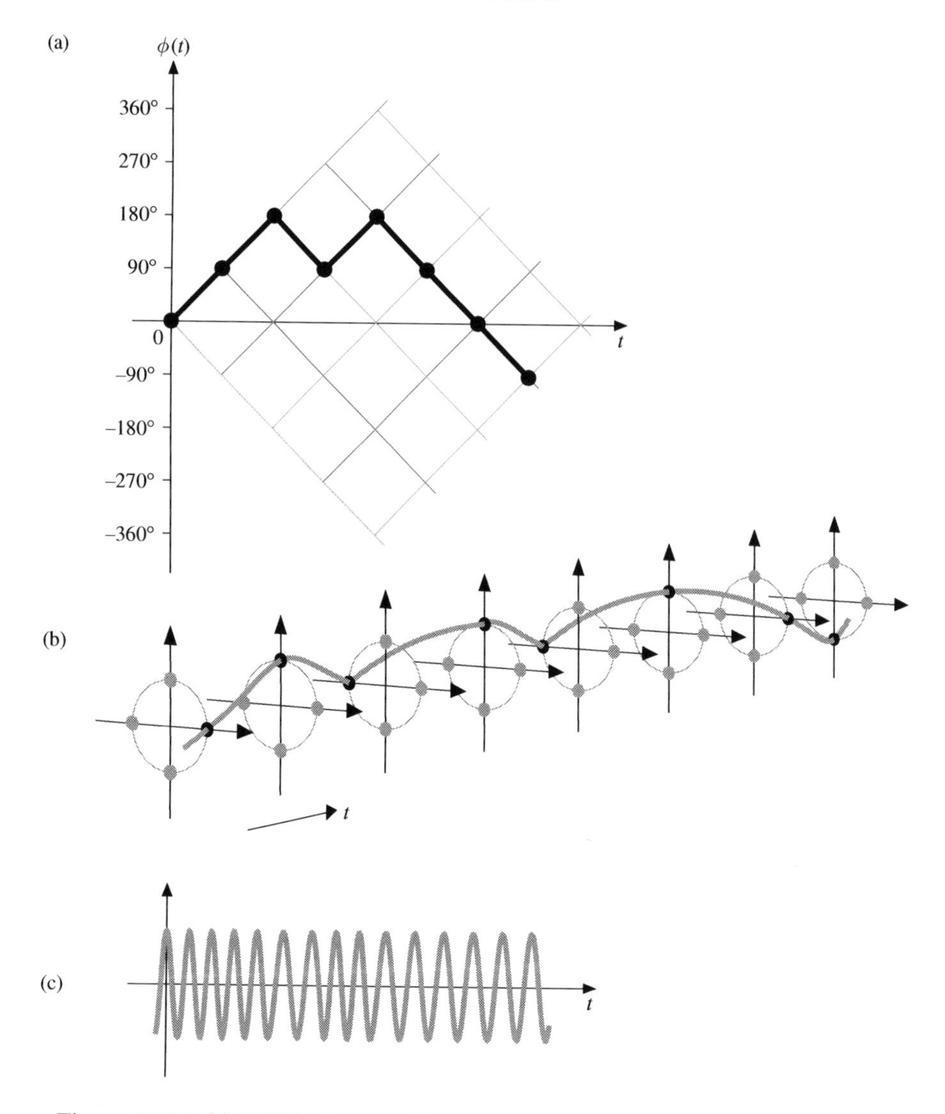

Figure 11.24 (a) MSK phase trellis with example trajectory shown in bold; (b) corresponding baseband signal; (c) modulated signal.

In practice the phase trajectory is filtered to make it smoother. The GSM (Groupe Spécial Mobile or Global System for Mobile Communications) mobile telephony standard uses a variant of this scheme called *Gaussian minimum shift keying* or *GMSK*.

Demodulating QAM signals

In principle it is simple to demodulate a QAM signal: we simply reverse the steps carried out in the modulator of Figure 11.16, giving the design shown in

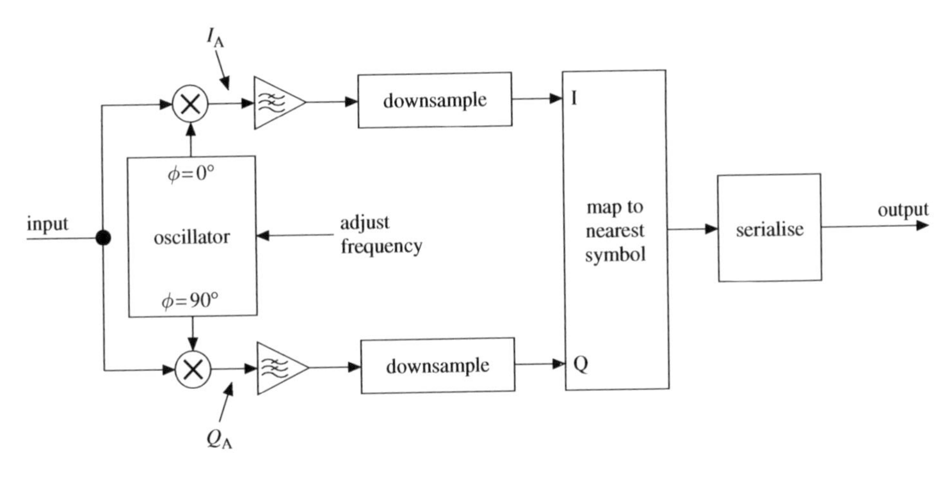

Figure 11.25 Basic principle of QAM demodulator.

Figure 11.25. The first section of the design shifts the spectrum of the signal down so that it is centred about zero frequency, thus recreating the baseband signal. We do this by multiplying the (real-valued) signal by a complex exponential $e^{-2\pi j f_c t}$, where f_c is the carrier frequency. Note the similarity of this step to the technique of heterodyning described above: the difference is that, because the result here is centred around zero frequency, the output of this stage must be complex-valued in order to prevent aliases between negative frequencies and positive ones. This technique of heterodyning with a zero intermediate frequency is called *homodyne detection*.

We now apply a low-pass filter to the baseband signal we have recreated to remove any frequency components that we might have received that are not in our channel of interest. The result can be safely downsampled to the correct symbol rate. We can now match up the complex samples of the baseband signal with the constellation diagram and decide which symbol was being transmitted at each sample time. The regular arrangement of the constellation diagram simplifies this decision. Finally we can reassemble the sequence of symbols into a stream of bits.

Unfortunately in practice things are not quite as simple as this. There are two main obstacles to implementing the above scheme.

The first obstacle is that the signal received by the demodulator will be corrupted by noise and interference. If we have a model for the noise and interference we can calculate the probability that any given baseband signal sample arose from any given constellation point, thus associating a set of probabilities with each baseband sample.

Simple demodulators just find the point in the constellation diagram nearest to the complex baseband signal sample and report that as the decoded symbol. This corresponds to using an additive white Gaussian noise model (see Section 6.3). As we saw in Chapter 6, the Gaussian model is not always an appropriate one to use, and of course more sophisticated receivers use more advanced models.

When a communications system is combined with an error-correction coding scheme it is common for the demodulator to output its set of probability estimates for each baseband sample rather than a single hard decision. These probabilities are then combined in the error-correcting decoder to determine the most likely original data stream. The algorithms used depend on the exact nature of the error-correction code and are outside the scope of this book.

The second obstacle is that the receiver is usually a separate device from the transmitter and so the two units do not have a common frequency reference: we say that the demodulation is *noncoherent*. This gives rise to three sub-problems: the receiver does not know in advance the correct times at which to sample the baseband signal in order to extract the sequence of symbols; the receiver does not know in advance the exact carrier frequency; and the receiver does not know the carrier phase.

Solving the first of these sub-problems is called *symbol synchronisation*; solving the second is called *carrier lock*; and the third sub-problem is solved using a technique called differential coding, which we shall describe later.

The presence of noise and interference on the received signal means that any techniques we use to try to solve these synchronisation problems must be robust.

Carrier lock

The IEEE 802.11 and IEEE 802.11b standards for wireless data transmission ('wireless Ethernet', or 'WiFi') specify that the transmitter must produce a carrier frequency with a maximum relative error compared to the nominal value (which is typically around 2.5 GHz, depending on the channel used) of $\pm 25 \times 10^{-6}$, or 25 ppm (parts per million). If we assume a similar tolerance in the local oscillator at the receiver, the transmitter may appear, as far as the receiver is concerned, to be operating at a frequency as much as ± 50 ppm away from its nominal value. This corresponds to an absolute frequency error of $\pm 50 \times 10^{-6} \times 2.5\,\text{GHz} = \pm 125\,\text{kHz}$.

A further contribution to apparent relative error in carrier frequency at the receiver can arise from the Doppler effect. If transmitter and receiver are moving towards or away from one another at a velocity v, the relative error introduced into the carrier frequency is equal to v/c, where $c = 3 \times 10^8\,\text{m/s}$ is the speed of light. For example, a receiver in a car moving at $30\,\text{m/s}$ ($108\,\text{km/h}$) will see a relative carrier frequency error of $30/(3 \times 10^8) = 10^{-7}$, or 250 Hz in 2.5 GHz.

Although the effect is probably negligible in this case, it can become an important factor in satellite communications.

Note that both the carrier frequency and the symbol rate are subject to error in this way; in some systems the carrier and symbol rate clocks are derived from the same master source in the transmitter (IEEE 802.11 includes this as an option), but we shall not discuss how this fact might be exploited in a receiver.

If we use a receiver design like the one shown in Figure 11.25, any error in the carrier frequency of the received signal will result in the shifted-down baseband signal we generate not being centred around zero frequency. Since the whole received spectrum is shifted by the same amount, *absolute* frequency error is preserved: in the example above, we would see the baseband spectrum offset by up to $\pm 125\,\text{kHz}$.

The frequency offset in the baseband signal is equivalent, as we have seen before, to its being multiplied by a complex exponential of the form $e^{2\pi j f_o t}$, where f_o is the frequency offset. Now, multiplying a signal by this complex exponential is the same as rotating it about the origin in the complex plane through an angle $2\pi f_o t$: in other words, the baseband signal at points I_A and Q_A in Figure 11.25 will look like the constellation diagram of Figure 11.22 but rotating slowly either clockwise or anticlockwise. If our receiver local oscillator frequency is too low, it will not shift the spectrum of the received signal down far enough, giving a positive frequency offset and anticlockwise rotation. If, on the other hand, our receiver local oscillator frequency is too high, it will shift the spectrum too far and the frequency offset will be negative and the rotation clockwise.

Figure 11.26(a) shows a plot of a short segment of a received BPSK signal that has been shifted to baseband with a small positive frequency offset. Each dot in the figure represents the complex value of one baseband sample; the lighter grey dots are samples earlier in the segment and the darker grey and black ones are later in the segment. The two constellation points can be seen (although they are not aligned with the real axis), as can some of the trajectories between the constellation points. The trails behind the constellation points arise from the rotation of the overall constellation pattern. Figure 11.26(b) shows a similar example for a QPSK signal. The fact that the points are scattered around the constellation points is a consequence of noise in the received signal.

We can detect the rotation of the baseband signal by continuously estimating the phase of the carrier. The only obstacle to this is that the phase of the signal we receive is constantly changing relative to the carrier as it is modulated by the symbol stream, and so we have to find a way to remove that element. For simplicity, we shall ignore the trajectories that the baseband signal takes between constellation points and assume that at any given time the signal is at one of the constellation points.

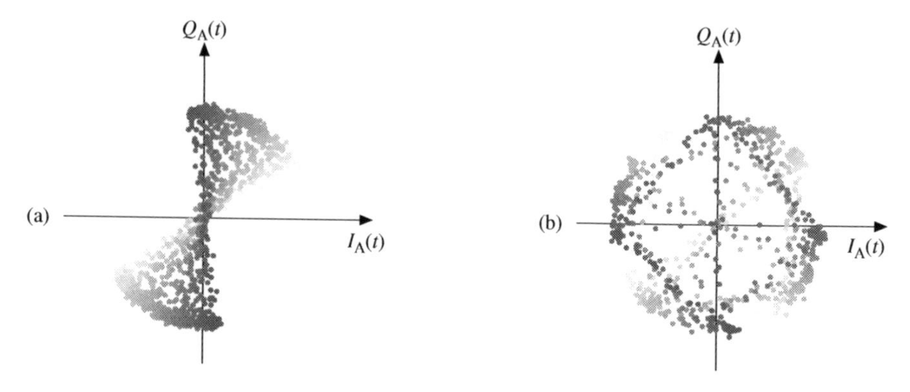

Figure 11.26 Trails of complex samples in a (a) BPSK and (b) QPSK receiver without carrier lock.

The Costas loop

Suppose that we are dealing with a BPSK signal and that the ideal constellation point at a given instant in time is u, which can be either $+1$ or -1. If the constellation is rotated anticlockwise through an angle θ because of carrier frequency error, the baseband sample we obtain will, ignoring noise, be $z = u\mathrm{e}^{\mathrm{j}\theta}$. Now, $u^2 = 1$ and so

$$\begin{aligned} z^2 &= u^2(\mathrm{e}^{\mathrm{j}\theta})^2 \\ &= (\mathrm{e}^{\mathrm{j}\theta})^2 \\ &= \mathrm{e}^{2\mathrm{j}\theta} \\ &= \cos 2\theta + \mathrm{j}\sin 2\theta. \end{aligned}$$

Thus z^2 is a complex number with magnitude 1 and argument 2θ. If θ is reasonably small $\sin 2\theta$ will be approximately equal to 2θ, or $\theta \approx \frac{1}{2}\sin 2\theta$. We can therefore estimate θ as half the imaginary part of the square of z, the baseband signal sample. If $z = x + \mathrm{j}y$,

$$\begin{aligned} \frac{1}{2}\mathrm{Im}\, z^2 &= \frac{1}{2}\mathrm{Im}\,(x+\mathrm{j}y)^2 \\ &= \frac{1}{2}\mathrm{Im}\,(x^2 + 2\mathrm{j}xy - y^2) \\ &= \frac{1}{2} \cdot 2xy \\ &= xy. \end{aligned}$$

In other words, we can estimate θ as the product of the real and imaginary parts of the baseband signal.

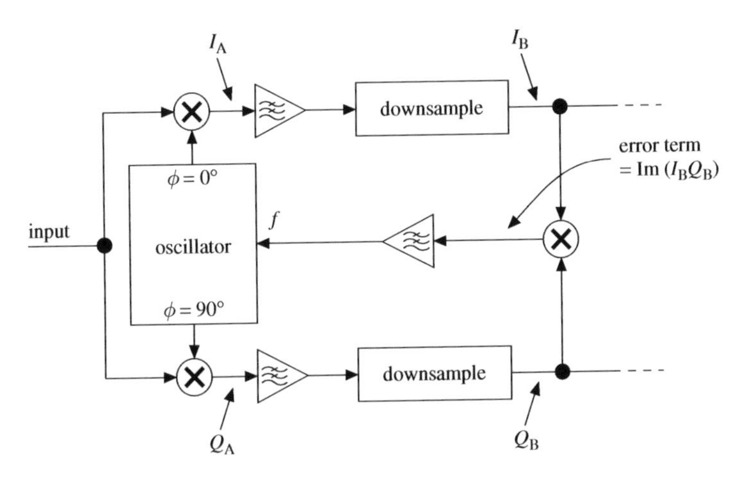

Figure 11.27 Costas loop for carrier lock.

We can use θ as an error signal in a feedback loop very similar to the phase-locked loop we discussed above. Since the error is derived directly from the signal, it will contain noise and a loop filter is required as before. Again, there are software packages available to help design a filter to suit the constraints of any given application.

Figure 11.27 shows how this feedback loop can be implemented: the filtered error signal is used to provide a fine adjustment to the local oscillator frequency. When the loop is locked the BPSK constellation points are aligned exactly with the real axis. Should they tend to rotate in either direction, the error signal will push the frequency of the local oscillator up or down slightly in order to bring them back into alignment. This method for achieving carrier lock is called a *Costas loop*.

Note that because the ideal BPSK constellation diagram looks identical when rotated through 180°, it is impossible for the above method (or, for that matter, any other carrier lock method) to distinguish between the case where the receiver local oscillator is exactly in phase with the carrier oscillator in the transmitter and the case where the oscillators are exactly 180° out of phase. In the latter case, the receiver would decode every bit incorrectly. The problem is solved by the use of differential coding, as we shall describe below.

A similar method to the Costas loop can be used for QPSK signals. In this case the ideal constellation points u are at $u = +1$, $u = +\mathrm{j}$, $u = -1$ and $u = -\mathrm{j}$. Now, as you can verify by squaring each of these numbers twice, $u^4 = 1$. The derivation above can be repeated using fourth powers rather than squaring; and the conclusion is that the error signal is one quarter of the imaginary part of the fourth power of the baseband signal. (Compare this with the BPSK result, where the error signal is one half of the imaginary part of the square of the baseband signal.) Now, as you can again verify,

$$\begin{aligned}\frac{1}{4}\operatorname{Im} z^4 &= \frac{1}{4}\operatorname{Im}(x+\mathrm{j}y)^4\\ &= \frac{1}{4}(4x^3y-4xy^3)\\ &= xy(x^2-y^2).\end{aligned}$$

This error signal can be filtered and used to adjust the local oscillator frequency as in the BPSK case.

The ideal QPSK constellation diagram is unchanged when rotated through 90°, 180° or 270°. There is therefore a fourfold ambiguity in the phase of the recovered carrier, analogous to the twofold ambiguity in the case of BPSK. Again, this problem is solved by the use of differential coding.

Decision feedback

An alternative method for achieving carrier lock that generalises more readily to QAM signals with more constellation points is called *decision feedback*. The idea is simple: if we knew the actual symbols transmitted, we could simply compare the baseband samples with the true constellation points and measure the angle between them. This would give an estimate of the phase angle between the local oscillator in the receiver and the carrier oscillator in the transmitter.

Unfortunately we do not, at the receiver, know the correct transmitted symbols; so instead, we proceed as shown in Figure 11.28. For each baseband sample we find the nearest ideal constellation point (this is the 'decision') and we pretend that this is the actual symbol transmitted. Figure 11.29 shows an example for the case of QPSK: the ideal constellation points at ± 1 and $\pm \mathrm{j}$ are shown by blobs, and an example baseband sample at $-0.5-0.8\mathrm{j}$ is shown as a cross. The nearest constellation point to the baseband sample is the one at $-\mathrm{j}$, and so the phase error is the angle marked θ in the figure.

The squaring and fourth-power methods, although simple to implement, have the disadvantage that when we square the baseband signal or raise it to the fourth power, we also square any noise on the signal or raise it to the fourth power. The decision feedback method is less sensitive to noise on the input signal, and is usually therefore preferred, especially for QPSK signals. Decision feedback also generalises more straightforwardly to use with constellation diagrams with more points, such as the one shown in Figure 11.15.

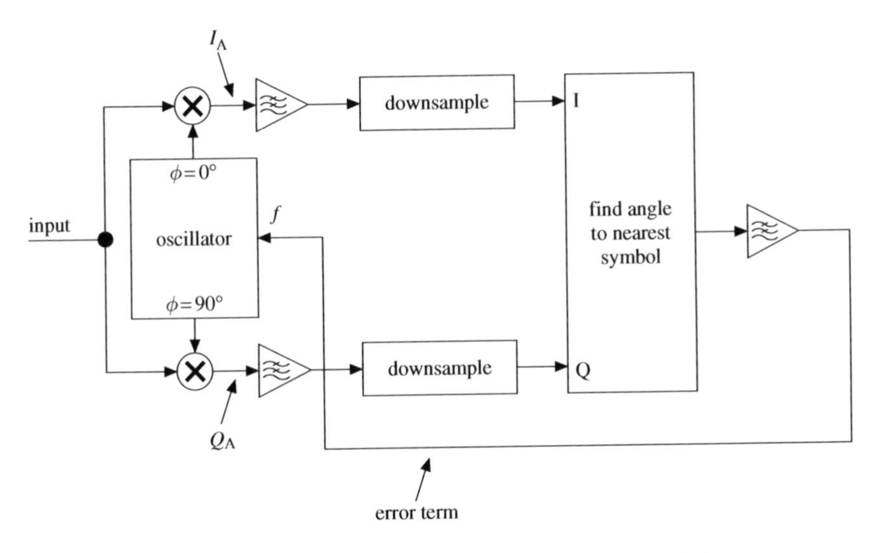

Figure 11.28 Decision feedback for QAM demodulation.

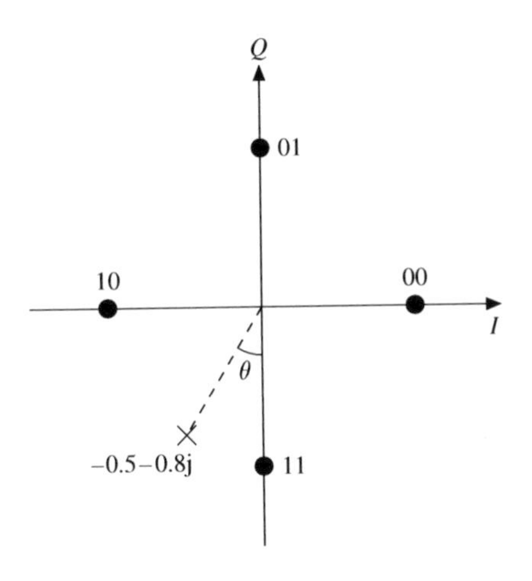

Figure 11.29 Decision feedback in QPSK.

Symbol synchronisation

Once we have recovered the original baseband signal, the next step is to decide exactly when to sample it to obtain the sequence of symbols.

Since the clocks in the transmitter and receiver are not necessarily operating at exactly the same frequency, the symbol rate will appear to be slightly different from its nominal value as far as the receiver is concerned. We will thus need another phase-locked loop, this time locked to the symbol clock.

The early–late gate

One approach to this problem is to arrange for the sample rate at the outputs of the downsamplers in Figure 11.25 to be a small multiple of the sample rate: three times the symbol rate works well. We also make the downsamplers *polyphase* interpolators, so that we can ask them to produce a downsampled version of the baseband signal with any phase offset we like. Let us consider the case of BPSK, where, since the constellation points and hence the baseband signal are nominally real, we only need to process the in-phase component. The downsampled baseband signal at three times the symbol rate might look like the example of Figure 11.30(a), where we can see that every third sample, marked by an arrow, is in the middle of a symbol. Alternatively, it might look like Figure 11.30(b), where our sampling is slightly early; or like Figure 11.30(c), where our sampling is slightly late.

Looking carefully at the figure, we see that when our sampling is early, the sample immediately before the arrowed sample, s_-, is smaller in magnitude than

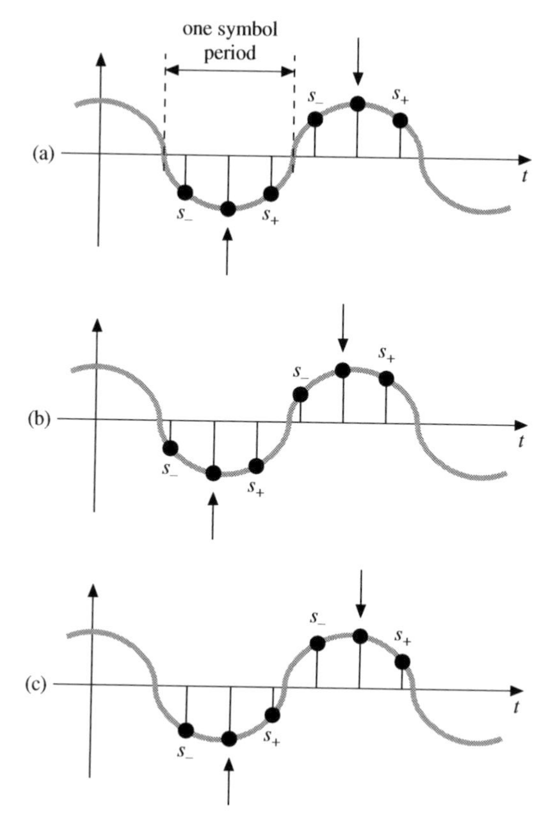

Figure 11.30 The early–late gate: (a) punctual sampling; (b) early sampling; (c) late sampling.

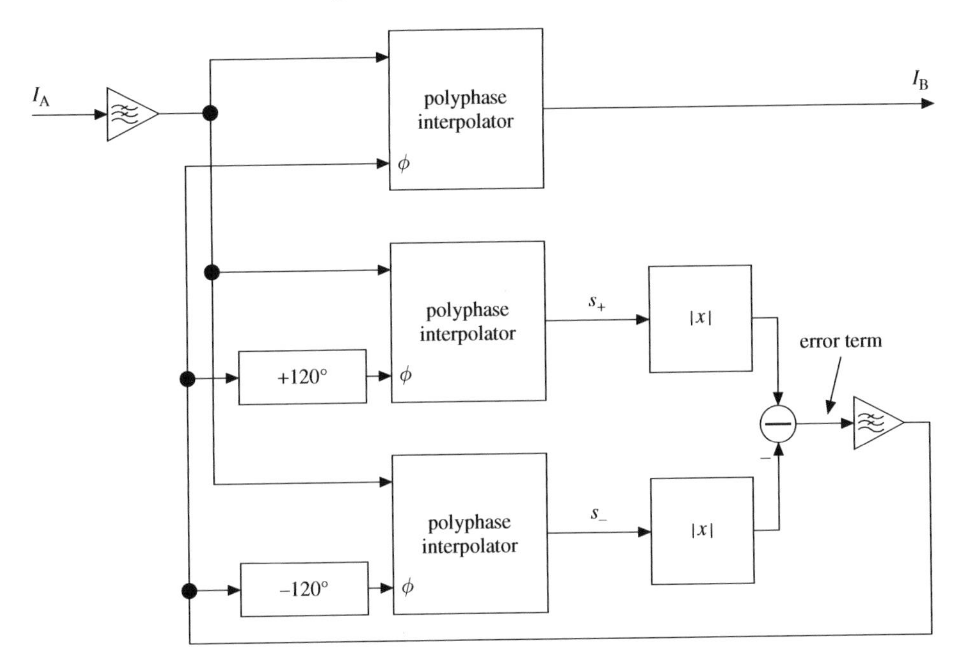

Figure 11.31 Early–late gate for symbol lock.

the sample after the arrowed sample, s_+; and when our sampling is late, s_- is larger in magnitude than s_+. This suggests a method for producing an error term: simply compute $|s_+| - |s_-|$. The result will be positive if our sampling is early, negative if it is late, and zero if it is punctual. This method is called the *early–late gate*. The overall arrangement is as shown in Figure 11.31, where we have shown the three samples per symbol generated using three separate polyphase interpolators with fixed phase offsets of $-120°$ and $+120°$.

It is not essential for the early and late interpolators to operate exactly one third of a sample ahead of and behind the main interpolator: other values will do, although the early and late samples should be equally spaced on either side of the main sample to ensure that the error term is zero when the sample point is correct.

For QPSK signals error terms are worked out separately for the I and Q channels and then summed before being passed to the loop filter.

Observe that if the signal is not changing, the early–late gate will produce a zero output: this is only reasonable, since if there are no edges in the signal there is no way to determine where the symbols start and finish. For this reason (among others) many communications systems include a *scrambler* in the transmitter which turns long runs of zeros or ones in the input data stream into patterns with more transitions to help the symbol synchroniser in the receiver. The receiver includes a complementary descrambler to restore the original data.

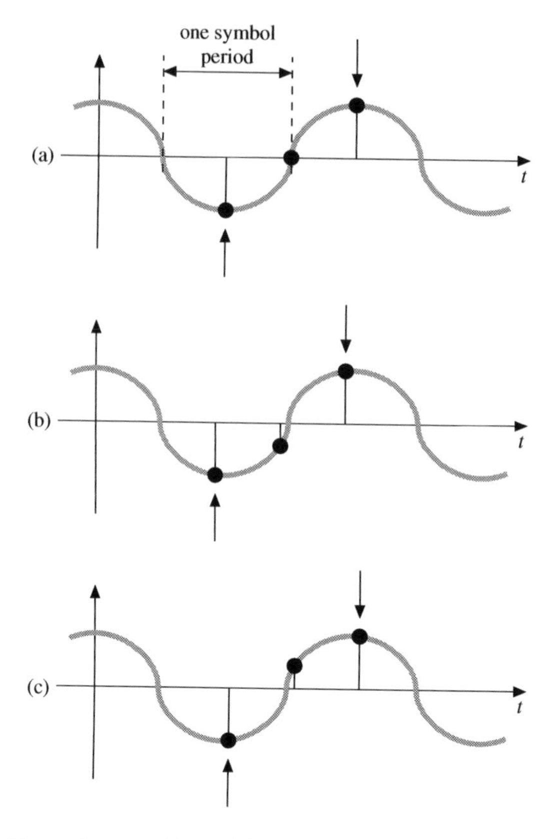

Figure 11.32 Transition tracking: (a) punctual sampling; (b) early sampling; (c) late sampling.

Transition tracking

An alternative approach to symbol synchronisation operates at exactly double the symbol rate. We shall consider the BPSK case.

We inspect the baseband signal at what we believe to be two successive symbol samples. If they have the same sign then we expect no transition between them and so we cannot obtain any symbol timing information from them. If they have different signs, however, then there is a transition between them: three possible situations are shown in Figure 11.32. In Figure 11.32(a) the symbol sampling time is correct, and the baseband signal is zero at the instant exactly mid-way between the two symbol samples. In Figure 11.32(b) the symbol sampling time is slightly early, and the baseband signal is negative mid-way between the two symbol samples. In Figure 11.32(c) the symbol sampling time is slightly late, and the middle sample is positive. Note that the sign of the middle sample would be reversed if the transition were negative-going.

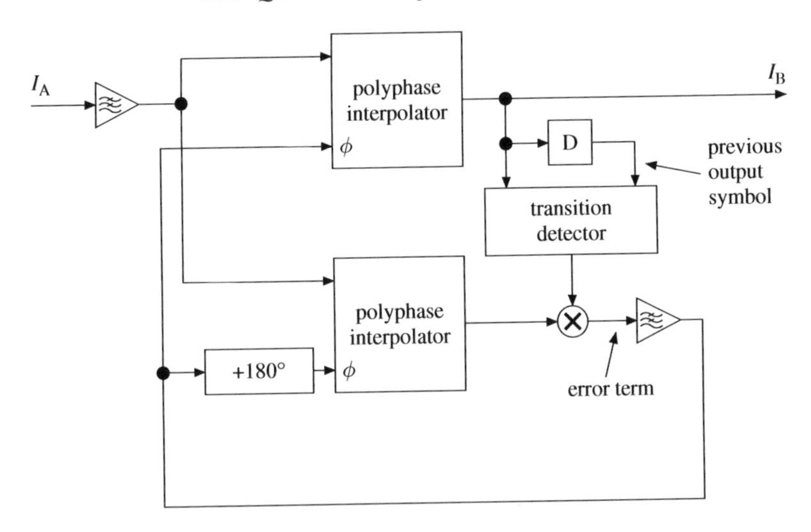

Figure 11.33 Transition tracking for symbol lock: the transition detector produces an output value of -1 on negative-going transitions, $+1$ on positive-going transitions, and zero if there is no transition.

The overall arrangement is as shown in Figure 11.33: we have shown the two samples per symbol being produced using a pair of polyphase interpolators at a fixed phase offset of 180°. The transition detector produces an output value of -1 on negative-going transitions, $+1$ on positive-going transitions. and zero if there is no transition. The error term is therefore equal to zero if there is no transition, equal to the middle sample value if there is a negative-going transition, and equal to minus the middle sample value if there is a positive-going transition.

For QPSK demodulation we can add the error terms from the I and Q channels together, as with the early–late gate.

Automatic gain control

We have assumed in the above discussion that the amplitude of the baseband signal is constant. In practice the amplitude of the received signal varies depending on the distance between the transmitter and the receiver, as well as other effects. We therefore need to employ an automatic gain control system along the lines described in Exercise 7.7. One strategy is to calculate the magnitude of the downsampled baseband samples, $\sqrt{I_B^2 + Q_B^2}$, and use the result in a feedback loop to control the amplitude of the input as shown in Figure 11.34.

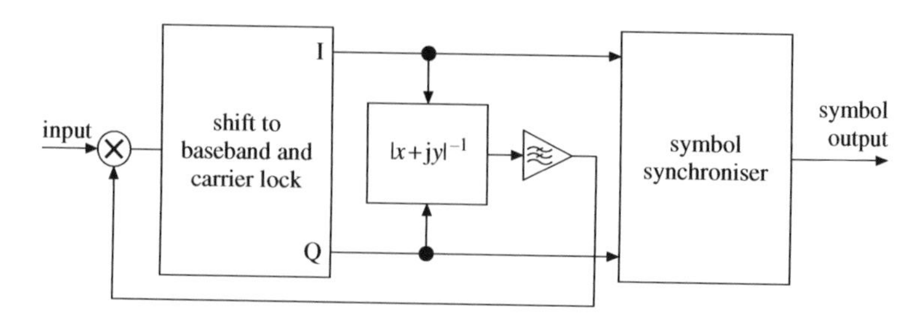

Figure 11.34 Automatic gain control for a QAM demodulator.

Differential coding

As we mentioned above, because of the symmetry of the constellation diagram it is impossible for a receiver to know whether it has locked on to the correct phase of the carrier. In the case of BPSK, there is an ambiguity of 180°: in other words, the constellation diagram at the receiver could be rotated by 0° or 180° relative to the diagram at the transmitter. In the case of QPSK, the rotation could be 0°, 90°, 180° or 270°. If the receiver locks on to the wrong carrier phase, every symbol it decodes will be wrong.

This problem is avoided using *differential coding*, where the data bits are encoded using the differences between consecutive symbols rather than their absolute values. For example, in a BPSK system, a '0' might be encoded by sending the same symbol as the previous symbol, or no phase change, and a '1' by sending the other symbol from the previous one, or a phase change of 180°. So if, for example, the data sequence runs 0011110010101, the sequence of symbols would be 00010100011001. Note that the encoded stream is one bit longer than the original data stream; we have assumed here that we start transmission with a zero symbol.

The receiver might see this transmitted stream correctly, as 00010100011001, or, with a carrier phase shift of 180°, as the complementary sequence 1110101 1100110. In either case it would decode the symbols by examining overlapping pairs of symbols: where they are the same, the data bit is a '0'; where they differ, the data bit is a '1'. You can verify that the two possible received streams give rise to the same result when decoded in this way.

A similar scheme can be used with QPSK modulation. A pair of bits is encoded by the difference in phase angle from each symbol to the next. A change of 0° might correspond to the bit pair 00, a change of +90° to the bit pair 01, a change of −90° (or, equivalently, +270°) to the bit pair 10 and a change of ±180° to the bit pair 11.

When necessary for clarity, BPSK modulation with differential coding is referred to as *DBPSK modulation*, and likewise QPSK modulation with differential coding is called *DQPSK modulation*.

11.4 Spread spectrum schemes

There are two main types of interference encountered in radio communications systems: impulsive interference (the kind that results in clicks and pops in AM and FM systems used for audio broadcast, for example when you turn on a light near to a radio receiver), and narrow-band interference, which causes continuous disruption to a channel. Narrow-band interference is radio-frequency energy concentrated at a single frequency within the communications channel, and can originate, for example, from digital circuits operating at fixed clock rates. As we saw in Chapter 4 signals with sharp transitions, like those encountered in digital electronic circuits, have frequency components that decrease in amplitude only slowly with increasing frequency. If one of these frequency components falls within a communications channel, it can easily prevent the receiver from being able to decode the transmission correctly: we say the channel is *jammed.*

In digital communications using the modulation schemes described so far, impulsive interference normally results in the loss of a few bits from the data stream. It can also cause loss of carrier lock and symbol synchronisation, which in turn causes a small number of consecutive bits in the stream to be decoded incorrectly. This type of error is called a *burst error*. Various error-protection techniques are available to deal with burst errors at the cost of a small reduction in the effective data rate.

Narrow-band interference, on the other hand, can result in a very large fraction of the received bits being incorrectly decoded. Using an error-protection scheme capable of correcting such a large number of errors would lead to a significant reduction in the effective data rate.

Complementary to the problem of narrow-band interference is that of *fading*. For example, if a signal simultaneously takes two different routes from transmitter to receiver – say directly as well as via a longer path involving a reflection off a passing aeroplane – there is a chance that the two copies will arrive at the receiver one half-cycle of the carrier wave out of phase. There will then be *destructive interference* between the two copies of the signal as they tend to cancel one another out. In the worst case the signal can be lost completely. Note, however, that for a given path length difference only certain frequencies will be affected: those where the path length difference is a whole number of wavelengths plus one half.

The more general case of fading when the transmitted signal takes several routes to the receiver, all of different lengths, is called *multipath distortion.*

A range of solutions to the problems of narrow-band interference and fading go under the umbrella term of *spread spectrum* methods. Broadly, their common feature is that they reorganise the use of a band of frequencies so that the effects of narrow-band interference are dispersed either in time or over a number of channels. The errors introduced by the interference are then either negligible or within the correction capacity of a simple error-correction code.

Frequency-hopping spread spectrum

The simplest spread-spectrum scheme is called *frequency hopping*. It is assumed that a number of independent data streams are to be transmitted within a certain band. In a conventional scheme each stream would be allocated permanently to a single channel occupying a narrow range of frequencies within the band, as shown in Figure 11.1. In a frequency-hopping spread spectrum scheme, however, the channels are reassigned pseudo-randomly among the streams at regular intervals. Thus a given data stream will spend a short while in one channel, where a few symbols will be transmitted; it will then hop to another channel and a few more symbols will be transmitted; and so on. An example hopping pattern is shown in Figure 11.35: as you can see, stream 0 (shaded) hops from frequency to frequency apparently at random among the six channels shown. (Strictly the scheme we have described is called *slow* frequency hopping, since several symbols are transmitted per hop; in *fast* frequency hopping, which we shall not cover here, the carrier frequency is changed several times per symbol.)

The pseudo-random hopping pattern is, as you might expect, pre-arranged between transmitter and receiver so that the receiver knows to which channel it must tune to receive the next few symbols of data. The transmitted signals need to carry some extra information to allow a receiver to synchronise with the hopping pattern. In some applications the pattern is kept a secret in the hope that this will make it harder for an eavesdropper to piece together the data corresponding to one stream. If the pattern is a simple repetitive one like the one illustrated in Figure 11.35, such 'security through obscurity' will only last as long as the secret does.

If a scheme like this is subject to narrow-band interference in one of the channels, the result will be disruption to a small proportion of the bits being

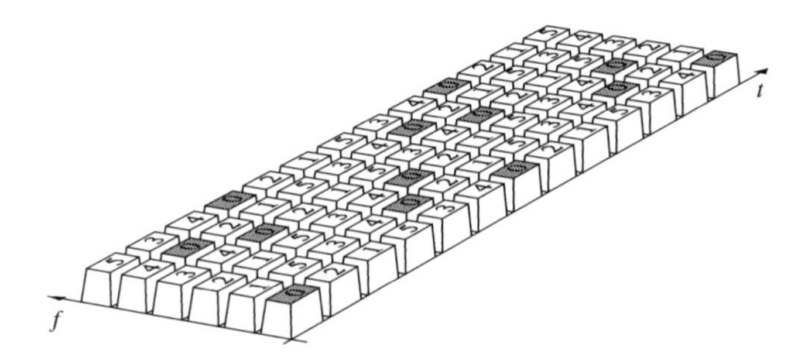

Figure 11.35 Example frequency-hopping pattern for six data streams numbered 0 to 5.

transmitted in each stream; with a suitable hopping pattern, all streams will be equally affected. The resulting short burst errors are easily correctable using suitable error-protection codes.

Note how the frequency-hopping scheme has spread out the effect of the narrow-band interference, degrading all channels to a small but manageable extent rather than degrading a single channel to the point where it becomes unusable.

The GSM mobile telephony standard used in many countries employs a frequency-hopping scheme of the type described above.

A disadvantage of slow frequency-hopping schemes is that it is usually necessary for there to be a gap in transmission at each point the carrier frequency changes, to allow receivers to retune and, depending on the details of the scheme, reacquire carrier lock and possibly also symbol synchronisation. This time, which we could otherwise be using for sending data, is effectively wasted.

Orthogonal frequency-division multiplexing

Although not usually classified as a 'spread spectrum' modulation technique, *orthogonal frequency-division multiplexing*, or *OFDM*, has many similarities to the frequency-hopping method described above. It is widely used for DVB-T (digital video broadcasting – terrestrial) digital television and DAB (digital audio broadcasting) digital radio in Europe. OFDM is also used in the IEEE 802.11a and IEEE 802.11g wireless Ethernet standards.

The figures given in the example below correspond approximately to those used in one variant of the DVB-T standard for terrestrial digital television. We will suppose for the sake of concreteness that we wish to transmit four video streams, each compressed using MPEG-2 to 4 Mbit/s, in a band 8 MHz wide.

A simple approach would be to divide the band into four 2 MHz channels and transmit each stream in one channel using QPSK. With a symbol rate of 2 MHz and 2 bits per symbol each channel provides the necessary capacity.

In OFDM, however, we divide the band into a very large number of channels, say 8000 channels each 1 kHz wide. Within each of these channels we transmit one symbol per millisecond using QPSK: each symbol represents two bits and each channel will have a capacity of 2000 bit/s. The total capacity of the 8000 channels is thus 16 Mbit/s as required. What benefit do we get in exchange for the apparent enormous leap in complexity in the transmitter and receiver? The point is that now we have a large number of channels, we can distribute the bits to be transmitted pseudo-randomly over them. If there is narrow-band interference or fading, only some of these channels will be rendered unusable at any given time, just as we described in the case of slow frequency hopping above. Thus only a few of the

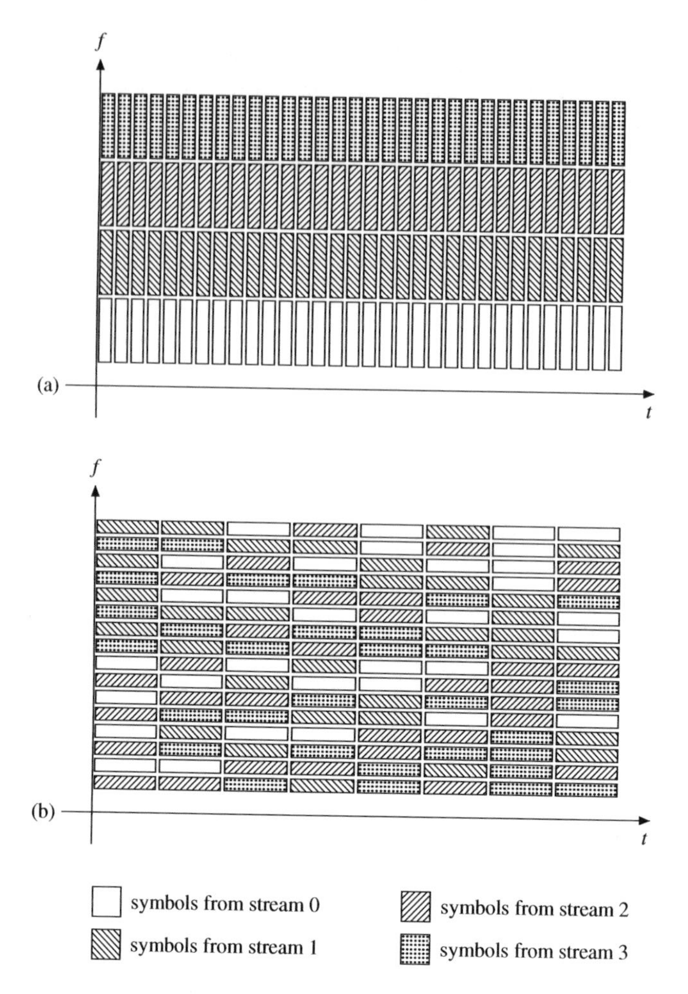

Figure 11.36 Allocation of frequency-time plane to symbols under (a) simple division into frequency bands; (b) OFDM.

transmitted bits received will be in error and we can again use an error-protection code to restore the original data.

Figure 11.36 contrasts the situations described in the last two paragraphs. Figure 11.36(a) shows the straightforward division of the band into four channels. As you can see, each symbol is comparatively brief in the time dimension and wide in the frequency dimension. Figure 11.36(b) shows the situation with OFDM, where each symbol occupies a comparatively long period of time but is narrow in frequency. The symbols corresponding to the different streams, distinguished by different hatching, are pseudo-randomly distributed over the available channels.

Because the use of suitable error-protection codes is essential to realising the benefits of OFDM, the method (with error-protection included) is sometimes called *coded orthogonal frequency-division multiplexing*, or *COFDM*.

In practical OFDM systems a few of the channels are dedicated to transmitting fixed signal patterns called *pilot signals*. These are designed to help the receiver achieve carrier lock and symbol synchronisation, and also provide a simple way for a receiver to measure the level of interference or noise present at certain frequencies. The receiver can build up information to decide how confident it can be in each of the symbols it has decoded: this will normally be a slowly varying function of the frequency of the channel on which the symbol was received. The confidence information, called *channel state information* or *CSI*, forms part of the overall noise model for the system and is fed into the error-correction mechanism to improve its performance.

When a signal suffers severe multipath distortion the transitions between symbols in the various channels may not line up perfectly in time. For this reason the symbols are lengthened slightly by an amount called the *guard interval* and receivers are designed to ignore a time period of that length around the symbol transition time.

Implementation of an OFDM transmitter and receiver

It may seem impractical to build a system which consists of several thousand QPSK transmitters, but implementation is surprisingly straightforward. The procedure is as follows: see the simplified block diagram in Figure 11.37. Take the set of (for example) 8000 symbols to be transmitted, one per channel in a given symbol period, and convert each to a complex number using the constellation diagram. Arrange the results as a vector of complex numbers of length 8000, and take an inverse Fourier transform. The result is the baseband signal to be transmitted!

Let us look in slightly more detail at what is happening. The result of the inverse Fourier transform is one period of a periodic signal which is a sum of complex exponentials, i.e., of baseband carrier waves. The carrier waves are linearly spaced in frequency, forming the 8000 individual channels. The amplitude and phase of each carrier is determined by the symbol at the corresponding input to the inverse Fourier transform, and thus the output in that channel is modulated in QAM fashion by that symbol.

Note that the modulating symbol in each channel is effectively fixed for its duration. It is as if zero-order interpolation had been used to produce the upsampled baseband signal for each channel, rather than smooth interpolation of the type illustrated in Figure 11.14: compare Figure 11.19. The contributions of the indi-

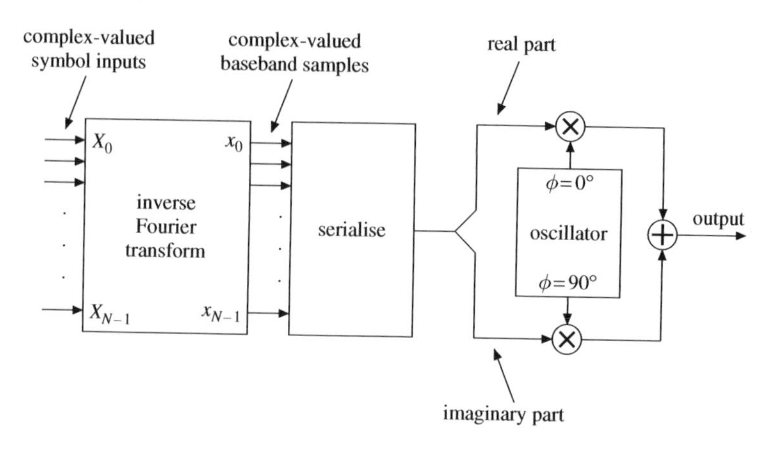

Figure 11.37 Simplified block diagram of an OFDM transmitter.

vidual channels therefore overlap: each channel leaks considerably into adjacent channels.

Demodulation of OFDM, as you might expect, can be done using a forwards Fourier transform. As we saw in Chapter 4, an inverse transform followed by a forwards transform exactly restores the original information: the leakage between adjacent channels in the transmitter can be exactly undone in the receiver.

The details of carrier lock and symbol synchronisation are complicated, and we shall not go into them here. In general the methods used are highly dependent on what provision is made in the overall scheme for pilot signals scattered among the available channels. Because the signals in the individual channels are generated together in a single inverse Fourier transform, only one carrier lock mechanism is required, rather than one per channel.

Direct sequence spread spectrum

The final class of spread-spectrum scheme we shall consider is called *direct sequence spread spectrum* or *DSSS*. In such schemes each symbol to be transmitted is expanded into a relatively long sequence of symbols, called *chips*, which are themselves transmitted using an underlying modulation scheme such as BPSK or QPSK. If there are n chips per symbol, the *chip rate* will be n times faster than the symbol rate. The mapping between symbols and chip sequences is fixed and adjacent symbols do not interact.

For example, the IEEE 802.11 wireless Ethernet standard includes a mode where the underlying modulation scheme is BPSK at 11 Msymbol/s: in other words, the chip rate is 11 MHz. Each data bit is represented by a sequence of 11 chips, and so the data rate is exactly 1 Mbit/s. The data bits 0 and 1 are represented by the complementary chip sequences $\mathbf{b} = (+1, -1, +1, +1,$

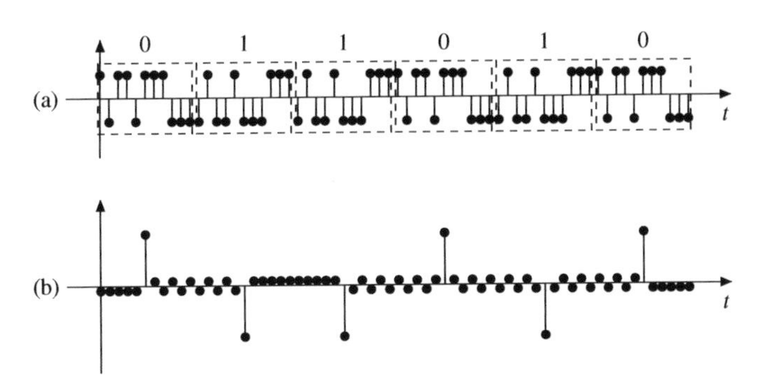

Figure 11.38 Direct-sequence spread spectrum: (a) data stream and resulting chip stream; (b) result of correlating (a) with chip vector.

$-1, +1, +1, +1, -1, -1, -1)$ and $-\mathbf{b} = (-1, +1, -1, -1, +1, -1, -1, -1, +1, +1, +1)$. Figure 11.38(a) shows an example data stream (after the differential encoding that would be required in practice) and the chip sequence corresponding to it. As you can see, the chip sequence consists of concatenated copies of **b** and $-\mathbf{b}$ under control of the data bits.

IEEE 802.11 also includes a mode based on QPSK, where the sequence **b** can be multiplied by any of 1, j, -1 or $-$j to produce one of four possible sequences of 11 complex-valued chips, encoding two bits. The overall data rate in this mode is thus 2 Mbit/s.

The sequence of chips in **b** is sometimes called the *pseudo-noise* or *PN* sequence because of its random-looking appearance. The particular sequence of chips given in this example is called a *Barker sequence* and it has a rather special property which you can verify for yourself if you wish: if you construct a vector that is a cyclic permutation of **b**, say $\mathbf{c} = (+1, +1, -1, +1, +1, +1, -1, -1, -1, +1, -1)$ (where we have moved two terms from the start of the vector to the end), $\mathbf{b}.\mathbf{c} = -1$. Here '.' is the vector dot product operator: pointwise multiply corresponding elements in the two vectors and add up the results. The result is the same whichever of the ten cyclic permutations you choose. Note that $\mathbf{b}.\mathbf{b} = 11$. The autocorrelation function (see Section 5.4) of **b** is thus equal to -1 except at zero time offset, where it has a spike of amplitude 11. All pseudo-noise sequences used in DSSS communications have an autocorrelation function with a strong spike like this.

We can now look at the effect of correlating the sequence shown in Figure 11.38(a) with **b**. (Remember from Section 5.4 that this is the same as applying an FIR filter whose impulse response is the reverse of the sequence **b**). The result is shown in Figure 11.38(b): it shows clear positive spikes where the signal corresponds to a 0 data bit and clear negative spikes where the signal corresponds to a 1 data bit. This signal could be decoded using the standard techniques we

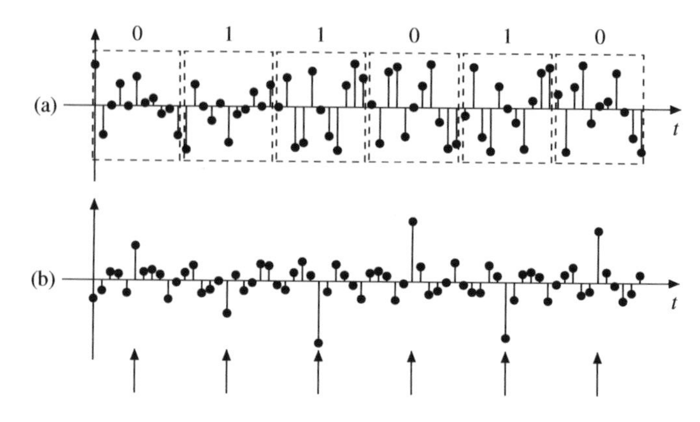

Figure 11.39 Direct-sequence spread spectrum: (a) chip stream with narrow-band interference added; (b) result of correlating (a) with chip vector.

described earlier in this chapter. The IEEE 802.11 standard requires that a fixed data sequence is transmitted at the start of each block, which can be used to help acquire carrier lock as well as chip and symbol synchronisation in the receiver. Once chip and symbol lock are acquired they can be maintained using variations on the methods described above. The early–late gate, for example, can be used on the signal *after* correlation with the pseudo-noise sequence.

Figure 11.39(a) shows the same signal as Figure 11.38(a) but with narrow-band interference added, with the amplitude of the interference set equal to that of the signal. Figure 11.39(b) shows the result of correlating with **b**: as you can see, the same positive and negative spikes are still recovered, while the interfering signal has been converted into lower-level noise. Figure 11.40 shows the same example, but with impulsive interference.

We can understand how DSSS provides this immunity to narrow-band interference more easily in the frequency domain. For simplicity, suppose that a steady stream of zero bits is to be sent using the IEEE 802.11 BPSK scheme. The resulting chip pattern will be periodic with period 11. The signal and its Fourier transform are shown in Figure 11.41(a): the frequency components have various arguments but all, apart from the one at DC, have the same magnitude.

First, for reference, we shall look at the situation when there is no interference. As we mentioned above, correlating with the Barker sequence is the same as convolving with the reverse of the sequence. The reversed sequence (keeping the sample at time $t = 0$ in place) and its Fourier transform are shown in Figure 11.41(b): the Fourier coefficients are the complex conjugates of the coefficients for the forwards sequence. We obtain the autocorrelation function by multiplying the two Fourier transforms together as shown in Figure 11.41(c): all

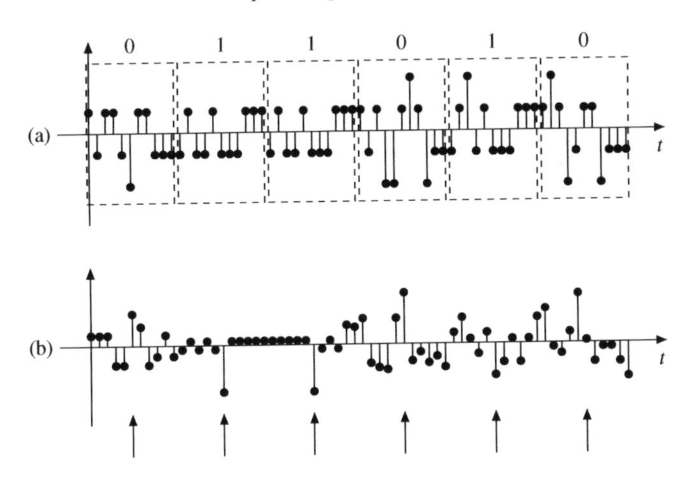

Figure 11.40 Direct-sequence spread spectrum: (a) chip stream with impulsive interference added; (b) result of correlating (a) with chip vector.

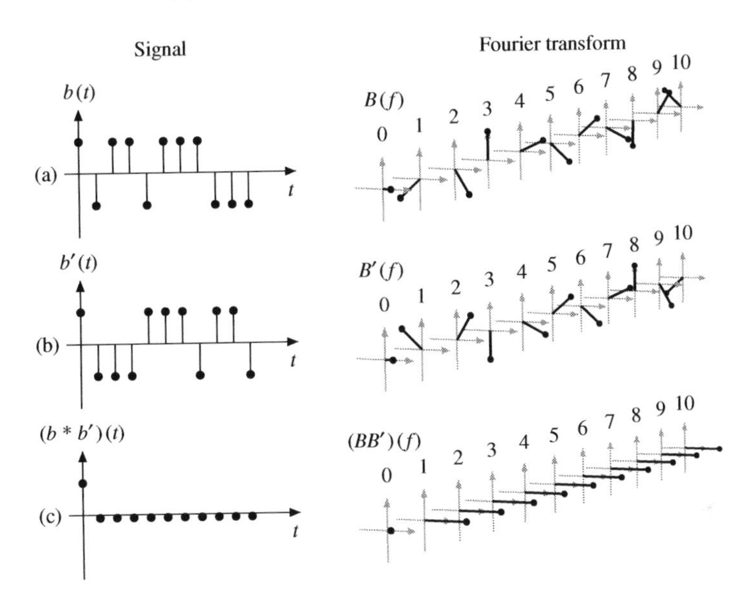

Figure 11.41 Barker code and its Fourier transform (a) forwards; (b) reversed; (c) convolution of (a) and (b).

the frequency components are neatly lined up along the real axis, and in the time-domain we have a spike at time zero, as mentioned above.

Figure 11.42(a), (b) and (c) show the situation when there is narrow-band interference, in this case a sine wave of amplitude 0.5 making three complete cycles in each period of the Barker sequence. As you can see by comparison with Figure 11.41, this only affects the third frequency component (and its partner, the eighth frequency component) on the right of Figure 11.42(a) and hence also

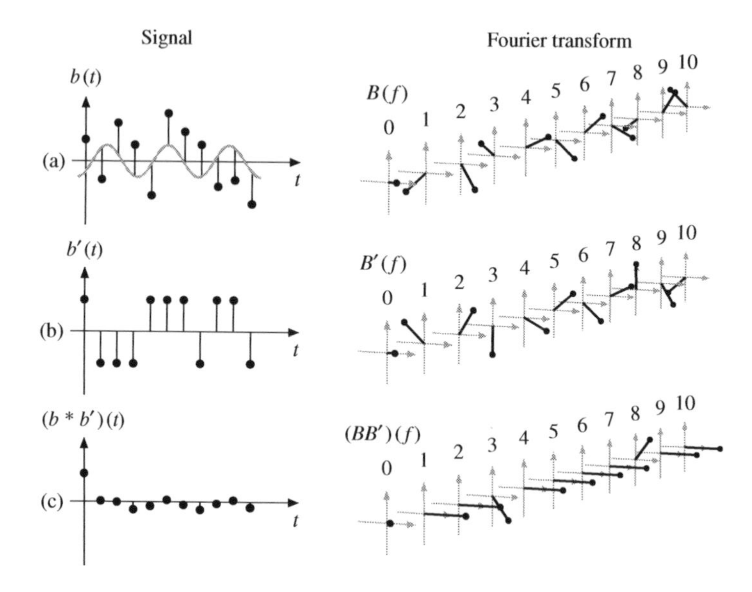

Figure 11.42 Barker code and its Fourier transform (a) forwards, with narrow-band interference added; (b) reversed; (c) convolution of (a) and (b).

on the right in Figure 11.42(c). When converted back into the time domain, on the left in Figure 11.42(c), the energy in the interfering signal is distributed over the 11 samples, while the energy from the desired signal is concentrated into the resulting sample at time zero.

Variations on DSSS modulation schemes are used in so-called 'third generation' mobile telephony systems, where resistance to narrow-band interference (and, equally, fading) is a particularly useful feature. The PN sequences used are usually rather longer than the 11-chip Barker sequence above. Typically a different PN code is allocated to each mobile station in a scheme called *code division multiple access*, or *CDMA*. A number of mobile stations will all communicate with the nearest base station in the same broad frequency band, but since they use different PN codes each appears as random noise to all the others. The PN codes must be carefully chosen to ensure that there are no unfortunate *cross-correlations*: suppose that one mobile station is assigned a code A and another is assigned a code B which, perhaps with a time offset, closely resembles A. Even after the received signal is correlated with A there might be a considerable residual effect from the transmission on code B. There is a great deal of theory on the subject of choosing good sets of PN codes to minimise cross-correlations.

Unwanted cross-correlations are a particular problem if the signal using code B is much stronger than the signal using code A, for example if one mobile station is much closer to the base station than the other. This is called the *near–far problem* and is solved by controlling the power of transmissions dynamically so as to equalise the strengths of the received signals.

Exercises

11.1 Why does the spectrum shown in Figure 11.2(a) have a large DC component?

11.2 Analogue superheterodyne receivers often retransmit from their antenna a low-amplitude frequency component directly from their local oscillator. You suspect your neighbour is using such a receiver with an intermediate frequency of 500 kHz. How could you tell what frequency he is listening to?

11.3 A digital QAM modulator is built using the design described in Section 11.3. The upsampling interpolation filters have an output sample rate eight times higher than the symbol rate and are implemented as FIR filters. What is the condition on the FIR filter coefficients to ensure that the system exhibits no inter-symbol interference?

11.4 A digital oscillator is constructed along the lines of the design described in Section 11.3, with the phase represented using eight-bit fixed-point arithmetic in 0Q8 format (see Chapter 7) as a fraction of a revolution. The output sample rate is 6 kHz and the desired carrier frequency is 2 kHz. What value should be used for the frequency setting? What is the actual carrier frequency in this case? Suggest a modification to the design of the oscillator that will produce a carrier frequency of exactly 2 kHz.

11.5 Show how to implement a QPSK modulator with the constellation diagram of Figure 11.20 using two BPSK modulators operating at the same carrier frequency.

11.6 An MSK transmitter using a phase trellis like the one shown in Figure 11.24 is transmitting a continuous stream of zeros using a carrier frequency of 1 MHz and a symbol rate of 2 kHz. Draw the corresponding phase trajectory. Describe the output of the transmitter as simply as you can. (Assume no differential coding is used.)

11.7 Show that if a scrambler takes as input an n-bit binary stream and produces another n-bit binary stream as its output, there must be an input sequence that gives an all-zero output. Does this mean there is never any point in using a scrambler? (**Hint:** No.) Why not?

11.8 A friend proposes a differential coding scheme for QPSK where the BPSK scheme described in the text is used separately on the I and Q channels in the transmitter and receiver. Why will his idea not work?

11.9 A friend is designing a highly sensitive radio receiver for a space probe designed to pick up transmissions from Earth in a channel of width 1 MHz centred on 2.25 GHz, using a digital signal processor with a clock frequency

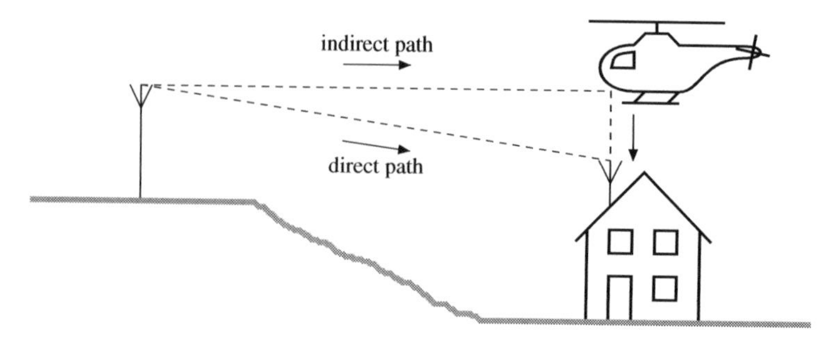

Figure 11.43 Fading resulting from multiple transmission paths.

of 250 MHz. Suggest a reason why this might not be a good idea, and recommend a modification to the design.

11.10 You are listening to a radio station broadcast using analogue FM with a carrier frequency of 100 MHz. A helicopter flies directly overhead in a straight line towards the transmitter. There are now two signals arriving at your receiver: the one directly from the transmitter and one reflected from the helicopter, as shown in Figure 11.43. Over the din of the helicopter you can just hear the radio signal fading in and then out into noise at a frequency of 5 Hz. At what speed is the helicopter approaching the transmitter? (Assume that the speed of light $c = 3 \times 10^8\,\mathrm{m/s}$ and that the distance from the helicopter to your receiver is constant.)

11.11 Show that the fact that the the non-DC frequency components of the Barker sequence all have the same magnitude is a consequence of the spiky form of its autocorrelation function.

11.12 Mobile telephones generally attempt to adjust their transmission power so that the signal level received at the base station is just strong enough for reliable demodulation. A friend suggests that you wrap your mobile telephone in aluminium foil to prevent the harmful radiation from it entering your brain. Aside from the practical difficulties of operating a telephone wrapped in aluminium foil, explain why his plan is not a clever one.

Should mobile telephone users who fear the harmful effects of radio waves campaign for or against the installation of base stations in their neighbourhood?

12
Implementations

This chapter looks at the implementation, and in particular the efficient implementation in software and in digital hardware, of some of the algorithms discussed earlier in this book. Signal processing algorithms are implemented on a huge range of different platforms: we will cover here only a tiny proportion of the possibilities to illustrate the general principles at work.

The design of a signal processing system usually takes place under a number of constraints. You might be required to:

(i) make a system as soon as possible;
(ii) make a system that is as flexible as possible;
(iii) make a system that consumes as little power as possible;
(iv) make a system that works as fast as possible;
(v) make a system that is as cheap as possible;

or indeed some combination of these and other requirements. Often there is conflict between the requirements and a compromise must be sought. Sometimes a requirement is 'hard': a signal processing system on a satellite will only be allowed to consume a given amount of power, and a digital television decoder must be capable of running in real time.

The last three items on the list above often boil down to the same thing: using available hardware as efficiently as possible. For most modern processors it is a reasonable approximation to say that every instruction executed consumes a certain amount of energy, so if the same task can be done using fewer instructions the system as a whole will probably consume less power. Consuming less power translates directly into increased battery life for portable devices, which is often an important selling point. Likewise, if you can replace a fast processor with a slower one by using a more efficient program to achieve the same result, the end product will probably be cheaper. Similar arguments apply to hardware implementations of signal processing algorithms, where, as we shall see later, there are several ways to make tradeoffs between cost, power consumption, and speed.

The first decision to be made when embarking on the design of a project involving signal processing is between hardware and software implementation. In many cases this decision will be determined by the nature of the application: for example an algorithm might have to run on an existing microcontroller system or in an existing programmable logic device. Otherwise the decision must be made by analysing the alternatives available, which we shall now examine in more detail.

12.1 Software implementations

Numerical simulation languages

At one end of the scale are applications which are practically unconstrained in terms of power, speed or cost, and where all that matters is getting a (possibly one-off) answer, for example to generate data for testing an implementation of an algorithm written in a different language. For these applications software implementation on an ordinary desktop computer is ideal, and programs can be written using the free and commercially available numerical simulation languages, which generally use double-precision floating-point arithmetic and which are often accompanied by comprehensive libraries including high-level functions to perform filtering, calculate Fourier transforms and so on. This can also be a good way to experiment with algorithms, although the high-level nature of the languages can mask significant amounts of underlying complexity. These languages are not ideal for more demanding real-time applications. Some of the simulation languages have a facility whereby they can compile code for use in embedded systems; unfortunately the languages do not encourage the expression of algorithms in ways that are efficient for such systems, and so the results often run slowly and require large amounts of memory.

It is also surprisingly common to find that one step in an algorithm cannot be neatly implemented using the particular set of functions provided, and the necessary work-around can be difficult to write and slow to execute.

General-purpose languages

A more flexible approach is to use a general-purpose programming language such as C. There are C compilers available for a wide range of machines from supercomputers down to the smallest embedded systems. It is relatively easy to keep control of memory usage within the language. Often C compilers for embedded systems include extra features to make it easy to control hardware directly or write interrupt service routines, which is useful when writing real-time applications. Almost all C compilers support floating-point as well as integer arithmetic. Some also offer extensions that directly support fixed-point arithmetic; failing that, it is always possible to implement your own fixed-point arithmetic routines.

Languages offering higher levels of abstraction than C are generally less suitable for embedded systems work. There is usually a considerable performance penalty associated with using these languages, and many implementations include features like just-in-time compilation and garbage collection that make it difficult or impossible to achieve reliable real-time operation.

Assembly language

For the highest level of performance it is necessary to program directly in the assembly language of the target device. For all the claims made on behalf of compilers, it is still the case that the code generated by a compiler is in general far from optimal. Sometimes this is a limitation of the language: a simple example is determining whether the process of adding two integers has caused an overflow.

Most processors have internal flags that indicate when such events occur whose status can be updated on every addition operation without any execution time penalty, but these flags are inaccessible to most high-level languages. In a high-level language one is thus forced to resort to a series of tests on the numbers being added, which can take considerably longer than the addition operation itself. In many high-level languages it is impossible to access the more significant half of the product of two integers, adding unnecessary complexity to many signal-processing algorithms which work with fixed-point arithmetic. Compilers are also generally very poor at taking advantage of special features in more complex processor architectures.

Assembly language can be tedious to write, however, and must be rewritten if the same algorithm is to be implemented on a new processor. Detailed knowledge of the workings of the processor is needed to write efficient assembly language. For this reason the commonest strategy for a high-performance signal-processing system is to implement a small number of time-critical functions at the core of the algorithm in assembler, possibly using hand-written library functions provided by the manufacturer of the processor or a third party, and write the rest of the algorithm in a language such as C. This gives a good compromise between the efficiency of the resulting code and the time taken to write it.

Assembly language is also used in very resource-constrained applications where the size of the program must be kept to an absolute minimum.

12.2 Processor architectures

In this section we shall look at the various types of processor on which you might want to implement signal processing algorithms. They fall broadly into three categories: microcontrollers, processors designed for desktop computers, and dedicated digital signal processors. As you might expect, the first two categories of processor are hampered by their general-purpose nature when executing signal processing algorithms: microcontrollers because of their simplicity and hence relative slowness; and desktop processors because of their complexity and hence high power consumption.

Many signal processing operations reduce to the basic functions of filtering or calculating Fourier transforms plus a small number of extra computations. These in turn reduce to sequences of multiplications and additions, and so it is the performance of a processor when doing these operations that largely determines its performance on signal processing algorithms.

Microcontrollers

Microcontrollers are low-cost processors designed for embedded control applications, often consisting of a single integrated circuit (or 'IC') which includes not only the processor, but also memory and peripherals. Microcontrollers are usually low-power devices and suitable for use in battery-powered products. For reasons of cost there is usually no hardware support for floating-point operations, which are consequently slow. Many microcontrollers are slow at multiplication: it is not unusual for a 16-bit by 16-bit multiplication to be 16 times slower than a 16-bit addition. They are therefore suitable for relatively undemanding applications where the sample rate is low or where little work needs to be done on each sample. The techniques mentioned below for minimising the number of multipliers in hardware implementations, including bit-serial and table look-up methods, can be useful.

Desktop processors

Processors for desktop computers are designed with performance on general tasks, rather than power efficiency, in mind. Since the processor is usually powered from the mains, power consumption is chiefly an issue with regard to the problem of removing heat, and so the processor is designed to operate as quickly as possible within the constraints of a practical cooling system (which itself also consumes power).

A subtle way in which desktop processors are different from, for example, typical microcontrollers, is in their use of deep pipelines. A *pipeline* is a scheme whereby a complicated operation, such as a multiplication, is divided into a number of simpler stages which are executed in sequence in different places within the processor, much like the way a complex device is assembled on a production line. Each stage in the pipeline takes one cycle of the processor clock. A multiplication might be divided into ten successive stages, and so would take ten clock cycles to complete. However, after the first cycle the first stage is ready to start another operation: in principle we could start a new multiplication on every clock cycle, the results appearing, one per cycle, with a ten-cycle delay. We say that the *throughput* in this example is one operation per cycle, but the *latency* is ten cycles. Figure 12.1 shows how the same calculation is carried out on three independent sets of data using a pipeline. The arrows show the flow of partially processed data from pipeline stage to pipeline stage, with the result in each case appearing at the output of the final stage ten cycles after the input data entered the first.

The example shown has a gap, called a *bubble*, between the second and third operations. The bubble flows through the pipeline just like an operation would.

The benefit of using a pipeline, from the point of view of the designer of the processor, is that the clock cycle can be made much shorter: in the example above, it only needs (in theory) to be long enough to get one tenth of the job done. The overall clock frequency of the processor can thus be increased, potentially by a factor of ten in this example.

The disadvantage of the pipeline, from the point of view of the programmer, is that to make the most efficient use of it a large number of *independent* operations

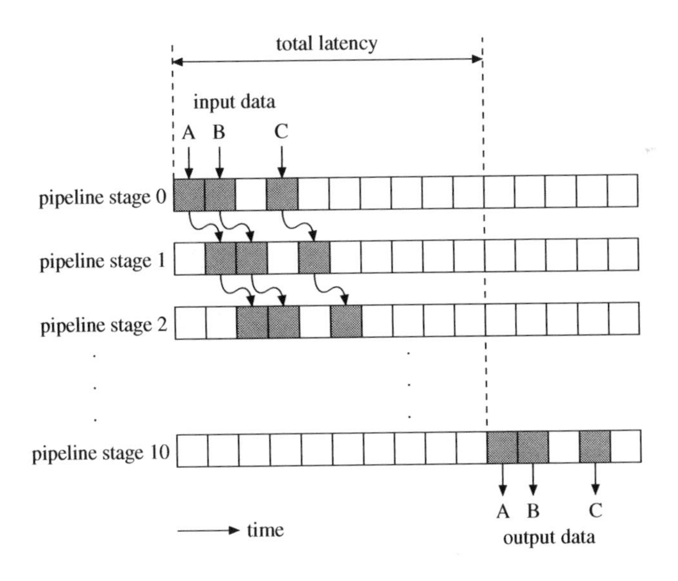

Figure 12.1 Pipelining.

need to be found. Suppose you wished to multiply all the entries in an array of 1024 numbers together using the example pipelined multiplier above. The simple approach of a loop which runs down the list calculating a running product would be very slow. At each iteration it would be forced to wait for the result of the previous operation to be calculated before it could be fed back into the beginning of the pipeline. At best, a multiplication could only be started once every ten cycles, the multiplier only being used at one tenth of its maximum possible throughput. Implementations of IIR filters can suffer from similar problems.

A better approach would be to multiply successive pairs of numbers in the list together: this is a total of $1024/2 = 512$ operations which are all independent of one another and which can therefore run through the multiplier at full speed in a total of 512 cycles. These 512 results can then be processed in the same way, using $512/2 = 256$ independent multiplication operations to reduce the list down to 256 numbers in 256 cycles. The process can be repeated until the list is reduced to a single number, which is the answer. A small number of bubbles needs to be introduced into the pipeline towards the end of the operation when the size of the list becomes smaller than the pipeline length, but overall the operation will keep the multiplier pipeline busy most of the time, and will therefore execute about ten times faster than the simple method.

Fortunately many signal processing operations, including the fast Fourier transform and filtering using long FIR filters, inherently involve large numbers of independent operations. Each stage of an N-point fast Fourier transform involves $N/2$ independent operations; the evaluation of a single output value of an N-point FIR filter involves N independent multiplications and $N-1$ additions, which, if the latency of the addition operation is great, can be done using an algorithm similar to the one we described above.

Many desktop processors have extra instructions to help with signal processing tasks, such as the ability to perform the same floating-point or integer operation simultaneously on several independent data values. This is called a *SIMD* (from *single instruction, multiple data*) operation. Some processors provide SIMD logical instructions which can speed up morphological operations, both greyscale and bilevel, considerably. Other useful features, such as saturation arithmetic (see Section 7.1), are also provided. Compilers are generally not capable of making the most of these extra instructions, and so assembler coding is necessary to get the best performance. The individual operations within a SIMD instruction usually do not interact, and so your processing must again be expressed as a number of independent calculations.

Because desktop processors are designed to run general-purpose programs which have not necessarily been written with detailed knowledge of the design of the pipeline, a considerable fraction of processor is dedicated to circuitry which analyses the code to be executed to find sets of independent operations, trying to arrange them into an order which maximises the performance of the processor core. As you might imagine, this is a complex process which accounts for a significant part of the power consumption of the device.

Desktop processors have on-chip caches, small areas of memory that hold copies of frequently accessed areas of main memory for quicker access. A complicated algorithm decides which parts of main memory are shadowed in this way at any given time, with the result that the time taken by a memory access is hard to predict. This makes desktop processors difficult to use in demanding real-time applications. Also, most operating systems used on desktop computers are not designed for real-time work: even on a lightly loaded machine, a program might be made to stop for a significant fraction of a second by the operating system so that other tasks can be carried out. This is one reason why (for example) DVD playback on a desktop computer is often considerably less smooth than on a dedicated player with much less processing power.

Dedicated signal processors

Dedicated digital signal processors usually have an architecture which is centred around a multiplier: a typical design is sketched in Figure 12.2. Often the multiplier is capable of doing an addition along with the multiplication, and so the basic operation is to calculate $a+b\times c$, called a *multiply–add* operation. Details vary from device to device, but often a is stored in a register, called an *accumulator*, and the result of the multiply–add is returned to this same register: in this case the operation is called a *multiply–accumulate*, or *MAC*. Multiply–accumulate operations are ideal for sum-of-products calculations, such as are involved in implementing FIR filters. Often the accumulator is an extra-long register capable of holding the sum of a large number of such products to simplify the implementation of long FIR filters. Some digital signal processors support floating-point arithmetic directly: these tend to be the more expensive and power-hungry devices.

The multiply–add unit is fed by units designed to fetch data from memory fast enough to keep the multiplier busy. These include address generators that automatically step through memory in various programmable patterns. The available patterns might include wrap-around of pointers to facilitate the use of circular buffers for input and output data, for applications where a continuous stream of samples is being processed; programmable pointer increment to simplify the implementation of polyphase filters and other algorithms; and a 'bit-reversed' addressing mode to accelerate fast Fourier transforms (see Exercise 4.15). The

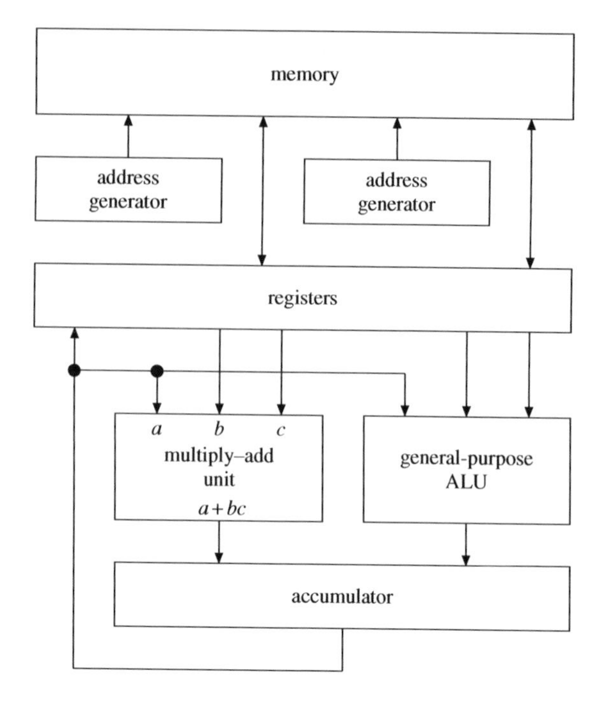

Figure 12.2 Block diagram of a typical dedicated signal processor.

memory is usually *dual ported*, which means that it can carry out two operations simultaneously: two reads, a read and a write, or two writes.

This whole system is pipelined so that the multiply–accumulate unit can start a new operation every cycle. Meanwhile the two operands for the next multiply–accumulate operation are being fetched from memory, and the address generators might be calculating the addresses of the operands for the next-but-one operation.

An n-point FIR filter can be implemented very efficiently in an architecture like this. One address generator is set up to read the filter coefficients from memory; the other to read a stream of consecutive sample values. The multiply–accumulate unit then calculates the sum of the products at the rate of one coefficient per cycle. One output sample can thus be generated in n cycles plus the time taken to set up the address generators and the time to store the result; for large n these overheads are negligible. However, in applications where n is small the overheads can be considerable, and it is worth experimenting to determine whether, for example, coefficients can be loaded once and kept in registers rather than being fetched from memory repeatedly for each output sample from the filter.

A general-purpose *arithmetic logic unit* or *ALU* is also provided to perform logical operations, shifts, and other miscellaneous tasks. For example, in a digital television receiver the incoming signal would be demodulated using signal processing algorithms mostly involving the multiply–accumulate unit; the general-purpose

ALU might be used for error correction and for parsing the resulting data stream; and then the multiply–accumulate unit would again be used to decompress the stream into a video signal.

Some signal processors include extra hardware to help accelerate tasks like error correction and parsing data streams. It is also common to include a range of other peripherals on the same IC, much like a microcontroller. A *direct memory access*, or *DMA*, controller is a popular peripheral: it allows you, for example, to read data into memory from an analogue-to-digital converter at a regular sample rate without software intervention. The samples might then be processed efficiently as a block, producing a block of output samples; these can then be output to a digital-to-analogue converter, again using DMA.

Other extensions include adding extra independent multiply–accumulate units, address generators and so on. As the number of functional units increases, so does the size of the instruction required to specify what each should be doing in each processor cycle. In *very long instruction word*, or *VLIW*, processors, each instruction can be hundreds of bits long.

The more sophisticated dedicated signal processors have caches, which, as with desktop processors, can make the time they take to perform a given function less predictable. Care is therefore required when using these processors in real-time applications.

12.3 Hardware implementations

In this section we will look at how signal processing algorithms can be implemented on digital hardware. We shall only consider *synchronous* hardware, where all operations are carried out in time with a single master clock signal. There are two main routes to hardware implementation: design an *application-specific integrated circuit* or *ASIC*, or use a programmable logic device such as a *field-programmable gate array* or *FPGA*. The former route involves very high initial costs but is cheaper overall if the final product is to be made in large quantities; the latter route has much lower initial costs but is more expensive per unit. ASICs can usually run at a higher speed than FPGAs and can fit more logic onto a single device.

As we mentioned in Chapter 7, hardware implementations of signal processing algorithms almost always use fixed-point arithmetic for reasons of efficiency.

To illustrate the techniques available when implementing signal processing algorithms in hardware we shall look at a concrete example: the calculation of an 8-point one-dimensional discrete cosine transform (see Section 9.5) using fixed-point arithmetic with eight-bit precision.

As we saw in Chapters 9 and 10, the one-dimensional discrete cosine transform can be used in the calculation of two-dimensional discrete cosine transforms in a digital video compression application. Suppose that we are dealing with a video signal with a resolution of 720 by 576 pixels at a frame rate of 25 Hz. Then each frame contains $720/8 \times 576/8 = 90 \times 72 = 6480$ eight-by-eight blocks of pixels, for a total of $6480 \times 25 = 162\,000$ blocks per second.

The two-dimensional transform involves eight one-dimensional row transforms followed by eight one-dimensional column transforms, a total of sixteen one-dimensional eight-point transforms. For real-time operation we therefore have to be able to calculate $16 \times 162\,000 = 2\,592\,000$ one-dimensional transforms per second, or just less than 400 ns per transform.

Referring back to Equation (9.1) we see that each of the eight output values produced by the discrete cosine transform is the sum of eight products of the input values with fixed coefficients. This suggests the simple hardware structure shown in Figure 12.3, involving a total of 64 multipliers and $8 \times 7 = 56$ adders.

It would probably be possible to implement the design shown on an ASIC to make it fast enough for real-time use. However, there is a number of improvements we can make which will make it smaller (and hence cheaper) and faster. As we shall see in Section 12.4, a faster design can potentially consume much less power than a slower one doing the same task.

There are two principal directions in which we can improve the design shown. The first is to make algorithmic changes: there are fast discrete cosine transform algorithms available, similar in nature to the fast Fourier transform. Another example occurs in implementing an FIR filter where two coefficient values happen to be the same (such as when the filter coefficients form a mirror-symmetric pattern): instead of multiplying each corresponding sample value by the coefficient separately and then adding the results, we can add the samples first and then multiply. This replaces two multiplications with one.

This kind of algorithmic change is independent of, and can be combined with, the lower-level hardware improvements we can make, which we shall now discuss.

Pipelining

The flow of signals in Figure 12.3 is entirely from left to right: there are no loops where a signal is fed back. This means that it is very simple to make the design pipelined (see Section 12.4). We can draw lines from the top of the diagram to the bottom at any horizontal positions we like in order to divide the design into individual pipeline stages: in Figure 12.4 we have shown a division into four stages numbered from 0 to 3. (We could go further and introduce pipeline stages

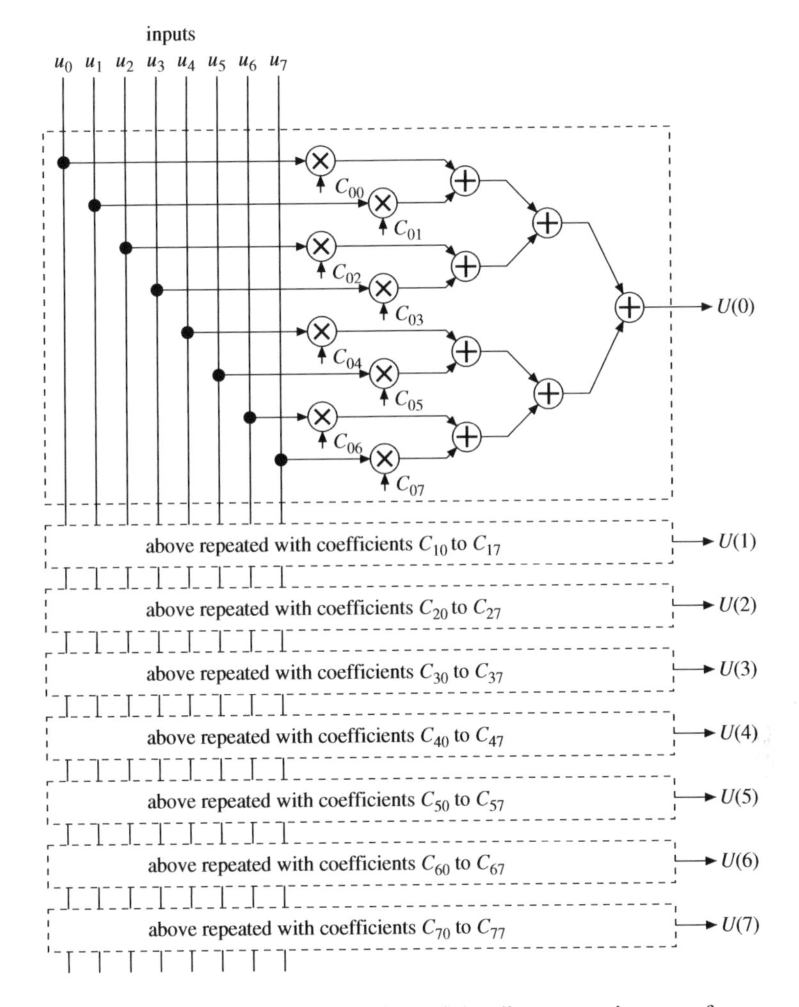

Figure 12.3 Direct implementation of the discrete cosine transform.

through signals in the middle of multipliers or adders, but in the interests of clarity shall not do so here.)

Wherever a line crosses a signal we introduce a pipeline register or *latch*, which transfers the value on its input to its output at the beginning of every clock cycle: the clock is common to all the latches. The latch holds its output steady for the duration of the clock cycle, even if the input subsequently changes. The calculation is thus carried out in a series of four steps, governed by the overall clock. The shortest clock period you can use is determined by the slowest path from an input of one stage to an output of the same stage: it thus makes sense to try to equalise these paths. If the paths have equal lengths then (ignoring the small overhead introduced by the pipeline registers themselves) the shortest clock period

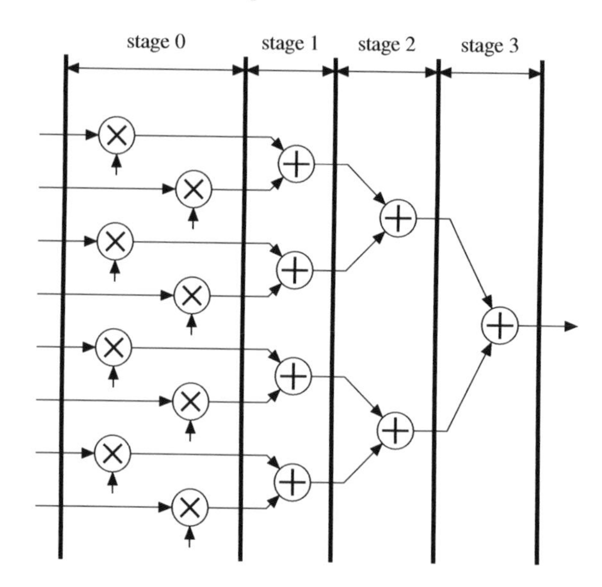

Figure 12.4 Pipeline stages in the direct implementation of the discrete cosine transform.

we can use will be four times shorter than if we had omitted the intermediate pipeline registers: the circuit can be run four times faster.

The pipeline can accept a new set of input values on every clock cycle, and produces the corresponding output values with a latency of four cycles. Note that at any time all the inputs to any adder or multiplier unit in the system will have passed through the same total number of pipeline registers (as you can verify by tracing back though the circuit to the left-hand edge counting the number of vertical lines you cross), and so the calculations commencing on different cycles are completely independent of one another.

Time-division multiplexing of hardware

Now suppose you wished to build a device that could simultaneously generate 64 audio waveforms to make a polyphonic synthesiser. One approach would be to design a waveform generator with a single output and then replicate the hardware for the other outputs. This would multiply the hardware cost by a factor of 64, which is clearly not desirable.

An alternative approach, called *time-division multiplexing*, shares the same hardware between the output channels. Generating one audio waveform is rarely a demanding task in hardware, and so it is easy to design a system that is capable of working at 64 times the speed that would be required for a single channel. The hardware might be controlled by loading in the parameters for the first channel, generating one sample of the waveform, loading in the parameters for the second channel, generating one sample of its waveform, and so on. A complete set of 64 samples, one for each channel, can be generated in one sample time.

Time-division multiplexed systems can often be pipelined efficiently as long as the calculations involved in different channels are independent of one another, as they would be in the case of the polyphonic synthesiser.

Multiplication by constants

In terms of the number of constituent logic gates, multipliers are generally much larger than adders with operands of comparable length (although the very fastest adder designs can be quite large). We should therefore concentrate on the multipliers if we wish to make our design more economical.

Note that in the design of Figure 12.3, as well as in fast discrete cosine transform designs, all the multipliers have one input that is a constant. Many FIR filter designs also use constant coefficients. This gives us a way to make them smaller.

First observe that if the constant is a power of 2, multiplying by it simply corresponds to a relabelling of the input bits to shift them up or down by the appropriate number of places. When shifting two's complement signed numbers down, care must be taken to preserve the sign bit: see the discussion of sign extension in Section 7.2.

Now consider, for example, a multiplier whose constant input is $\sqrt{2}$. In binary, this is equal to $1.01101010000\ldots$, so $1.0110101_2 = 181 \times 2^{-7}$ is a very good approximation: the relative error is about one part in ten thousand, or better than 13 bits of precision. For clarity, we shall work with binary integers rather than binary fractions. The problem is thus equivalent to multiplying a number by $10110101_2 = 181$. Since $181 \approx 2^7\sqrt{2}$, the result of the integer multiplication will be the answer we want shifted up by seven places. Now,

$$10110101_2 = 2^7 + 2^5 + 2^4 + 2^2 + 1,$$

so we can implement multiplication by 10110101_2 using the scheme shown in Figure 12.5. The boxes marked '$\times 2^k$' are multiplying by powers of 2, and so are simply relabellings of their input bits as noted above. We have therefore replaced what might have been a 13-bit multiplier with four adders. Exercise 12.4 shows how we can go further, and do the job with just three adders.

Still further savings are possible when we share work between two calculations. For example, in a discrete cosine transform, we might want to multiply the same number x by both $\cos\pi/16$ and $\cos 7\pi/16$. Now, to 10 bits of accuracy,

$$\cos\frac{\pi}{16} = 0.1111101100_2$$

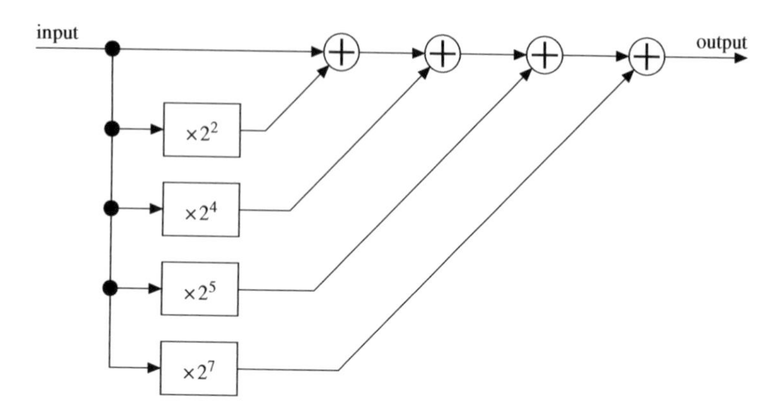

Figure 12.5 Approximate multiplication by $2^7\sqrt{2}$.

and, to 12 bits of accuracy,

$$\cos\frac{7\pi}{16} = 0.001100011111_2.$$

In integer terms, we want to multiply by the twelve-bit numbers 111110110000_2 (4016_{10}) and 001100011111_2 (799_{10}); the results we get will be the correct answers shifted up by twelve places. Spotting some similarities between these two bit patterns, we let $u = 11_2 x = 3x$ and $v = 11111_2 x = 31x$. Then

$$\begin{aligned} 111110110000_2 x &= (11_2 \times 2^4 + 11111_2 \times 2^7)x \\ &= 2^4 u + 2^7 v \end{aligned}$$

and

$$\begin{aligned} 001100011111_2 x &= (11_2 \times 2^8 + 11111_2)x \\ &= 2^8 u + v. \end{aligned}$$

This idea is illustrated in Figure 12.6. We calculate v using a shift and a subtraction, because $11111_2 = 31 = 2^5 - 1$, and u using a shift and an addition; as $11_2 = 3 = 2^2 - 1$ we could equally well have calculated it like v using a shift and a subtraction. These two intermediate values are then combined according to the equations above to produce the two results, using a total of four adders.

This idea can be extended to cases where there are more than two different constant multiplying factors, although it is tedious to find efficient structures by hand; for larger examples, exhaustive search by computer is the only practical method, and even that can be time-consuming.

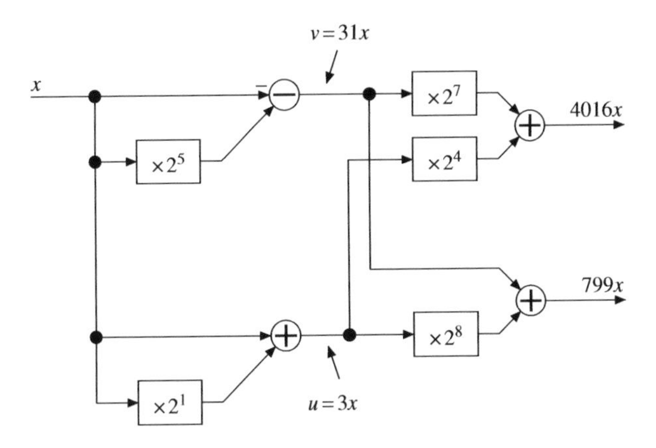

Figure 12.6 Efficient simultaneous multiplication by two different constants: $799 \approx 2^{12}.\ \cos\frac{7\pi}{16}$, $4016 \approx 2^{12}.\ \cos\frac{\pi}{16}$.

12.4 Bit-serial arithmetic

A case of time-division multiplexing that is particularly useful in resource-constrained applications is *bit-serial arithmetic*. Here, instead of having elements such as adders operate on all bits of their operands simultaneously, we build (for example) a one-bit adder and reuse it for each bit of the operand.

The designs we look at will work equally well when operating on unsigned numbers or on signed numbers using two's complement representation with appropriate sign extension. Sign extension mechanisms have been omitted from diagrams in the interests of clarity.

Bit-serial addition

A *full adder* (see Figure 12.7) is a unit with three inputs and two outputs. The inputs are two bits to be added, A and B, and a carry input C_{in}; it adds up its three inputs and produces two outputs, a sum bit S and a carry output C_{out}. The function it implements is one column of an ordinary binary addition sum where A and B are two bits of equal significance from the operands, C_{in} is the carry brought across from the previous (next less significant) column, and C_{out} is the carry to take forward to the subsequent (next more significant) column. The truth table for the full adder (a list defining its outputs for every possible set of inputs) is shown in the figure.

It is possible to build a parallel adder, one that operates on all the bits of its operands simultaneously, by cascading a series of full adders, connecting the carry output of each one to the carry input of the next: see Figure 12.8. The carry input of the least significant full adder is set to zero (or the full adder can be replaced with a *half adder*, which is just a full adder missing the carry input).

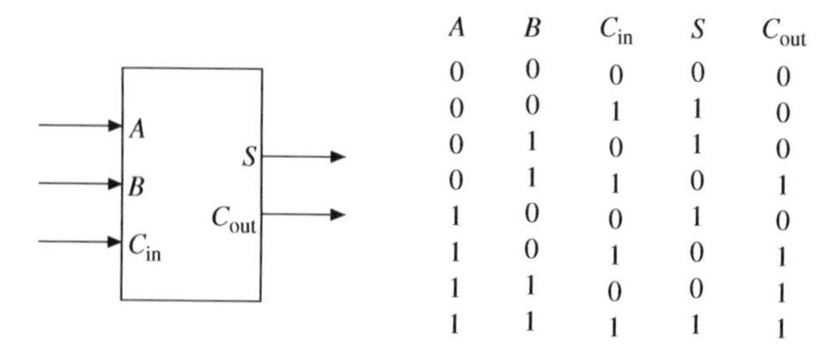

A	B	C_{in}	S	C_{out}
0	0	0	0	0
0	0	1	1	0
0	1	0	1	0
0	1	1	0	1
1	0	0	1	0
1	0	1	0	1
1	1	0	0	1
1	1	1	1	1

Figure 12.7 A full adder and its truth table.

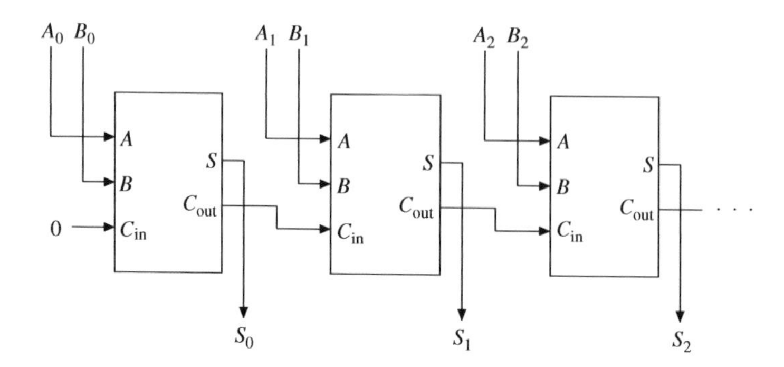

Figure 12.8 A cascade of full adders forming a parallel adder.

If we wish to time-division multiplex a single full adder to replicate the function of Figure 12.8, we will need to present the bits of A and B in order from least significant to most significant. Processing data serially in this order is called using a *little-endian* convention: this is almost always more convenient in practice than the reverse *big-endian* convention.

All that remains is to deal with the carry. The carry out from one bit has to feed in to the carry input of the next bit, which will arrive one clock cycle later. We must therefore delay the carry output by one cycle in a latch before connecting it back to the carry input. The whole arrangement is shown in Figure 12.9. Operands A and B are presented serially at the inputs, and the sum appears serially at the output. A new addition can be started at any time. Note, however, that we have to make provision to clear the carry output latch when starting a new addition: otherwise, the carry coming out of the most significant bits of one addition will enter into the least significant bit of the next.

Bit-serial multiplication

We shall describe a method of bit-serial multiplication where one input is presented serially and the other in parallel: the method is used in many smaller

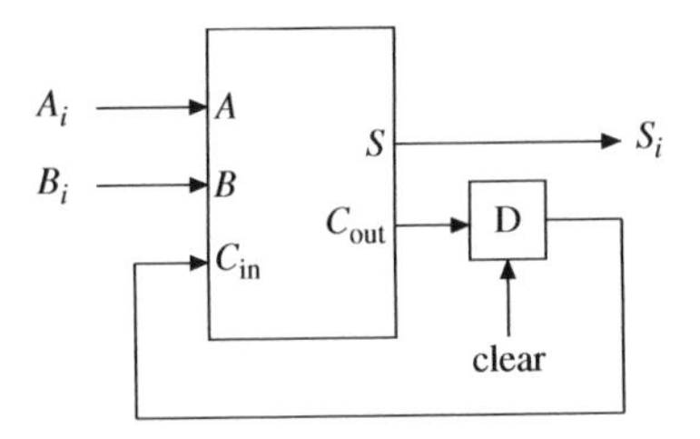

Figure 12.9 A bit-serial adder.

				1	0	0	1
			×	1	1	0	1
				1	0	0	1
+			0	0	0	0	
+		1	0	0	1		
+	1	0	0	1			
	1	1	1	0	1	0	1

Figure 12.10 Binary long multiplication.

microcontrollers to provide a multiply instruction without incurring the cost of a large amount of hardware dedicated to multiplication.

Consider the long multiplication of $9_{10} = 1001_2$ (the multiplicand) by $13_{10} = 1101_2$ (the multiplier). Calculating the result by hand, we would proceed as shown in Figure 12.10. At each row we add in a new *partial product*, which is either zero (where the corresponding multiplier bit was a zero) or a copy of the multiplicand shifted up by the appropriate amount (where the corresponding multiplier bit was a one). The total of the partial products is the final answer. Notice that only the first partial product affects the least significant bit of the answer; only the first two partial products affect the second least significant bit of the answer; and so on.

We can construct a multiplication unit that works like this where the multiplier is presented serially, from least significant bit up to most significant bit, and the multiplicand is presented in parallel. We also generate the answer serially from least significant bit to most significant bit.

One possible arrangement is shown in Figure 12.11. In this and subsequent figures in this section we distinguish a bit-parallel data path from a bit-serial one by marking the former with a short diagonal line. A selector generates the partial products, choosing between zero and the multiplicand on the basis of the current bit of the multiplier. At each cycle the least significant bit of the running total of partial products is output as an answer bit; the sum is then divided by two (i.e., shifted down) and held in a latch before the next partial product is added in on the next cycle. Shifting the running total down in this way is equivalent to

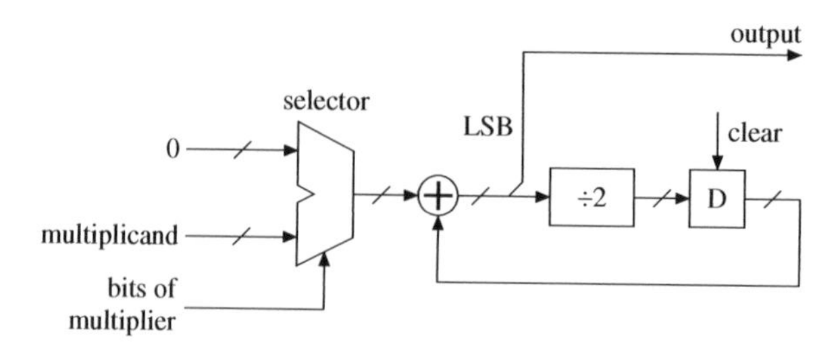

Figure 12.11 Bit-serial multiplication: parallel data paths marked by short diagonal line.

shifting the individual partial products up, and makes the generation of answer bits more convenient.

As with the carry in the adder, the latch holding the running total of partial products needs to be cleared before each new operation.

Bit-serial multiplication by a constant

It is also possible to implement structures for multiplication by a constant, such as the one shown in Figure 12.6, using bit-serial arithmetic. The adders can be constructed as shown in Figure 12.9. Multiplication by a power of 2, say 2^k, simply involves delaying the sequence of bits by k clock cycles. The block diagram for the system thus looks exactly the same.

Note that multiplication by negative powers of 2 is not possible in bit-serial arithmetic as it would require a negative delay. The problem is avoided by recasting the arithmetic in terms of integers (as we did previously); this is equivalent to adding extra pipeline delays into the system.

Distributed arithmetic

Suppose we wished to construct a three-point FIR filter using bit-serial adders and multipliers like the ones described above. One way would be to build three multipliers and sum the results from them as shown in Figure 12.12. The top multiplier multiplies one sample by the coefficient c_0; the middle one multiplies the next sample by coefficient c_1; and the bottom one multiplies the third sample by coefficient c_2. The design could of course be extended to longer FIR filters.

You will notice that in these three multipliers the partial products generated at one instant all have the same significance, since they all correspond to the same bit index in the input samples. (This means that care is required with the design if the bit-serial adders have non-zero latency.) We could modify the design to add the three partial products together immediately after they are generated,

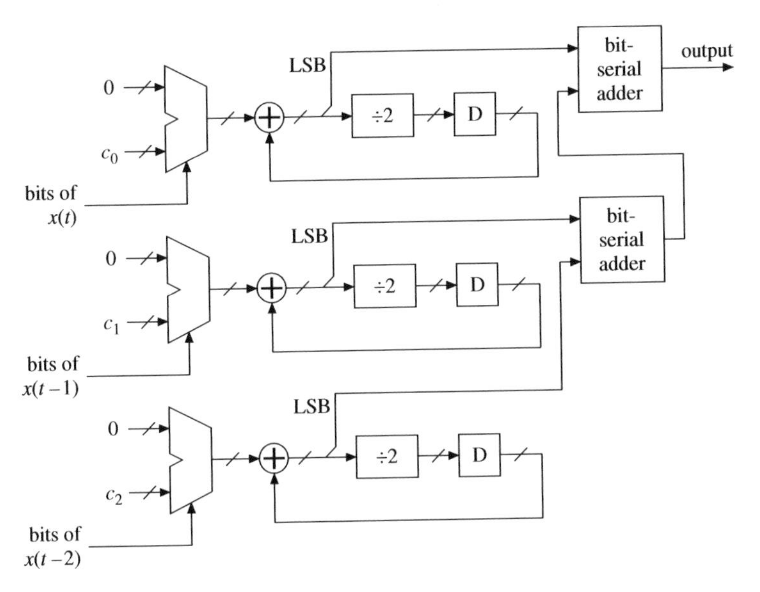

Figure 12.12 Bit-serial FIR filter (first three coefficients shown).

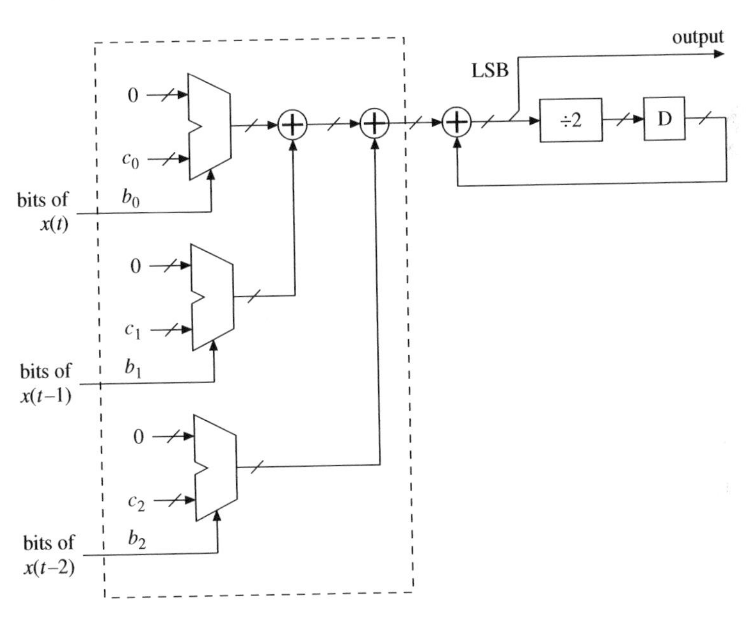

Figure 12.13 Alternative structure for three-point bit-serial FIR filter.

rather than adding the outputs of the multipliers: the result of this transformation is shown in Figure 12.13.

If we now inspect the part of the system within the dotted box, we see that at any given time it calculates a fixed function of its three input bits, b_0, b_1 and b_2. There are therefore only at most $2^3 = 8$ possible different values it can produce at

Table 12.1 *Look-up table for distributed arithmetic FIR filter.*

b_2	b_1	b_0	Table entry
0	0	0	0
0	0	1	c_0
0	1	0	c_1
0	1	1	c_1+c_0
1	0	0	c_2
1	0	1	c_2+c_0
1	1	0	c_2+c_1
1	1	1	$c_2+c_1+c_0$

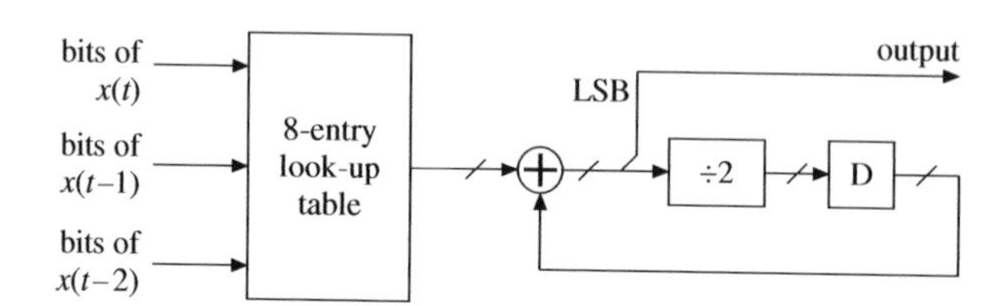

Figure 12.14 A three-point bit-serial FIR filter using distributed arithmetic.

its output, and the whole box could be replaced by a look-up table. The contents of the look-up table would be as shown in Table 12.1.

The resulting architecture is shown in Figure 12.14. Doing the addition part of the FIR calculation in a table in this way is called *distributed arithmetic.*

Distributed arithmetic can also be used for other sum-of-products operations. For example, we can redesign the discrete cosine transform structure shown in Figure 12.3 to use distributed arithmetic: see Figure 12.15. For each output of the discrete cosine transform there will be an arrangement like the one shown in Figure 12.14; since there are eight input values to the discrete cosine transform, each look-up table will receive eight input bits on each cycle. On the first cycle, each table will receive the least significant bits of each of the eight input values; on the next cycle, each table will receive bit 1 from each of the eight input values; and so on. Since each table has eight inputs, it will have $2^8 = 256$ entries. Its output is a sum of between zero and eight of the constants feeding the multipliers in the corresponding part of Figure 12.3, as directed by those bits; the pattern is just the natural extension of Table 12.1.

Another way to think about distributed arithmetic is to imagine the process of inspecting the input bit by bit as rewriting the original eight-component input vector $\mathbf{x}$ as a sum of eight-component vectors $\mathbf{x} = 2^0\mathbf{x_0} + 2^1\mathbf{x_1} + 2^2\mathbf{x_2} + \cdots$ where the eight components of each $\mathbf{x_i}$ are only allowed to be zero or one. The look-up

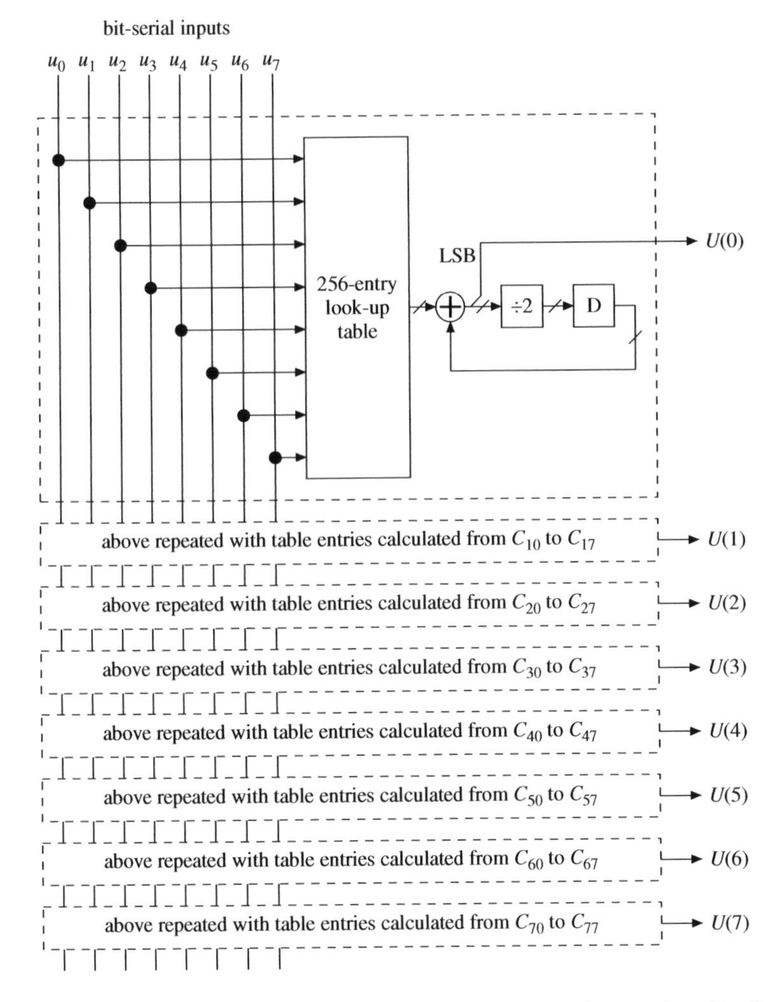

Figure 12.15 Implementation of the discrete cosine transform using distributed arithmetic with bit-serial inputs and outputs.

tables store the discrete cosine transforms of all the 2^8 possibilities for the $\mathbf{x_i}$ vectors. The transform of $\mathbf{x}$ is then calculated by calculating the transforms of the $\mathbf{x_i}$ and adding together the results with the correct significances, taking advantage of the fact that the transform is linear.

Distributed and bit-serial arithmetic are particularly well suited to implementation on FPGAs. A bit-serial adder can often be implemented in a single logic cell (of which the device might contain many thousands), and the pipelining of the carry in the adder avoids introducing long carry chains like the one in Figure 12.8, which often limit the maximum clock speed at which a design can run. Indeed, using bit-serial arithmetic it is usually possible to design a system whose speed approaches the theoretical maximum for the device.

Some FPGAs include small programmable look-up tables, which are ideal for helping to implement filters and discrete cosine transforms using distributed arithmetic. The resulting design is usually much smaller and faster than if full multipliers had been used.

Clock speed and power

We mentioned earlier that a faster design can consume less power than a slower one doing the same task. This may seem counterintuitive: surely a faster design would tend to consume more power?

In the CMOS (complementary metal-oxide semiconductor) technology used in most modern ICs, the majority of the energy is used when signals *change* logic level. Signals that do not change state use very little energy. The power consumption of a CMOS device is approximately given by

$$P = \alpha V + \beta f V^2,$$

where α and β are constants, f is the frequency at which the IC is clocked, and V is the supply voltage. For all but the most advanced CMOS devices, α is relatively small, and we shall for now assume that

$$P \propto f V^2.$$

This suggests, as we might expect, that to minimise power consumption we should reduce f as far as possible, which implies performing as few calculations as possible. This means:

(i) processing signals at the lowest practical sample rate by downsampling early;
(ii) reducing the precision of arithmetic to the minimum level consistent with requirements on accuracy;
(iii) using 'fast' algorithms where functions such as the Fourier transform or discrete cosine transform are to be computed.

The *gate delay* (the time it takes for a change in the input levels to a logic gate to propagate to a change in its output) of a CMOS device depends on the process technology used to manufacture it; for a given process technology, gate delay increases as the supply voltage is reduced. For a given circuit design, the maximum clock frequency achievable is inversely proportional to the gate delay, and thus maximum clock frequency decreases with decreasing supply voltage.

If we have a specified task, we could execute it slowly on a slow device, or quickly on a fast device. The faster device would consume more power, but over a shorter period of time, and the *average* power consumption would be roughly the same. Equivalently we can say that the total energy associated with a particular task is largely independent of how fast we do it.

Let us look at a (rather simplified) concrete example. Suppose we have a design for a circuit function whose longest path from input to output is thirty gate delays. The minimum clock period at which this design will work is also thirty gate delays and the maximum clock frequency is the reciprocal of this period. Suppose this frequency is just fast enough for real-time operation in our application.

Now consider modifying the design by dividing it into three pipeline stages with a maximum of ten gate delays between each stage. The minimum clock period at which this design will work is (to a first approximation) ten gate delays, three times smaller than the unpipelined design, and the maximum clock frequency is three times faster. However, instead of increasing the clock frequency, we could instead reduce the supply voltage: as we said earlier, this would increase the gate delay. In the pipelined design we have the freedom to reduce the supply voltage to the point where the gate delay is increased by a factor of three while still achieving real-time operation; we do not have this freedom in the unpipelined version. The relationship between voltage and gate delay is not a linear one, but in this example we might be able to reduce the supply voltage by as much as a factor of 2. Since power is proportional to the square of the voltage, this would correspond to a factor of four reduction in power consumption.

Exercises

12.1 Write a program in a general-purpose high-level language such as C to add up the integers from 1 to 1 000 000 using 32-bit unsigned integers throughout. (You will have to find out the name of the data type corresponding to 32-bit unsigned integers.) Measure how long the program takes to execute; if it is too fast to measure, add an outer loop to repeat the calculation a suitable number of times.

Work out how to determine whether the result of the addition of two unsigned integers has overflowed. Implement your test at each iteration of the above loop and hence find how many times the addition overflows during the calculation. Measure how long your new program takes to execute; how many times slower is it when the test is included?

12.2 Write a program in a general-purpose high-level language such as C to add two 64-bit unsigned integers using only operations on 32-bit unsigned integers. (On most 32-bit processors, this operation can be done in just two assembly language instructions.)

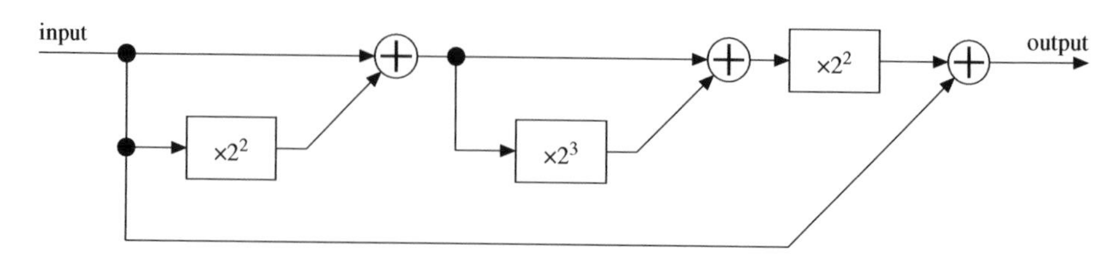

Figure 12.16 Approximate multiplication by $2^7\sqrt{2}$: alternative method.

12.3 Show that $ab = \frac{1}{4}\left((a+b)^2 - (a-b)^2\right)$.

You wish to implement an 8-bit by 8-bit multiplier in a programmable logic device that includes a 512-entry by 18-bit RAM module. Show how to use the RAM as a look-up table in conjunction with three adders or subtractors to implement the multiplier.

12.4 Explain how the circuit whose block diagram is shown in Figure 12.16 performs the same calculation as the one shown in Figure 12.5.

12.5 Draw a block diagram of a system along the lines of Figure 12.16 for multiplying numbers by $21845 \approx \frac{1}{3} \cdot 2^{16}$. How many adders do you need?

12.6 Design a system along the lines of Figure 12.6 for multiplying numbers simultaneously by $21845 \approx \frac{1}{3} \cdot 2^{16}$ and $13107 \approx \frac{1}{5} \cdot 2^{16}$. How many adders do you need?

Answers

1.1 Three (one time dimension and two space dimensions); four (altitude as well).

1.2 Domain: three; range: one.

1.3 You will most likely get a machine precision of either about one part in 10^7 or one part in 10^{15}: see the discussion of floating-point arithmetic in Section 7.3 for more details. These correspond to variations in length in the capacitor plate of 10^{-10} m and 10^{-18} m, or one atom and ten billionths of an atom respectively.

1.4 You could represent each bearing as a complex number with magnitude 1 and argument θ, add up the complex values and take the argument of the result. To weight the readings, you could give the numbers different magnitudes.

2.3 The first alias of a 400 Hz sine wave sampled at 2 kHz is at 1600 Hz. The lowest-frequency alias of a 400 Hz sine wave sampled at 500 Hz is at 100 Hz.

2.4 To reproduce frequencies of up to 20 kHz faithfully a sample rate of at least 40 kHz is required.

2.5 The apparent motion is at a frequency of -8 spokes per second and the sample rate is 24 Hz. The actual motion could therefore be at $24-8=16$ spokes per second, $2\times24-8=40$ spokes per second, $3\times24-8=64$ spokes per second and so on. One spoke corresponds to one eighth of a revolution, which is $4/8=0.5$ m on the ground. The possible speeds are thus $16\times0.5=8\,\text{m/s}$, $40\times0.5=20\,\text{m/s}$, $64\times0.5=32\,\text{m/s}$ and so on.

2.6 The candidate frequencies are $75k\pm10$ Hz for integer k. In the range specified, the possibilities are 290 Hz, 310 Hz, 365 Hz, 385 Hz, 440 Hz, 460 Hz and 515 Hz.

2.8 (a) Yes; yes; yes; yes; 20 Hz. (b) Aliasing would make frequencies of (for example) 124 Hz and 126 Hz indistinguishable. (c) 60 Hz; $86\frac{2}{3}$ Hz.

3.3 Yes, 3-bit logarithmic quantisation is a reasonable thing to do: if we assume one bit is used to represent the sign of the number, the other two bits can be used to represent four possible magnitudes which can be logarithmically spaced.

3.6 The aperture time of the converter is equal to one period of the 100 MHz sine wave. The average value of the sine wave over the aperture period will therefore always be zero. Requirements on aperture are more demanding when sub-Nyquist sampling is used.

3.8 The 20 kHz sine wave is given by $x(t) = \sin(2\pi \cdot 20\,000t)$ where t is in seconds. The 2^{16} quantisation steps are linearly spaced over the amplitude of the wave from -1 to 1, and so one quantisation step (ignoring rounding) corresponds to a change in amplitude of $2/2^{16} = 2^{-15}$. Solving $\sin(2\pi \cdot 20\,000t) = 2^{-15}$ we get $t = 2.43 \times 10^{-10}$ s, or about a quarter of a nanosecond. For a 200 Hz sine wave the answer is 100 times bigger.

CD players include a buffer to store samples read from the disc before conversion. The number of samples stored in the buffer increases and decreases to absorb the variation in speed of the disc.

4.1 3; 1.

4.11 There are two $N/2$-point transforms, $N/2$ complex multiplications and N complex additions and subtractions involved: $u(N) = 2u(N/2) + N/2 + N = 2u(N/2) + 3N/2$. $u(1) = 0$; $u(2) = 3$; $u(4) = 12$; $u(8) = 36$; $u(16) = 96$; $u(32) = 240$; $u(64) = 576$; $u(128) = 1344$; $u(256) = 3072$; $u(512) = 6912$; $u(1024) = 15\,360$; $u(2048) = 33\,792$. In general, $u(N) = \frac{3}{2}N \log_2 N$.

For a straightforward implementation of Equation 4.1, we have $v(N) = 2N^2$. $v(1024) = 2\,097\,152$. Going purely on operation counts, you would expect the fast Fourier transform algorithm to be $2\,097\,152/15\,360 \approx 137$ times faster.

4.12 A complex multiplication can be done using a total of six real operations: four multiplications, one addition and one subtraction. A complex addition or subtraction takes two real operations.

The 4-point Fourier transform using the fast algorithm requires four complex multiplications and eight complex additions or subtractions, or $4 \times 6 + 8 \times 2 = 40$ real operations.

All the multiplications are by $+1$, -1, j or $-$j and can thus be eliminated. The 4-point Fourier transform can thus be calculated using just eight complex additions or subtractions, for a total of $8 \times 2 = 16$ real operations.

4.15 The right-hand column of the table is the reverse of the left-hand column.

4.16 Write $r(a)$ for the number whose binary expansion is the expansion of a with the bits in reverse order. Loop over the values of a from 0 to $N-1$ and exchange $z(a)$ with $z(r(a))$ if $a < r(a)$. The condition ensures that each exchange only occurs once.

5.1 11; $m + n - 1$.

5.3 7.

5.4 Applying Filter A twice is the same as multiplying by its frequency response twice in the frequency domain.

5.6 (a) (2, 8, 7, 3, 0, 0, ...). (d) Some of the values in the final result are ten or more. These need to be processed like the carries in the final addition of a long multiplication in order to obtain the correct answer. (e) The long multiplication requires 1024^2 individual digit-by-digit multiplications and about 1024^2 individual digit additions, for a total of about two million operations. Using the Fourier transform method we take three transforms (two forwards and one inverse) and perform 2048 pointwise multiplications. The total operation count in this case is thus $3 \times 33\,792 + 2048 = 103\,424$, but note that these are operations on complex numbers, not single digits.

5.7 If the length of $x(t)$ is less than or equal to 258 it can be split into two parts of length at most 129. The convolution of each of these with $h(t)$ will be at most 256 samples long and can thus be done with the library function. Three forwards transforms (of $h(t)$ and of the two halves of $x(t)$) and two inverse transforms are required. If the length of $x(t)$ is 259 samples then an extra forwards and an extra inverse transform are needed.

5.8 Add the two samples receiving the same weighting from the filter together before the multiplication.

5.10 Reverse the order of the coefficients.

5.11 For example, the signal (... 0, 0, 1, 1, 0, 0, ...).

5.12 The downsampling step will destroy frequency components between 8 kHz and 16 kHz which will not be restored by the upsampling step.

5.13 Use the first four bits after the binary point, considered as an integer in the range 0 to 15. You may wish to round this value (see Section 7.1).

5.15 For example, $a = 1$, $b = 2$.

6.2 The four equally likely times consistent with the pattern you can see are 2:34, 3:34, 7:34 and 12:34. The probabilities are therefore 1/4, 1/4, 0, and 1/2. $(\frac{6}{20} + \frac{3}{20})/(\frac{6}{20} + \frac{3}{20} + 0 + \frac{3}{20}) = \frac{9}{12} = \frac{3}{4}$.

6.3 (a) Yes. (b) No: the filter does not have a flat frequency response.

6.4 Alliteration.

6.5 $\Lambda = 22.05$.

6.7 (a) $\Lambda = 8.342$. (b) $\Lambda = 5.458$. (c) $\Lambda = 4.113$.

7.1 0011.10010_2.

7.5 No; no; yes; yes; no. $u + 2v \leq 32$.

7.8 0.0101011_2. $1.0000001_2 \neq 1$. $n = 1, 2, 4, 8, \ldots 128$.

7.9 Shift the remainder up by k places and divide by the divisor. The quotient gives k fractional bits of the fixed-point result.

7.12 For example, 0.0001001_2; 0.0000111_2; 0.0001001_2.

7.14 0; $2^{16}x$; $2^{16}y$; $2^{16}(x+y)$ (modulo 2^{32}).

7.15 The absolute value of the most negative number representable in a two's complement system is not representable in that system.

7.16 $(2^{15} - 1)(-2^{15}) = -(2^{30} - 2^{15})$; $(-2^{15})(-2^{15}) = 2^{30}$; $2n$. (If it were not for the product $(-2^{n-1})(-2^{n-1}) = 2^{2n-2}$, $2n - 1$ bits would do.)

7.17 For example, $a = 10^{-10}$, $b = 10^{10}$, $c = -10^{10}$.

7.18 You could convert all the numbers to fixed-point format with enough bits to represent all the possible floating-point values exactly, and then add them. A faster, but less accurate, approach is to add up the numbers in increasing order of magnitude.

8.1 $2 \times 44\,100 \times 16 = 1\,411\,200$ bits; approximately 783 million bytes. CD-ROMs use more error protection and contain more cataloguing information than audio compact discs.

8.2 $1\,411\,200 : 128\,000$, or approximately $11 : 1$.

8.3 Have the compressor check whether it can compress the data. If it can, have it output a '1' bit followed by the compressed data; otherwise, have it output a '0' bit followed by the uncompressed data. The resulting compression ratio averaged over your collection of files will then be approximately 9 : 1.

8.5 One possible set would be $\pm\frac{2}{3}\cdot 2^k$ (rounded to integers) for $k = 1$ to $k = 16$. A number can be quantised to the nearest of these delta values by finding the position of the most significant '1' bit in the binary representation of its absolute value.

8.6 For example, the signal of period 3 where $s(0) = 0$, $s(1) = 1$ and $s(2) = 2$.

8.7 Examine the sample values either side of the zero-crossing and work out where the straight line joining them crosses the time axis. (More sophisticated approaches are possible.)

9.2 Group the rows of the image into pairs and transform them two at a time using the method of Exercise 4.9, calling the library function 512 times. You can now transform columns 0 to 512 inclusive using a further 513 calls to the library function, and then reconstruct the answer for columns 513 to 1023 using the symmetry property of the two-dimensional transform of real data. The total number of calls to the library function is then $512 + 513 = 1025$. (You can shave one further call off this total, bringing it down to 1024, by observing that columns 0 and 512 of the result of the first batch of transforms are real, and so can be transformed together.)

9.3 $1024 \times 15\,360 = 15\,728\,640$ operations.

Using direct calculation, approximately 1024^4 multiplications and 1024^4 additions, for a total of $2 \times 1024^4 \approx 2 \times 10^{12}$ operations are required. Using the Fourier transform method, we need three transforms (two forwards and one inverse) plus 1024^2 pointwise multiplications, for a total of approximately 50 million operations.

9.13 Convolve the bilevel image and the structuring element as if they were greyscale images. The dilation is the set of points where the result is non-zero and the erosion is the set of points where the result is equal to the number of set pixels in the structuring element.

9.15 No.

10.1 They probably find it appears flickery, since the field rate is only 50 Hz rather than 60 Hz. (They may also have some comments on the quality of the programmes.)

10.2 Every six crosses or five dots, because $576 : 480 = 6 : 5$.

10.4 (a) $720 \times 576/16^2 = 1620$. (b) $1620 \times (2 \times 15 + 1)^2 = 1\,556\,820$. (c) $1\,556\,820 \times 16^2 = 398\,545\,920$. (d) $398\,545\,920 \times 25 = 9\,963\,648\,000$, or about one every 100 ps.

10.5 The motion estimator is finding incorrect motion vectors in the blurred faces of the crowd, where there is little reliable detail to use. In such circumstances the compressor may simply use a zero vector in the interests of improving the compression ratio.

10.6 13: the previous two B-pictures and the whole of the rest of the group beginning with the corrupted I-picture might be affected.

10.7 Each group of 12 pictures is compressed to $140\,000 + 3 \times 40\,000 + 8 \times 20\,000 = 420\,000$ bytes and lasts $12/25 = 0.48$ s. The data rate is thus $420\,000/0.48 = 875\,000$ bytes per second or $7\,000\,000$ bits per second.

The original signal contains $720 \times 576 \times 25 = 10\,368\,000$ pixels per second, and so the stream represents the signal using $7\,000\,000/10\,368\,000 \approx 0.675$ bits per pixel.

If every picture were coded as an I-picture, the data rate would be $140\,000 \times 25 = 3\,500\,000$ bytes per second, or $28\,000\,000$ bits per second. This corresponds to $28\,000\,000/10\,368\,000 \approx 2.7$ bits per pixel.

10.8 The B-picture two pictures before an I-picture (the picture labelled '−2' in Figure 10.17) cannot be decoded until it has been completely received, $140\,000 + 20\,000 = 160\,000$ bytes after the start of transmission of the I-picture. The I-picture is displayed two picture times later, and so the total delay is $160\,000/875\,000 + 2/25 \approx 0.263$ s. You can check that this is the worst case.

10.9 A pattern of vertical stripes moving horizontally will appear slanted when using a scanning sensor.

11.1 The signal is given an offset to ensure that it is always positive. This gives rise to a large DC component in its spectrum.

11.2 You could try to detect the reradiation from the local oscillator, measure its frequency, and add or subtract 500 kHz.

11.3 All coefficients whose time indices are multiples of eight (except the one at time zero) must be zero.

11.4 .01010101; approximately 1992 Hz; change the frequency setting to .01010110 on every third increment, or, more simply, replace the phase accumulator with a modulo-3 counter.

11.5 Ensure the oscillators in the two BPSK modulators are operating 90° out of phase and add the outputs of the modulators.

11.6 The transmitter output will be a sine wave at 1 000 500 Hz.

11.7 The sequence that scrambles to all-zeros will probably be much less commonly transmitted than the all-zeros sequence, and so there is overall benefit in using the scrambler. Some higher-level protocols ensure that when a sequence needs to be retransmitted because of an error it will be scrambled differently.

11.8 It will not work correctly if the constellation diagram in the receiver is rotated by 90° or 270°.

11.9 The Fourier transform of the (periodic) clock signals used in the processor may well have significant peaks at integer multiples of 250 MHz, including $9 \times 250\,\text{MHz} = 2.25\,\text{GHz}$. This will probably cause interference in the radio receiver.

The clock frequency should be changed to a nearby value such as 240 MHz or 260 MHz which has no multiples near 2.25 GHz.

11.10 The wavelength of the signal is $3 \times 10^8/100 \times 10^6 = 3\,\text{m}$. The helicopter is moving at a speed of 5 wavelengths per second, or 15 m/s.

11.12 The telephone will increase its transmit power to compensate for the effect of the shielding.

They should campaign for the installation of base stations so that their mobile telephones can transmit at lower power.

12.3 Suppose $a > b$ and that entry k in the look-up table contains the value of k^2. Calculate $a+b$ (which may be 9 bits long) and $a-b$. Use the look-up table to obtain $(a+b)^2$ and $(a-b)^2$. Subtract these, producing $u = (a+b)^2 - (a-b)^2$ and shift u down two places to obtain ab. Note that since we discard the two least significant bits of u (which will always both be zero), the look-up table can be trimmed to 512 entries each of 16 bits.

12.5 Let x be the number to be multiplied. Calculate in turn $x + 2^2 x = 5x$, $5x + 2^4 \cdot 5x = 85x$ and $85x + 2^8 \cdot 85x = 21\,845x$, using a total of three adders.

12.6 Let x be the number to be multiplied. Calculate in turn $x + 2^4 x = 17x$, $17x + 2^8 \cdot 17x = 4369x$, $4369x + 2 \cdot 4369x = 13\,107x$ and $4369x + 2^2 \cdot 4369x = 21\,845x$, using a total of four adders.

Index

65143355R00193

Made in the USA
Lexington, KY
02 July 2017